Satellite Thermal Control for Systems Engineers

Satellite Thermal Control for Systems Engineers

Robert D. Karam
Gaithersburg, Maryland

Volume 181
PROGRESS IN
ASTRONAUTICS AND AERONAUTICS

Paul Zarchan, Editor-in-Chief
Charles Stark Draper Laboratory, Inc.
Cambridge, Massachusetts

Published by the
American Institute of Aeronautics and Astronautics, Inc.
1801 Alexander Bell Drive, Reston, Virginia 20191-4344

To my wife,
Maggie

Progress in Astronautics and Aeronautics

Editor-in-Chief
Paul Zarchan
Charles Stark Draper Laboratory, Inc.

Table of Contents

Preface

This book is about the theory and methods of controlling the temperature of a satellite. It is intended to give spacecraft systems engineers a background for directing and advising during the evolution of a thermal design, but there is also enough information here to invite specialists to compare and exchange knowledge and ideas.

Satellite thermal control has its foundations in heat transfer, a subject about which very many books have been written. Fortunately, only the basics from this extensive and complex field are needed for the development of thermal control, and because all treatments present more or less the same basics it was a matter of personal preference that *Fundamentals of Heat Transfer*, by Frank P. Incropera and David P. DeWitt (Wiley, New York, 1981), was selected as the main reference. The buildup, however, as it applies to satellites, is derived from actual involvement and contacts and from some of the innumerable number of publications that have appeared in various journals and compilations since the launch of Sputnik in 1957. A careful review of all these works would be a lifelong project, and, indeed, only a very modest sampling was consulted and credited. But, in any case, the influence cannot stop here. In fact, it was sobering to discover how many vital aspects in the development of thermal control were never reported in general publications and remain lodged in stored boxes of interoffice communications, notes from meetings, skimpy records of telephone conversations, and memories from informal discussions. Although some of this background is implicit in this work, there was no practical way for citing the sources or their authors in terms of accessible references.

As this book is one man's unsolicited view of what constitutes the basis for appreciating satellite thermal control, there will undoubtedly be disagreements on the range and depth of the material. Perhaps worse, it is expected that there will be serious refutation of some of the statements and numerical data, particularly those pertaining to hardware performance. In many respects this has always been part of the nature of thermal control. The variety of requirements to sustain potential extremes in a satellite's environment and performance makes any restrictions fragile and open to legitimate criticism. On the other hand, to come up with generalizations or a complete classification would certainly require more than this one volume, as well as the participation of a great many thermal engineers, each with different experience and perceptions. In this regard, it is hoped that this work will induce comments, corrections, and additions in anticipation of a future, more encompassing edition.

Having spent a whole professional life at Fairchild Space Company, often in close association with the NASA Goddard Space Flight Center, I was fortunate to work and learn with some of the most talented people in aerospace. I pay a special tribute to the memory of Ralph Hall and Bill King who, perhaps more than anyone, were responsible for my approach to technical presentations. Among the many others who helped make this book possible, my colleagues and friends Michael Coyle, Robert Eby, Jack Hunter, William Kelly, and Han Whangbo had the greatest influence. I also acknowledge with thanks the valuable assistance of Jodi Glasscock and Rodger Williams at the American Institute of Aeronautics and Astronautics. Finally, a very special appreciation of my wife Maggie for her patience and encouragement during the many long days and nights I spent on this work.

Robert D. Karam
July 1998

Nomenclature

A	=	area, m^2
A	=	albedo (reflected solar radiation off Earth), W/m^2
b	=	dimension (width), m
C	=	convection conductance, W/K
c	=	specific heat (solids), W-h/kg K (also J/kg K)
c_o	=	speed of light in vacuum, 3.0×10^8 m/s
c_p	=	specific heat at constant pressure (fluids), J/kg K
c_v	=	specific heat at constant volume (fluids), J/kg K
D	=	diameter, m
D_h	=	hydraulic diameter, m
E	=	modulus of elasticity (bending of tubular elements), N/m^2; voltage, V
E	=	Earth [infra red (IR)] radiation, nominally 236 W/m^2
Ec	=	Eckert number
e	=	coefficient of expansion (composites; bending of tubular elements), /K
F	=	shape factor; radiation factor, W/K^4; also W/m^2 K^4; friction factor (heat pipe), N/W m^3
$\mathbf{F}$	=	radiation blockage factor (measure of ability to view)
F_A	=	shape factor
F_E	=	emissivity factor
f	=	friction factor (Fanning); factor relating multilayered insulation properties to radiator's effective optical properties
f	=	albedo factor, nominally 0.33
G	=	radiation G factor, W/K^4
Gr	=	Grashof number
g	=	gravitational acceleration, m/s^2
H	=	dimension; length (heat pipe in a radiator); height, m
HTF	=	heat transport factor (heat pipe), W-m
h	=	convection coefficient, W/m^2 K; Planck's constant, 6.626×10^{-34} J s
I	=	radiation intensity, W/m^2 μm; moment of inertia, m^4
K	=	conduction conductance, W/K; conductance, W/m^2 K; also W/K /m
k	=	conductivity, W/m K
$\mathbf{k}$	=	conductivity matrix (components k_x, k_y, k_z in W/m K)
L	=	length, m
l	=	length; characteristic length, m
M	=	Mach number
M	=	mass, kg; heat pipe figure of merit or transport factor, W/m^2; thermal moment of inertia, N m
Mc	=	thermal mass (mass times specific heat), J/K

$\dot{m}$	= rate of mass flow, kg/s
N	= integer; number; number of conductances; number of parallel planes; number of grooves in a heat pipe
Nu	= Nusselt number
P	= pressure, N/m^2; perimeter, m; thermal load (louvers bimetallic), N
Pr	= Prandtl number
p	= percent perforation of surface material
Q	= heat rate, W
Q^*	= total heat generated in a volume, W
q	= heat rate, W/m^2
q^*	= volumetric heat generation, W/m^3
R	= radiation conductance, W/K; thermal resistance, K/W; electrical resistance, Ω; radius, m; principal radius of curvature of meniscus (heat pipe), m
R_E	= Earth radius, average 6340 km
R_v	= gas constant of the vapor in a heat pipe; J/kg K
Ra	= Rayleigh number
Re	= Reynolds number
r	= radius, m
S	= solar flux or solar constant, nominally 1350 W/m^2
s	= distance connecting two locations, m
T	= temperature, K or °C
T_∞	= temperature of far fluid (in convection); sink temperature (in radiation), K
t	= time, s; thickness, m
t_c	= time constant, s
U	= constant heat flux, W/m^2
U	= total conductance (combined modes of heat transfer), W/K
u	= speed (fluids), m/s
u_s	= local speed of sound (fluids), m/s
W	= dimension; length; width, m
We	= Weber number
w	= width; heat pipe groove width, m
X	= distance (from air inlet inside fairing at launchpad), m
x, y, z	= position or distance; dimension in x, y, z coordinate system, m
α	= absorptivity (average for all wavelengths and directions)
α_n	= normal absorptivity (measured at normal angle of incidence)
α^s	= solar absorptivity (average in the solar spectrum)
β	= volumetric coefficient of expansion, per K; dimensionless parameter in conducting–convecting or conducting–radiating fins; beta angle (angle between solar vector and orbit plane)
Γ	= vapor pressure gradient from radial flow in a heat pipe, N/m^3

γ	=	ratio of specific heats, c_p/c_v; dimensionless parameter in conducting–radiating fins
Δ	=	difference; change
Δ or ΔT	=	temperature difference, K
ΔP	=	pressure difference causing flow in a pipe, N/m^2
δ	=	thickness; deflection (tubular elements), depth (heat pipe groove); m
ε	=	emissivity (total hemispherical)
ε_n	=	normal emissivity (measured at normal angle of incidence)
ε^*	=	effective emissivity of multilayered insulation (variation on ε_{ins})
η	=	dimensionless heating input
θ	=	angle (from vertical in radiation exchange); dimensionless time; angular displacement (louvers bimetallic); louvers blade angle from closed position
(θ, φ)	=	direction of radiation (θ is in-plane angle and φ out-of-plane)
κ	=	Boltzmann's constant, equal to 1.38×10^{-23} J/K; proportionality factor
λ	=	wavelength, μm; latent heat of two-phase system (heat pipe working fluid), J/kg
μ	=	viscosity, N s/m^2
ν	=	ratio of distance from a sphere surface to its radius (ratio of orbit altitude to planet radius)
ξ	=	dimensionless distance (x/L in a longitudinal fin)
ρ	=	density, kg/m^3; ratio of internal to external emissivity (tubular element)
ρ	=	reflectivity
σ	=	Stefan–Boltzmann constant, 5.67×10^{-8} W/m^2 K^4
σ	=	surface tension, W/m
τ	=	transmissivity
τ	=	dimensionless temperature
ϕ	=	angular position on a cylinder
$\phi(t)$	=	heating input from external sources (thermal energy equation), W/m^2
φ	=	angle (in-plane in radiation exchange; from vertical position in spherical coordinates; around cylinder; sun angle off louvers plane)
ψ	=	angle (in-plane in spherical coordinate; sun angle in louvers plane; angular extent of Earth shadow)
ω	=	solid angle
$\Im$	=	radiation emission factor
$\Im_{ij}$	=	radiation exchange factor (script F) between surfaces (or nodes) i and j

xvi

Subscripts

a	=	adiabatic section (heat pipe); antenna; aluminum (louvers)
av	=	average
b	=	black; back of solar array; louvers blade; vapor bubble (heat pipes)
bl	=	bond line (heat pipes)
bp	=	baseplate (electronics box)
c	=	closed (louvers); coolant; condenser (heat pipe); capillary (heat pipe); core (honeycomb)
cold	=	under cold case conditions
D	=	daytime (during sunlit portion of orbit)
e	=	evaporator (heat pipe)
eff	=	effective
f	=	fiber (composites); fluid
fs	=	face sheet (honeycomb)
hot	=	under hot case conditions
hp	=	heat pipe
htr	=	heater
i	=	location; node; inner
ij	=	from node, body, or surface i to node, body, or surface j
ic	=	inner cover (innermost sheet of multilayered insulation)
$i \rightarrow j$	=	from i intercepted by j (radiation)
ins	=	multilayered insulation
j	=	location; node; junction (electronics)
L	=	location at length L (fin end)
l	=	liquid; louvers
m	=	mean
max	=	maximum
min	=	minimum
N	=	nighttime (during eclipse in orbit)
nb	=	nonblack
o	=	open; outer; optimum
oc	=	outer cover (insulation)
p	=	panel (solar array)
r	=	radiator; location at radius r
s	=	space; saddle (heat pipes); sun shield (shielded louvers); surface (convection)
sa	=	solar array
ss	=	steady state
t	=	transverse (across thickness); at time t
u	=	unshielded louvers
v	=	vapor
w	=	width; wick (heat pipe); wall (shroud); white paint (on louvers blades)

x, y, z = Cartesian coordinates; position (m) from initial point
θ or $\theta - \varphi$ = directional (radiation)
λ = spectral (radiation)
λ, θ = spectral directional (radiation)
0 = initial position (fin stem); initial time (transients)
12 = from node, body, or surface 1 to node, body, or surface 2

Superscripts

A = associated with albedo
a = associated with absorbed radiation
c = associated with heat convection
d = associated with dissipated heat
E = associated with emitted Earth radiation
e = associated with radiation emission
env = associated with surrounding environment
fmh = free molecular heating
i = associated with incident radiation
k = associated with heat conduction
r = associated with radiation
S, s = associated with solar radiation
s = saturation conditions (change of phase)
0-g = zero gravity (assumed condition in space)

Chapter 1

Satellite Temperature and Thermal Energy Management

I. Introduction

S ATELLITES perform better and last longer when their components remain within certain temperature limits, usually, but not always, near the level at which they are assembled. Satellite thermal control deals with the theory and practice by which these temperatures are produced, and the function of the thermal engineer is to determine the influencing factors and manage them within the constraints of the satellite as one system. The process involves unique methods of analysis and test and often requires the use of some highly specialized hardware.

Thus, the problem of thermal control must begin by establishing the temperature specifications under which the satellite is to exist during various stages of its life. The thermal design should then be such that the specified values, especially those in orbit, are not exceeded. This is asserted during development by analysis, similarity studies, and tests. Additional confidence in long-term operation is gained when the satellite and its separate components are shown to maintain their integrity and perform satisfactorily when subjected to short-term temperatures that go beyond those expected in the course of the mission.

II. Specifications

Some of the immediate questions are: What are these temperatures? How are they established, and where exactly do they apply? Normally, the interest is in locations where temperature is known to affect the operation. When it is not possible to monitor the precise location, adjacent but more accessible regions are designated and related by analysis or test to those under consideration.

But because of the diversity of spacecraft applications, questions concerning actual values of temperature and the rationale behind their selection cannot always be answered in nonrestrictive terms. This is especially apparent when components of the same heritage are used in scientific, military, and commercial satellites. Cultures differ among these orders, and the level of acceptable risk varies, making variations in the required temperature limits not uncommon. Differences may also exist in the same industry, depending on the frequency of a component's activity, its exposure to space, and the duration of the mission.

A list of specifications for which there seems to be some general agreement is given in Table 1.1. But even here there may be a need for narrower management, if only to be in concert with today's economic realities that commit aerospace companies to the building of satellites with extended useful lives.

Table 1.1   Specifications for Orbital Temperature

Battery (mounting platform average): 0 to 20°C
Battery (temperature difference among multiple units): ±5°C
Power regulating unit (mounting platform average): 0 to 40°C
Remote interface unit (component baseplate average): −5 to 60°C
Tape recorder (component baseplate average): 0 to 32°C
Tape recorder (average tape temperature, from analysis): 0 to 35°C
Transponder (component baseplate average): 0 to 50°C
Attitude control wheel (mounting points): −5 to 45°C
Attitude control electronics (component baseplate average): 0 to 40°C
Optical bench (average bench temperature): −5 to 45°C
Optical bench (temperature differential): ±3°C
Earth sensor (mounts): −5 to 50°C
Experiments mounting platform (average): 0 to 35°C
Hydrazine propulsion electronics (local mounting platform): 5 to 45°C
Hydrazine propellant elements (wet points or surfaces): 10 to 50°C
Solar array (low-Earth-maximum and -minimum cell temperature): −100 to 85°C
Antenna dish (low Earth orbit): −80 to 60°C
Antenna elements (including extendable booms): −100 to 150°C
Viscous dampers (at deployment): −50 to 60°C
Viscous dampers (after deployment): <70°C
Inactive structure: −100 to 100°C

Because a satellite's performance and durability are both enhanced by a more benign and stable temperature, stricter requirements emerge and established ranges are frequently modified.

A.   The Role of Thermal Control

Clearly, temperature specifications must be developed early in a program. And because their selection cannot be arbitrary, immediate involvement by thermal control is mandatory to determining practical ranges. The main effort here lies in conducting concepts and parametric studies as background to negotiating the requirements and margins for an acceptable thermal design. Among the tasks in this connection is a thermal analysis of the satellite electronics which relates their temperature to that of the mounting platform. The major factors influencing this analysis are dissipated heat, the manner of packaging, mounting procedures and locations in the satellite. The predictions should be reported in documents with wide distribution, such as a satellite resources and margin management report, and periodically updated to reflect the expected temperatures of the evolving design.

Tradeoff studies are also conducted on improving temperatures at the expense of added weight, specialized hardware and heater power. This effort is normally coordinated with structures and power, and it includes determining the orbital temperature profiles of the solar array panels. In addition to being a guideline for writing the solar array thermal specifications, this information is required for assessing batteries operation and sizing and shaping the panels. The shape and weight of satellite arrays are among the first data needed by attitude control.

Other tasks include support for establishing launch time lines and procedures for orbit insertion and deployments. These data are used in writing the temperature specifications for mechanisms associated with deployable solar arrays, antennas and booms. The models for these analyses are also used to determine if special orbital maneuvers or spacecraft orientations will be required to evade unmanageable heating of a particularly sensitive instrument or component.

Although perhaps of a lesser consequence, it should nonetheless be remarked that it is thermal engineering, more than any other factor, that defines the appearance of a satellite. Satellite exterior surfaces are shaped, colored and oriented almost exclusively for the purpose of achieving the needed temperatures.

B. Definitions

In quoting temperatures, modifiers are used to distinguish various applications. The definitions given here are the most common. For dissipating electronics, the reference is almost always to the average temperature in the region of the mounting structure, where a component makes contact. For solar arrays, antennas, and nondissipating components, reference is usually made to regions with maximum and minimum local temperatures.

1) Allowable flight temperature limits are the highest and lowest values specified for normal operations in flight. They encompass the range within which a component is assumed to perform most reliably. Flight allowables are the baseline to which temperature margins are added and subtracted.

2) Predicted flight temperature limits are the maximum and minimum values obtained from mathematical models that incorporate positive and negative tolerances on the factors that affect temperature. Generally, the thermal design is said to be verified by analysis when the predicted highest and lowest flight temperatures are at or within the allowable flight limits.

3) Acceptance temperature limits are the extremes that flight equipment is subjected to during thermal testing. Acceptance tests are usually conducted in a vacuum and can be designed to demonstrate both workmanship (for example, by temperature cycling) and compliance with performance requirements at temperatures with margins beyond the expected normal operating limits. The margins are selected based on component history and mission duration, but specific criteria may differ depending on the origin of the program.

4) Qualification temperature limits are the extremes that qualification or protoflight units are subjected to during thermal testing, which is frequently performed in vacuum. Qualification tests demonstrate short-term performance at temperature levels well beyond those expected during normal operation. As in acceptance testing, the margins of qualification are selected in accordance with the background of a component and its planned operation.

5) Design temperature limits are the maximum and minimum values assumed for material selection and for thermal stress analysis. Design limits can, and often do, exceed the qualification limits, but it is not standard practice to test assembled hardware at these levels.

6) Nonoperating temperature limits are the extremes acceptable for a dormant component. Associated with this definition are storage temperatures, transportation temperatures, and startup temperatures for the electronics. The cold startup temperature is the lowest permitted on a component when it is activated.

Example: TOPEX/Poseidon Temperature Specifications

An example of how temperature specifications are negotiated and established is found in the TOPEX/Poseidon program,[1-3] which was contracted by NASA's Jet Propulsion Laboratory (JPL).

JPL's culture centers mainly around planetary missions (usually with reduced activity until encounters) and their methods for screening and qualifying components are more in line with the requirements of Military Standard Specification MIL-STD-1540-B (Ref. 4), which generally exceed the levels required by the NASA Goddard Space Flight Center (GSFC). In addition, JPL recommends that qualification units of satellite electronics be tested in extended operation at or near their design temperature; this follows from statistical evidence of a high probability of revealing flaws and operational deficiencies under such conditions. NASA GSFC, on the other hand, prefers qualification by cycling at somewhat lower temperatures and for shorter times, but reserves the option to also cycle flight units at least three times (and more often eight) at 20°C above and below flight allowables.

The dilemma facing the TOPEX/Poseidon program was that it was a JPL project using components derived mostly from GSFC's Multimission Modular Spacecraft (MMS). MMS heritage was well established and, in certain respects, differed considerably from the JPL tradition. A resolution would have been to requalify MMS under JPL's requirements, but this was deemed costly and complicated in that no alternatives could be devised within program constraints in the event of a failure of a component already qualified by Goddard's procedures.

Final decisions were made based on the argument that because the satellite was to operate in an Earth orbit, which is the operational domain of NASA's MMS, MMS requirements should form the basis for qualification. But, concurrently, JPL's concerns were to be alleviated by conducting analyses (not necessarily backed by test confirmation) to verify that the components would, in fact, meet JPL's standards. It was also agreed that all new, nonheritage components be qualified by GSFC methods but at their design temperature levels.

The resulting specifications for TOPEX/Poseidon are given in Table 1.2.

III.   Thermal Energy

A specified temperature field is produced by appropriate thermal energy management. This means that energy in the form of heat that is generated, received, and rejected by a satellite is directed (by shielding, routing, reflecting, etc.) in such a manner that its balance (the first law of thermodynamics) occurs at the desired temperature.

Satellite heating is classified under two categories. The first is environment heating, which refers to energy interacting directly or indirectly (through intervening structure, reflector, etc.) with the satellite's external surface. Sources of this energy include sunlight, radiation from other heavenly bodies, and effects from surrounding media such as air (in continuum or free molecular form) and charged particles found in radiation belts around planets. Management of environment heating is accomplished by orienting or conditioning and treating the surface.

The second category is dissipation, which is heat generated by operating electronics, heaters, shunts, and other, mostly I^2R, losses. Unlike environment heating, which is expansive and acts on the outside, dissipation is usually released locally within enclosures. Its management is accomplished by promoting the paths that distribute then reject it to the surroundings.

Table 1.2   TOPEX/Poseidon Temperature Specifications

1) Dissipating components with MMS heritage:

Maximum design temperature[a]: an acceptable performance at 85°C to be verified by analysis.

Maximum nonoperating temperature: 75°C (except for batteries[b]) soak for 100 h, verified by test.[c]

Maximum qualification[d] temperature: 20°C (except for batteries) above allowable flight, verified by test.

Maximum acceptance[e] temperature: 10°C (except for batteries) above allowable flight, verified by test.

Maximum allowable flight temperature: 40°C (batteries, 25°C; tape recorders, 35°C), verified in balance test.

Maximum predicted flight temperature, from mathematical models that include combined hot case biases.

Analysis and thermal balance test shall confirm compliance
with allowable flight temperatures

Minimum predicted flight temperature, from mathematical models that include combined cold case biases.

Minimum allowable flight temperature: 0°C to be confirmed in thermal balance test.

Minimum acceptance temperature: 10°C below allowable flight to be verified by test.

Minimum qualification temperature: 20°C (except for batteries) below allowable flight to be verified by test.

Minimum nonoperating temperature: −20°C (except for batteries), startup to be confirmed after warm-up.

Minimum design temperature, acceptable reduced performance at −30°C, verified by analysis.

2) Newly designed dissipating components[f]

Maximum design temperature, an acceptable performance at 85°C, verified by test on qualification units.

Maximum qualification temperature: 75°C for at least 250 h.

Maximum allowable flight temperature: 50°C.

Analysis and thermal balance test shall confirm compliance
with allowable flight temperatures

Minimum allowable flight temperature: 0°C.

Minimum qualification temperature: −25°C for at least 250 h.

Minimum design temperature, an acceptable performance at −30°C, verified by test on qualification units.

---

[a]Temperatures are average values at the mounting surface.

[b]Batteries are qualified by similarity to previous and concurrent MMS applications.

[c]Test verification refers to confirming operational performance in vacuum.

[d]Component qualification given at 24 cycles with 12-h dwells at high and low temperatures. Levels for nondissipating structures are ±40°C beyond flight allowables. Wet hydrazine components must remain at 10°C or higher in all tests. The solar array is considered a dissipating component.

[e]Component acceptance given at 8 cycles with 6-h dwells at high and low temperatures. Levels for nondissipating structures are ±30°C beyond flight allowables. This includes deployment mechanisms, with exceptions to be negotiated.

[f]Two major components were considered newly designed on the TOPEX/Poseidon program: the instrument module interface unit (IMIU) with flight allowables of −5 to 55°C, and the propulsion module electronics (PME) with flight allowables of 0 to 45°C.

Ultimately (the second law of thermodynamics) all heating experienced by a satellite is transferred to the surface (radiators), where it is rejected to the environment.

IV. Relation Between Heating and Temperature: Method of Thermal Control

The connection between heating and temperature comes from the knowledge that heat transfer between bodies is the result of temperature differences and that surfaces above absolute zero (0 K) radiate heat at rates proportional to their temperature raised to the fourth power. Symbolically,

$$Q_{a-b} = K_{a-b}(T_a - T_b)$$

and

$$Q^e = FT^4$$

Q_{a-b} (usually watts) is heat transferred from body a to body b, and Q^e (also watts) is heat radiation emitted by a surface at temperature T (in Kelvin, K). The proportionality factors K_{a-b} (W/K) and F (W/K^4) depend on physical constants, material properties, surface conditions, geometry, and possibly temperature.

Basic thermal control can now be illustrated, as shown in Fig. 1.1, by considering the main body of a satellite as an idealized thermal system with one dissipating interior region (say, location 1 representing the electronics) exchanging heat with

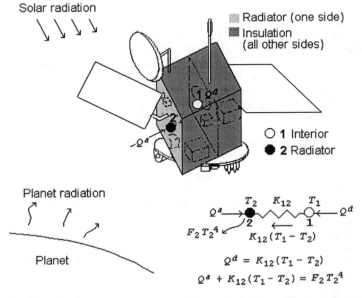

Fig. 1.1   Method of satellite thermal control: in analysis, T_1 and T_2 are calculated for given Q^a, Q^d, K_{12}, and F_2; in design, Q^a, K_{12}, and F_2 are manipulated to achieve the desired T_1 and T_2 for a given Q^d.

a boundary (radiators represented by point 2) exposed to the environment. Energy conservation (balance) in steady state requires that heat dissipated in the interior equals heat transferred to the radiator, and that, at the boundary, heat from internal dissipation plus heat absorbed from outside equals heat rejected to space. Denoting interior dissipation by Q^d and the absorbed portion of environment heating by Q^a, the energy balance is written

$$Q^d = K_{12}(T_1 - T_2)$$

and

$$Q^d + Q^a = F_2 T_2^4$$

where K_{12} and F_2 are heat exchange factors. Hence, if heat dissipated and heat absorbed from the environment are known and if the functional form of the proportionality factors can be determined, then there will be two equations that, theoretically, can be solved for the two unknowns: the temperature of the satellite interior and that of the radiator.

What is also revealing here is that the temperature of the satellite can be made to vary by changing the proportionality factors. This, in essence, is the method of thermal control. For if these factors are indeed functions of geometry and surface and physical properties, then the desired temperature, particularly that of the interior, can be obtained by the judicious selection of materials and configurations.

V.   Thermal Interface Requirements

On its surface, the principle of thermal control appears simple. But, aside from physical limitations on the availability of devices and materials with favorable thermal characteristics, controlling temperatures is only one facet in the building of a satellite and, as such, it cannot have a consuming impact on overall design, cost, or schedule. As it happens, much of thermal engineering involves negotiating acceptable interactions with other satellite subsystems. These are known as thermal interface requirements and they include: temperature and heat transfer requirements specifying the bounds of thermal influence among the various subsystems, including payload, power, attitude control, telemetry and command, structure, propulsion, and the launch vehicle; limitations on available electrical power for thermal control; a weight budget that could restrict thermal enhancements through added material; requirements to use only space-qualified hardware and materials; and a satellite envelope confining radiators size and location.

In many cases the negotiations fall short of producing within budget and schedule an ideal thermal interface for the entire satellite. But experience has shown that acceptable temperatures for all components will invariably result if thermal involvement is initiated at concepts definitions and continued throughout the program.

Example: Thermal Interface of Satellite Electronics

One of the more crucial thermal interfaces in the design of satellites is the region where a component (a box) containing dissipating electronics makes contact with a mounting platform. Thermal control of these platforms is directed toward providing

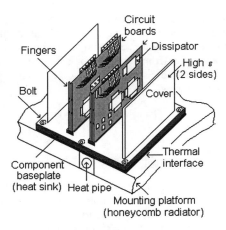

Fig. 1.2   Electronics box mounted to a satellite platform.

an efficient heat dump for the dissipators to keep them from overheating during operation.

Generally, the electronic parts within a component are packaged rather densely because of real estate limitations and the need for short-signal paths. Figure 1.2 shows an arrangement with parallel circuit boards placed edgewise in contact with a relatively thick plate, called the heat sink or component baseplate. Most of the heating in the integrated circuits is made to flow to the component baseplate, which must be placed in intimate contact to the mounting platform with the desired temperature. Despite a concerted effort to facilitate heat flow, however, packaging requirements and the nature of materials used in circuit boards make heat passage from component parts to the baseplate generally quite tortuous.

The traditional measure of the resistance to heat flow from a junction j to the component baseplate bp is the theta factor θ_{j-bp} (Kelvin per watt). It is a combination of resistances from where the heat is generated, to the board, to conductive rails, to board mount, across adhesives and bonds, and, finally, to the baseplate. The value is usually specified by the manufacturer on the component data sheet and, depending on package size and design, can be in the range of 10–100°C/W (Ref. 5).

By definition, heat transfer from junction to baseplate Q_{j-bp} is related to the difference between the junction temperature T_j and the baseplate temperature T_{bp} by the equation

$$T_j - T_{bp} = \theta_{j-bp} Q_{j-bp}$$

and because θ_{j-bp} is a characteristic of material and system properties (and possibly temperature), the thrust of the electronics thermal design is to make selections and assemblies with reduced θ_{j-bp} so that, for a given Q_{j-bp}, the junction temperature will be more effectively moderated by that of the baseplate. A good practice in this direction is to place high dissipating elements nearer to the baseplate and where heat can flow with relative ease. Also, a thicker, heat conductive baseplate will tend to smear out the heating and dampen potential hot spots. The limitations here are mainly weight and, to a lesser degree, transverse temperature variations. The goal in components with silicone transistors and diodes is to keep junction temperatures below 110 or 120°C.

But, in addition to a reduced θ_{j-bp}, the component baseplate must reject the dissipated heat at the appropriate temperature. This is done by coupling well to the satellite's mounting platform, which is designed to remain at moderate temperatures. A number of boxes may have to be mounted on the same platform, and thermal control must devise the means by which their dissipation (most of which ending up on the platform) is distributed and transferred efficiently at the specified interface temperature.

If Q^d (W) is the total heat generated within a particular box, A (m^2) the contact area, and K (W/m^{2}K) a measure of the heat flux per unit temperature difference (called conductance) across the baseplate/platform interface, then a steady-state heat balance on the baseplate (of a box) gives

$$Q^d = \sum_j Q_{j-bp} = K A(T_{bp} - T_{platform})$$

or

$$T_{bp} = T_{platform} + Q^d / K A$$

Substituting the junction temperature from the earlier formula,

$$T_j = T_{platform} + Q^d / K A + \theta_{j-bp} Q_{j-bp}$$

These last two equations show that the higher the value of conductance K, the nearer both T_{bp} and T_j are to the platform temperature.

On bulk basis, the temperature of the platform itself depends on the dissipation of the components and on its own interaction with spacecraft radiators that eventually reject the heat to space. The value of K enters the picture only insofar as the temperature of the electronics differs from that of the mounting platform. Thus, a sloppy mount (low K) will result in high electronics temperatures even as the platform remains cool. On the other hand, one cannot assume that the platform will be an infinitely tolerant heat sink. Most spacecraft platforms are honeycomb panels with limited capability to transfer heat in the lateral direction, and localized hot spots and reduced performance can readily develop as a result of high-density heating.

The rules for guaranteeing a satisfactory interface between the electronics and the platform are given in interface control documents (ICDs) consisting of drawings and requirements. A high value of K is generally one of the requirements listed in these documents as well as in the specifications of the thermal control system. A sample statement may read in part: Provisions shall be made to ensure that heat conductance across the interface between a dissipating component and the satellite mounting platform shall be greater than 300 W/m^{2}K.

Values of K in the order of hundreds might not be possible to achieve by just cleaning and smoothing the surfaces in contact, nor even by increasing the contact pressure with tighter mounting bolts or screws.[6-8] But large values are almost always obtained when a filler is placed between the baseplate and the platform. A series of silicone room-temperature-vulcanized compounds (RTVs) are sometimes used for this purpose. These can be cured in place at room temperature, and they generally fill out most interface gaps formed by surface roughness or deformation caused by bolting down the components. Electrical grounding with RTVs has

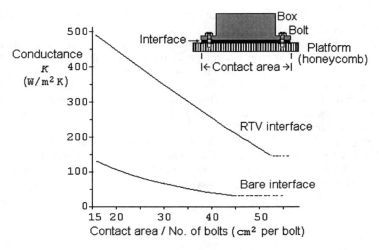

Fig. 1.3   Conductance K across a mounting interface with and without a filler.[8]

been mostly mastered by enriching the mold with electrically conductive particles and providing localized direct contacts. Component removal after curing remains a chore, however, often requiring the dismantling of harnesses and neighboring units for access to pry the component free.

A comparison between a bare and RTV-filled interface conductance is shown in Fig. 1.3. Other filler materials include CHO-THERM (a composite used as a gasket) and thin films of silicone thermal greases (which, however, have lost favor due to their outgassing and uncontrollable migration).

For heritage electronics, the vendor supplies the physical description of the component, including the pattern of mounting holes and a detailed footprint of the power profile. NASA requires that vendors also supply a mathematical thermal model for interface analysis. Thermal control then determines the integration requirements based on analyses that connect the electronics model to the satellite thermal model. The results may lead to recommendations for placing the electronics at specific locations on the platform, where, for instance, there are heat pipes or heaters. Sometimes adaptors or extensions to the baseplate with new hole patterns may be needed. Also, it may sometimes be necessary to enhance lateral heat transfer by incorporating a conductive doubler. It is rare, however, that vendors are directed to make modifications on the heritage package itself.

Design of nonheritage components must take into account the limitations of the satellite mounting platform. Of importance are size and power distribution. Thermal control system specifications often include a requirement similar in tone to the following: Dissipators mounted on a satellite platform shall have power density no greater than 0.05 W/cm^2 of contact area without heat pipes and no greater than 0.30 W/cm^2 of contact area with heat pipes. Of course, the actual values are the result of interface analysis and negotiations. In fact, it is not uncommon that a spacecraft platform is modified or designed specifically to accommodate a new component or experiment.

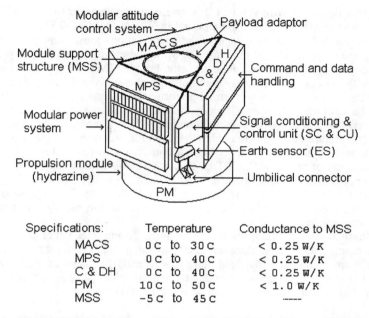

Specifications:	Temperature	Conductance to MSS
MACS	0 c to 30 c	< 0.25 w/K
MPS	0 c to 40 c	< 0.25 w/K
C & DH	0 c to 40 c	< 0.25 w/K
PM	10 c to 50 c	< 1.0 w/K
MSS	−5 c to 45 c	-----

Fig. 1.4   NASA's MMS interface specifications; C&DH, MACS, and MPS measure 1.194 × 1.194 × 0.457 m.

Example: Structure Interface Specifications

Similar considerations apply to nondissipators and coupled structures. For example, a satellite may consist of a group of subsystems (power, communications, etc.) supported by a basic structure (usually called a bus) with individual units built and tested by different contractors, and, because contractual obligations and satellite integration are simplified when the thermal interaction among the separate systems is small, the specifications often include a requirement to limit the heat exchange at the coupled interfaces with instructions to design and test individual subsystems with adequate margins.

NASA's MMS is typical. As illustrated in Fig. 1.4, MMS is an assembly of individual modules that provide the necessary services to a general payload that mounts on an adaptor on the carrier structure. For the most part, the thermal performance of MMS has been consistent with the listed data.

Limiting heat flow between coupled structures that can have different temperatures requires an interface of low-heat conductance. Conventional attachments made from materials with great resistance to heat flow and having small contact areas do not generally meet the requirements to sustain the mechanical loads at launch. The thermal–structural compromise for attaching the MMS modules consisted of an upper and a lower preloaded titanium bolt, a two-axis corner restraint, and a three-axis corner restraint. The average thickness-to-length ratio of the bolts was kept as small as structure requirements would permit, and this combined with titanium's inherent low-heat conduction to yield conductances somewhat below the upper limits given in Fig. 1.4. For interested readers, details on design and installation of MMS modules are reported in NASA's MMS S-700 series.[9,10]

As noted before, the language in the specifications must convey the desired conditions on both sides of the interface while preserving as much as is practical the independent effort of each subcontractor. For the MMS, the approach has been to specify hot case and cold case assumptions to be included in analyses and tests to contain all eventualities of thermal performance on either side of the interface.

In the cold case, the thermal analysis of a subsystem attached to a structure considers a heat loss based on the arithmetic product of the interface conductance by a temperature difference related to the minimum allowable temperature on the structure side (generally $-5°C$ for MMS). Thus, from the data in Fig. 1.4, the cold case analysis of the MACS, C&DH, or the MPS must include a heat loss from the module at the support structure of $(0.25)[T_{cold} - (-5)]$ W, where T_{cold} is the module's cold case temperature being calculated. Similarly, for hot case analysis either an adiabatic interface or one that supplies maximum heating into a module must be assumed.

In thermal tests of individual MMS modules, an interface panel (a cold plate) with a controllable temperature serves as simulated attached structure. Generally, the attachment points at the cold plate have exactly the same characteristics as flight hardware, and the test is conducted with the panel temperature maintained constant at the corresponding hot or cold test requirements.

References

[1] *Journal of Geophysical Research* (Oceans), Vol. 99, No. C12, 1994.

[2] *Journal of Geophysical Research* (Oceans), Vol. 100, No. C12, 1995.

[3] Cleveland, P. E., and Braun, C. E., "Thermal Design of the TOPEX/Poseidon Instrument Module," Proceedings of the Twentieth Intersociety Conference on Environmental Systems, 1990, pp. 29–37.

[4] "Military Standard, Test Requirements for Space Vehicles," U.S. Air Force, MIL-STD-1540B, Oct. 1982.

[5] Fink, D. G., and Christiansen, D. (eds.), *Electronics Engineers Handbook*, 2nd ed., McGraw–Hill, New York, 1982, pp. 15-45–15-56 and 28-21–28-24.

[6] Gluck, D. F., "Mounting and Interfaces," *Satellite Thermal Control Handbook*, edited by D. G. Gilmore, The Aerospace Corp. Press, El Segundo, CA, 1994, pp. 4-17–4-72.

[7] Peterson, G. P., and Fletcher, L. S., "Heat Transfer Enhancement Techniques for Space Station Cold Plates," *Journal of Thermophysics and Heat Transfer*, Vol. 5, No. 3, 1991, pp. 423–428.

[8] Bevans, J. T., Ishimoto, T., Loya, B. R., and Luedke, E. E., "Prediction of Space Vehicle Thermal Characteristics," Air Force Flight Dynamics Lab., TR AFFDL-TR-65-139, Aug. 1965.

[9] "Multimission Modular Spacecraft (MMS) Thermal System Specification," NASA S-700-12, 1981.

[10] "Multimission Modular Spacecraft (MMS) Mechanical System Specification," NASA S-700-13, 1979.

Chapter 2

Heat Transfer

I. Introduction

A N understanding of satellite thermal control is not possible without a satisfactory knowledge of the principles of heat transfer. It is in using these principles that the thermal engineer determines the means of regulating the heating of a satellite to produce the desired temperatures.

There are three modes of heat transfer: radiation, conduction, and convection. Conduction and radiation are always present and account for heat exchange among satellite components in a vacuum with final rejection by radiation to space. Convection occurs mainly on the ground, during ascent, and in heat transfer from fluids in sealed containers. Situations involving energy transfer by satellite outgassing and venting into space, or interaction with atomic and electrically charged particles, are specialized and require separate analysis.

II. Radiation

Radiation is heat exchange by electromagnetic energy (wavelength range from about 0.1 to 100 μm) between a surface and its surroundings. It is the most complex of heat transfer modes, and its mathematical treatment must invoke many simplifying assumptions to make it tractable. The next chapter in this book, Chapter 3, is an extended discussion with emphasis on the connection to thermal control.

Briefly, all surfaces at temperatures greater than absolute zero emit radiation at rates proportional to their temperature to the fourth power. A black surface (referred to in physics as a blackbody) is by definition the most efficient emitter. Its radiation emission, q_b^e (W/m^2), at temperature T (K) is given by the Stefan–Boltzmann law

$$q_b^e = \sigma T^4$$

or

$$Q_b^e \text{ (W)} = A^r \sigma T^4$$

where $\sigma (= 5.67E-8$ W/m^2 K^4) is the Stefan–Boltzmann constant and A^r(m^2) is the radiating area.

Radiation from a nonblack surface of the same area and temperature is

$$Q_{nb}^e = \varepsilon A^r \sigma T^4$$

where $\varepsilon (< 1)$, known as total hemispherical (or average) emissivity, is the ratio of energy emission by the nonblack surface to emission if it were black and at the same temperature. Reference to this property in thermal control is so frequent that the term has been abbreviated simply to emissivity. By way of contrast, there is also the normal emissivity (actually, total normal emissivity) ε_n, which is emission of a surface in the normal (perpendicular) direction divided by the value if the surface were black. Direct measurements are usually made with infrared reflectometers designed to determine normal emissivities, which are then converted to hemispherical values by using special equations and charts (Chapter 3, Sec. IV.C).

Radiation exchange among black surfaces depends only on their temperature and how they view one another. The geometric dependence is in terms of the shape factor (also known as view factor or configuration factor), which is defined as the fraction of radiation leaving a surface that is intercepted by another. It is shown in Chapter 3 that net radiation from a black surface element dA_1 at temperature T_{dA_1} to another interacting black surface element dA_2 at temperature T_{dA_2} is

$$d^2 Q^r_{dA_1-dA_2} \text{ (black surfaces)} = \left(\sigma T^4_{dA_1} - \sigma T^4_{dA_2}\right) \cos \theta_1 \cos \theta_2 \, dA_1 \, dA_2 / \pi s^2$$

which gives the net radiation from A_1 to A_2 as

$$Q^r_{A_1-A_2} \text{ (black surfaces)} = \int_{A_1} \int_{A_2} \left(\sigma T^4_{dA_1} - \sigma T^4_{dA_2}\right) \cos \theta_1 \cos \theta_2 \frac{dA_1 \, dA_2}{\pi s^2}$$

where θ_1 and θ_2 are the angles that the normals to surface elements dA_1 and dA_2 make with the line (length s) connecting them and the superscript r indicates net exchange, as opposed to emission alone which is designated by the superscript e.

When each one of the two interacting surfaces is isothermal, the equation can be written in the familiar form

$$Q^r_{12} = F_{12} A_1 \left(\sigma T^4_1 - \sigma T^4_2\right)$$

where Q^r_{12} is net radiation from surface 1 and F_{12} the shape factor calculated from

$$F_{12} = (1/A_1) \int_{A_1} \int_{A_2} \left(\frac{\cos \theta_1 \cos \theta_2}{\pi s^2}\right) dA_1 \, dA_2$$

If N surfaces are involved, the net radiation from surface i to the others is

$$Q^r_i = \sum_j F_{ij} A_i \left(\sigma T^4_i - \sigma T^4_j\right)$$

with the summation taken over N.

The rule of manipulating the indices must be consistent with the conservation of energy, which requires that (in equilibrium) the net heat into surface 1 is the net heat leaving surface 2. Thus,

$$F_{ij} A_i = F_{ji} A_j$$

Also, from physical considerations,

$$\sum_j F_{ij} = 1.0$$

in an enclosure.

Radiation exchange among real surfaces also depends on temperature and geometric aspects, but, in addition, it can be a function of the spectrum (wavelength) and direction, as well as surface material, smoothness, and curvature. An analysis that will account for all effects would be extremely complicated and may not even be useful due to a lack of experimental data. Fortunately, most satellite surfaces, especially those requiring radiation exchange analysis, have very nearly diffuse gray characteristics, which allows an approximation of the heat exchange by introducing a determinable script F ($\Im$ with referenced indices) emulating the shape factor in black surface radiation. Thus, the exchange between two isothermal surfaces is written

$$q_{12}^r = Q_{12}^r/A_1 = \Im_{12}(\sigma T_1^4 - \sigma T_2^4)$$

and net radiation from surface i to N surfaces is

$$q_i^r = \frac{Q_i^r}{A_i} = \sum_j \Im_{ij}(\sigma T_i^4 - \sigma T_j^4)$$

with j summed over N.

Within this framework, script F portends all underlying mechanisms of exchange including geometry, extent of departure from black surface behavior, temperature, and any other effect bearing on the problem of radiation.

In a number of practical cases, script F between two surfaces is found to be a relatively simple product of the shape factor (here designated by F_A) and a so-called emissivity factor F_E that contains the emissivities and geometric ratios. Net radiation from isothermal surface i to isothermal surface j is then written

$$q_{ij}^r = Q_{ij}^r/A_i = (F_E F_A)_{ij}(\sigma T_i^4 - \sigma T_j^4)$$

and, for exchange between two surfaces each with its own temperature distribution,

$$Q_{A_1 - A_2}^r \text{ (diffuse gray)} = \int_{A_1} \int_{A_2} F_{E(dA_1 - dA_2)}(\sigma T_{dA_1}^4 - \sigma T_{dA_2}^4)$$

$$\times \cos\theta_1 \cos\theta_2 \frac{dA_1 dA_2}{\pi s^2}$$

In a collection of finite isothermal surfaces, the net radiation from surface i to the rest (j) is

$$Q_i^r = \sum_j (F_E F_A)_{ij} A_i (\sigma T_i^4 - \sigma T_j^4) \equiv \sum_j G_{ij}(T_i^4 - T_j^4)$$

where the quantities $G_{ij} \equiv \sigma(F_E F_A)_{ij} A_i (\equiv \sigma \Im_{ij} A_i$ when script F is used) have units of W/K^4 and are known as radiation factors or G factors. They are a distinctive part of the input data fed into thermal analyzers for solving heat transfer problems.

Some configurations with simple expressions for F_E are listed in Table 2.1. The formulas for interacting finite planes are especially useful when the emissivities are high because then the lower and upper bounds approach one another resulting in

$$(F_E)_{12} \approx \varepsilon_1 \varepsilon_2$$

A radiating surface in equilibrium with its surroundings must be absorbing energy at the same rate at which it is emitting it. This gives rise to the concept of absorptivity $\alpha(< 1)$, which is defined as the fraction of incident radiation (also

Table 2.1   Shape factors and emissivity factors

From small surface 1 completely enclosed by a large surface: $F_A = 1$; $F_E = \varepsilon_1$
From surface 1 enclosed by comparable surface 2: $F_A = 1$; $F_E = [1/\varepsilon_1 + 1/\varepsilon_2 - 1]^{-1}$
Case between the preceding two: $F_A = 1$; $\varepsilon_1 > F_E > [1/\varepsilon_1 + 1/\varepsilon_2 - 1]^{-1}$
From infinite plane 1 to infinite parallel plane 2: $F_A = 1$; $F_E = [1/\varepsilon_1 + 1/\varepsilon_2 - 1]^{-1}$
Parallel surfaces with ratio of length to distance apart $\gg 1$ can be approximated by infinite
  planes
From surface element dA_1 to A_2: F_A from formulas; $F_E = \varepsilon_1\varepsilon_2$
From inner to outer concentric spheres, radii $r_1 < r_2$: $F_A = 1$;
  $F_E = [1/\varepsilon_1 + (r_1/r_2)^2(1/\varepsilon_2 - 1)]^{-1}$
From inner to outer concentric long cylinders, $r_1 < r_2$: $F_A = 1$;
  $F_E = [1/\varepsilon_1 + (r_1/r_2)(1/\varepsilon_2 - 1)]^{-1}$
Finite planes 1 and 2: F_A from formulas; F_E between $\varepsilon_1\varepsilon_2$ and $[1/\varepsilon_1 + 1/\varepsilon_2 - 1]^{-1}$

called irradiation and here given the symbols q^i and Q^i), and which is absorbed by the surface. Hence, the equation defining absorptivity is

$$q^a = Q^a/A^i = \alpha Q^i/A^i = \alpha q^i$$

where q^a (or Q^a) is absorbed energy and A^i the normal area exposed to incident energy.

Kirchhoff's law asserts that when emission and irradiation are of the same spectrum (or integrated spectra) then the absorptivity of a surface must equal its emissivity. But the two properties need not be equal when viewed separately in terms of incoming and emitted energies of different spectra. Hence, a surface may be conditioned (shined, anodized, coated, painted, etc.) to make it react differently to each spectrum. This practice is used to optimize the performance of satellite radiators exposed to the sun. Because the spectrum of solar radiation is concentrated in the range of about 0.2–3.0 μm whereas over 90% of emission at usual satellite temperatures spans the much wider range of 0.5–35 μm, a surface may be treated to make its solar absorptivity α^s (defined as the absorbed fraction of impinging solar radiation) relatively low, while its emissivity is simultaneously high. The sun's heating impact is thus reduced, and satellite dissipation is rejected at a relatively cooler temperature.

The relation between heating and temperature of a general surface in equilibrium in a solar–space environment is obtained from the conservation of energy in the form

$$\int_{A^s} \alpha^s q^S dA^s = \int_{A^r} \varepsilon\sigma T^4 dA^r$$

where A^s and A^r are the areas associated with incident and radiated energies and q^S the normal component of incident solar energy. The equation provides a basis for defining the average temperature in the sun as

$$\sigma T_{av}^4 \equiv \int_{A^s} \alpha^s q^S dA^s \bigg/ \int_{A^r} \varepsilon \, dA^r$$

which applies to nonisothermal as well as isothermal surfaces. For example, if side one (α^s, ε_1) of a flat plate is exposed to uniform solar radiation while both side one and side two (ε_2) are radiating to space, then the average temperature of the plate is defined by

$$T_{\text{av (flat plate)}} \equiv \left[\frac{\alpha^s q^S}{\sigma(\varepsilon_1 + \varepsilon_2)}\right]^{\frac{1}{4}}$$

For a cylinder with uniform exterior properties

$$T_{\text{av (cylinder)}} \equiv \left[\frac{\alpha^s q^S}{\pi \sigma \varepsilon}\right]^{\frac{1}{4}}$$

and for a sphere

$$T_{\text{av (sphere)}} \equiv \left[\frac{\alpha^s q^S}{4\sigma \varepsilon}\right]^{\frac{1}{4}}$$

Associated with absorptivity are reflectivity ρ and transmissivity τ, referring to the fractions of incoming energy that are reflected by and transmitted through a surface. From energy conservation, $\alpha + \rho + \tau = 1.0$.

Reflectivity must be known to determine the irradiation from reflected energy. It is also an easier surface property to measure than absorptivity, and knowing it for an opaque surface ($\tau = 0$) gives an immediate value of α. Transmissivity, on the other hand, rarely enters the analysis because radiation in satellites is almost exclusively among opaque surfaces, so much so, in fact, that there is seldom an occasion for an explicit reminder. Along these lines, it is generally assumed in thermal control that surface radiation activity takes place at an infinitessimally thin layer of the surface (hence the term black surface instead of the more common blackbody).

Application: Temperature Distribution in a Cylindrical Shell in the Sun

Consider a cylindrical shell heated by the sun as shown in Fig. 2.1. For long cylinders (length much greater than radius) edge effects and longitudinal temperature variations may be ignored, and, by assuming the cylinder to be sufficiently thin, circumferential heat conduction and temperature change through the thickness also become negligible. Under these assumptions, the steady-state temperature at a location on the shell's circumference is determined from an energy balance that equates, at that point, the heat absorbed from the sun to radiation both from outside to space and from inside to the surrounding enclosure.

Referring to Fig. 2.2a, an element situated at circumferential position ϕ on the shell's surface has an area

$$dA = R \, d\phi \, dy$$

where R is radius and y distance along length L. As noted in Fig. 2.2b, the normals of two elements situated at ϕ and ϕ' make equal angles with their connecting line s. Hence, from Fig. 2.2c,

$$\theta_1 = \theta_2 = \theta = \left(\tfrac{1}{2}\right)[\pi - (\phi' - \phi)]$$

$$\cos \theta = \sin[(\phi' - \phi)/2]$$

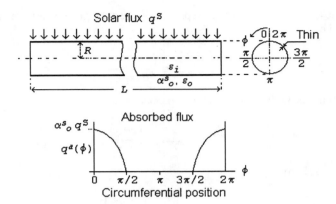

Fourier expansion:

$$q^a(\phi) = \frac{\alpha^s_o q^S}{\pi}\left[1 + \frac{\pi}{2}\cos\phi - 2\sum_{n=1}^{\infty}\frac{(-1)^n\cos 2n\phi}{4n^2-1}\right]$$

Fig. 2.1   Cylindrical shell in the sun.

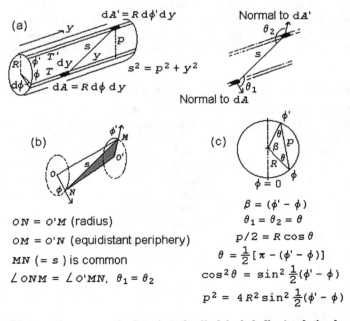

$ON = O'M$ (radius)

$OM = O'N$ (equidistant periphery)

$MN\ (= s)$ is common

$\angle ONM = \angle O'MN,\ \theta_1 = \theta_2$

$\beta = (\phi' - \phi)$

$\theta_1 = \theta_2 = \theta$

$p/2 = R\cos\theta$

$\theta = \frac{1}{2}[\pi - (\phi' - \phi)]$

$\cos^2\theta = \sin^2\frac{1}{2}(\phi' - \phi)$

$p^2 = 4R^2\sin^2\frac{1}{2}(\phi' - \phi)$

Fig. 2.2   Elemental areas on the interior of cylindrical shell: a) relative location of two elements, b) normals make equal angles with connecting line, and c) p and θ in terms of radius and relative angular position.

and

$$s^2 = 4R^2 \sin^2[(\phi' - \phi)/2] + y^2$$

Substituting into

$$F_{dA-dA'} = \cos\theta_1 \cos\theta_2 \, dA \, dA'/\pi s^2$$

$$(F_A)_{dA-dA'} = \frac{\sin^2 \frac{1}{2}(\phi' - \phi) \, dA \, R \, d\phi' \, dy}{4\pi R^2 \sin^2 \frac{1}{2}(\phi' - \phi) + \pi y^2}$$

Therefore, with the elements at temperatures T and T',

$$d^2 q^r_{dA-dA'} \equiv \frac{d^2 Q^r}{dA} = (F_E)_{dA-dA'} \frac{\sin^2 \frac{1}{2}(\phi' - \phi) \, dA \, R \, d\phi' dy}{4\pi R^2 \sin^2 \frac{1}{2}(\phi' - \phi) + \pi y^2}(\sigma T^4 - \sigma T'^4)$$

and, because there is no temperature variation along y, the total radiation per unit area from ϕ to the rest of the curved interior of the cylinder is obtained by the integration

$$q^r_{\phi-\text{inside}} = \int_0^L \int_\phi^{\phi+2\pi} (F_E)_{dA-dA'}(\sigma T^4 - \sigma T'^4)$$

$$\times \frac{\sin^2 \frac{1}{2}(\phi' - \phi)}{4\pi R^2 \sin^2 \frac{1}{2}(\phi' - \phi) + \pi y^2} R \, d\phi' \, dy$$

For large L (as when $L > 1$ m and $R < 1$ cm) the complexity of integration due to the presence of F_E can be resolved by noting that the final integrated result must contain the term ε_i (emissivity of the interior) so that the effective emissivity from a point to its surrounding enclosure is the emissivity of the point itself (Table 2.1). Moreover, using the formula

$$\int_0^L \frac{a}{b + cy^2} \, dy = \frac{a}{(bc)^{1/2}} \tan^{-1}\left[\left(\frac{c}{b}\right)^{1/2} y\right]\Bigg|_{y=0,}^L \qquad b, c > 0$$

it is found that in the limit

$$\lim_{L\to\infty} \int_0^L \int_\phi^{\phi+2\pi} \frac{\sin^2 \frac{1}{2}(\phi' - \phi)}{4\pi R^2 \sin^2 \frac{1}{2}(\phi' - \phi) + \pi y^2} R \, d\phi' dy = 1.0$$

which affirms the physical fact that the shape factor from an interior point to its whole surrounding is 1.0. Hence, in the limit,

$$q^r_{\phi-\text{inside}} = \varepsilon_i \sigma T^4(\phi) - \varepsilon_i \int_0^\infty \int_\phi^{\phi+2\pi} \frac{\sigma T^4(\phi') \sin^2 \frac{1}{2}(\phi' - \phi)}{4\pi R^2 \sin^2 \frac{1}{2}(\phi' - \phi) + \pi y^2} R \, d\phi' dy$$

Integrating first with respect to y,

$$q^r_{\phi-\text{inside}} = \varepsilon_i \sigma T^4(\phi) - \varepsilon_i \int_{\phi}^{\phi+2\pi} \tan^{-1} \frac{y}{[4\pi R^2 \sin^2 \frac{1}{2}(\phi' - \phi)]^{1/2}} \bigg|_{y=0}^{\infty}$$

$$\times \frac{R\sigma T^4(\phi') \sin^2 \frac{1}{2}(\phi' - \phi)}{[4\pi R^2 \sin^2 \frac{1}{2}(\phi' - \phi)]^{1/2}} \, d\phi'$$

which gives

$$q^r_{\phi-\text{inside}} = \varepsilon_i \sigma T^4(\phi) - \frac{\varepsilon_i}{4} \int_{\phi}^{\phi+2\pi} \sigma T^4(\phi') \sin \frac{1}{2}(\phi' - \phi) \, d\phi'$$

Now, every exterior point (emissivity ε_o) sees space with a shape factor of 1.0. Hence, radiation from ϕ to surrounding deep space (0 K) is

$$q^r_{\phi-\text{outside}} = q^e_{\phi-\text{outside}} = \varepsilon_o \sigma T^4(\phi)$$

Therefore, in a long, thin cylindrical shell with no discernable temperature difference across its thickness, the total net radiation (watts per square meter) from a point on the shell to the whole of its surroundings is

$$q^r_{\phi-\text{whole surroundings}} = (\varepsilon_i + \varepsilon_o) \sigma T^4(\phi) - \frac{\varepsilon_i}{4} \int_{\phi}^{\phi+2\pi} \sigma T^4(\phi') \sin \frac{1}{2}(\phi' - \phi) \, d\phi'$$

As indicated in Fig. 2.1, the sun's radiation coming from a great distance impinges as a heating flux vector (normal component near Earth is nominally $S = 1350$ W/m^2). Hence, if the solar absorptivity of the outer surface α^s_o is constant around the circumference, then absorbed solar energy $q^a(\phi)$ is represented by the rectified cosine function shown. Fourier series expansion gives

$$q^a(\phi) = \frac{\alpha^s_o q^S}{\pi} \left[1 + \frac{\pi}{2} \cos \phi - 2 \sum_{n=1}^{\infty} \frac{(-1)^n \cos 2n\phi}{4n^2 - 1} \right]$$

Since σT^4 (just as T) is also periodic (period 2π) in ϕ, it can be represented by

$$\sigma T^4(\phi) = c + \sum_n (a_n \cos n\phi)$$

where c and a_n are constant coefficients. With circumferential conduction neglected, the local solar input is equal to local heat radiated and, hence,

$$\frac{\alpha^s_o q^S}{\pi} \left[1 + \frac{\pi}{2} \cos \phi - 2 \sum_{n=1}^{\infty} \frac{(-1)^n \cos 2n\phi}{4n^2 - 1} \right] = (\varepsilon_i + \varepsilon_o) \left(c + \sum_{n=1}^{\infty} a_n \cos n\phi \right)$$

$$- \frac{\varepsilon_i}{4} \int_{\phi}^{\phi+2\pi} \left(c + \sum_{n=1}^{\infty} a_n \cos n\phi' \right) \sin \frac{1}{2}(\phi' - \phi) \, d\phi'$$

Equating the coefficients and using the relations

$$\int_{\phi}^{\phi+2\pi} \sin\frac{1}{2}(\phi' - \phi)\,d\phi' = 4$$

$$\int_{\phi}^{\phi+2\pi} \cos n\phi' \sin\frac{1}{2}(\phi' - \phi)\,d\phi' = \frac{4\cos n\phi}{1 - 4n^2}$$

it is found that

$$\sigma T^4(\phi) = \frac{\alpha_o^s q^S}{\varepsilon_o \pi}\left[1 + \frac{\pi\cos\phi}{2\left(1 + \frac{4}{3}\rho\right)} - 2\sum_{n=1}^{\infty}\frac{(-1)^n \cos 2n\phi}{[1 + \rho + \rho/(16n^2 - 1)][4n^2 - 1]}\right]$$

where $\rho = \varepsilon_i/\varepsilon_o$. The summation converges very rapidly for all angles ϕ.

The thin cylindrical shell is a reasonable model for tubular extendable elements (TEEs) carried by spacecraft as gravity gradient stabalizers, dipole antennas, instrumentation extenders, and for other uses. A discussion on applying the preceding results to finding the thermal distortion of a tube due to solar heating is given as an example in Chapter 5, Sec. IV.B.

III.   Conduction

Conduction is transmission of heat (vibratory kinetic energy) from material particle to particle with no discernable displacement of matter. The process is stated by Fourier's law

$$q^k = -k\frac{\partial T}{\partial n}$$

in which q^k is conduction heat rate per unit area, which is proportional to the gradient of the temperature in a direction n normal to the area. The proportionality factor k (W/m K) is the thermal conductivity, commonly called conductivity, and the negative sign indicates heat flow from higher to lower temperature.

Conductivity is a material property that generally depends on temperature. But it is almost constant in the range of temperatures usually encountered in satellites. When dependence is taken into account, the variation is often assumed in the form

$$k = k_0 + \gamma(T - T_0)$$

where k_0 is reference conductivity at T_0 (usually 300 K) and γ a constant. It has been noted in satellite solid materials that γ is negative except for tungsten, and it appears only slightly negative for aluminum and its alloys in the neighborhood of 300 K.

One-dimensional heat conduction Q^k from a region at temperature T_h to one at a lower T_l in a homogeneous material with constant properties and no other heating is given by

$$Q^k = (kA^k/L)(T_h - T_l)$$

where A^k is the area normal to heat flow and L the distance between the two regions. The quantity

$$K = kA^k/L$$

(in W/K) is known as conduction conductance (often called conductance) and is an important quantity in discrete numerical thermal modeling. Conductance is not a material property and must be defined by the associated area and length in the coordinate system representing the geometry of heat flow. Thus, the radial conductance across a cylindrical shell of length L and radii $r_2 > r_1$ that accounts for the variation in the area normal to heat flow is given by

$$K_{\text{cylinder}} = \frac{2\pi L k}{\ell_n(r_2/r_1)}$$

so that

$$Q^k_{\text{cylinder}} = \frac{2\pi L k}{\ell_n(r_2/r_1)}(T_{r_1} - T_{r_2})$$

with T_r the temperature at r. Similarly, for a spherical shell of radii $r_2 > r_1$,

$$K_{\text{sphere}} = \frac{4\pi k r_1 r_2}{r_2 - r_1}$$

so that

$$Q^k_{\text{sphere}} = \frac{4\pi k r_1 r_2}{r_2 - r_1}(T_{r_1} - T_{r_2})$$

The analogy to an electric network in which temperature difference represents a potential difference and heat flow an electric current gives rise to the quantity known as conduction thermal resistance R^k (K/W), which may be viewed as the inverse of conductance. For heat flow across N resistances,

$$Q^k = \frac{T_1 - T_N}{R^k_{\text{total}}}$$

where R^k_{total} is the effective resistance of the thermal network that obeys the algebra of combinations (Sec. V of this chapter) and depends on whether the conductors are working in collusion to aid the flow of heat (parallel resistances) or hinder it (series resistances).

Data on the conductivities of materials of engineering interest are found in handbooks[1] and appendices of standard books on heat transfer. The proceedings of the International Thermal Conductivity Conferences[2] are a rich source, often featuring methods of predictions and measurements. A list (k, together with density ρ and specific heat c) for some common satellite materials is given in Table 2.2.

When dealing with materials (such as composites) in which the heat conducted depends on direction, Fourier's law is generalized to read

$$q^k_i = -k_{ij} T_{,j}$$

where q^k_i is heat conduction along the ith direction, $T_{,j}$ is temperature gradient (K/m) with summation over j, and $k_{ij} (= k_{ji})$ are the components of thermal conductivity $\mathbf{k}$, which is now treated as a second-rank tensor.

Table 2.2 Conductivity k, density ρ, and specific heat c of some satellite materials

	k, W/m K(°C)	ρ, kg/m³	c, J/kg K	c, W-h/kg K
Aluminum 2024-0 (extensive use)	185 (20 ± 20) 190 (150)	2769	936	0.26
Aluminum 2024-Tx; 5052-H38 (extensive use)	120 (20 ± 20) 125 (150)	2769	936	0.26
Aluminum 6061-T6 (extensive use)	150 (20 ± 20) 155 (150)	2769	936	0.26
Stainless steel; 200-series (high temperature; strong isolator)	16 (20 ± 20) 17 (150)	7884	936	0.26
Beryllium copper; CDA-series (extendable elements)	58 (−100) 83 (0) 93 (100)	8252	792	0.22
Titanium alloys (strong, light isolator)	7 (20 ± 20)	4430	540	0.15
Copper; 80,000 series (conduction enhancer)	360 (20 ± 20)	8861	360	0.10
Fiberglass G4 (isolator material)	0.17 (20 ± 20)	4430	540	0.15
Kapton (high-temperature isolator)	0.16 (20 ± 20)	1410	1080	0.28
Air (air conditioning)	0.024 (0) 0.026 (20) 0.027 (40)	1.3 1.2 1.1	1006 1007 1008	0.26
Ammonia (saturated) (heat pipe working fluid)	0.50 (0) 0.49 (20) 0.45 (40)	639 612 580	4608 4690 4896	1.28 1.30 1.36

Example: Honeycomb Transverse Conductance–Test Verification

Honeycomb constructions (see Fig. 2.3) offer great advantages in weight and strength, but their ability to transport heat by conduction is limited by the large ratio of void to solid material. On occasion, the thickness and materials of core and face sheets are selected to enhance heat transfer, but, generally, the design of honeycomb panels follows mainly from structural and weight considerations.

The thinness and extended length of a corrugated honeycomb core makes the contribution to lateral (parallel to surface) heat transfer small in comparison to conduction by the generally thicker face sheets. Calculations involving the total lateral conductance are, therefore, usually based only on the conductivity and thickness of the face sheets.

By contrast, transverse (through the honeycomb) heat transfer is along much shorter paths. But it is an intricate process that involves mechanical and thermal interactions between face sheets and core and between structure material and bonding agents. The standard approach is to consider a honeycomb slab thermally consisting of two sections (nodes) with centers of mass located at the face sheets. The mass allocation may vary, depending on applied coatings and other substrates, but the distance between the nodes is usually taken as the core's thickness. Heat

Honeycomb construction

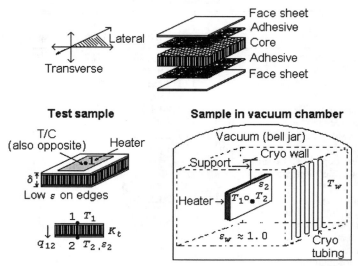

Fig. 2.3   Honeycomb mounting panels, test for determining transverse conductance.

conduction between the two nodes is then based on a core "conductivity" k_c estimated from the relation

$$k_c = k_m(\rho_c/\rho_m)$$

where ρ is density and c and m refer, respectively, to the core and the material from
which it is made. Thus, a core of density 40 kg/m^3 made from 2024 aluminum
(conductivity 185 W/m K and density 2769 kg/m^3) has a core conductivity

$$k_c = 185(40/2769) = 2.67 \, \text{W/m K}$$

For aluminum honeycomb, k_c ranges from about 1 W/m K (very light core) to
5 W/m K (heavy core).

Honeycomb unit transverse conductance K_t is defined as the rate of heat conduction per unit panel area per unit honeycomb thickness. Hence, $K_t = k_c/\delta$,
where δ is thickness, with units of W/K per m^2 of panel surface. For example,
the unit transverse conductance of a 3.5-cm-thick aluminum honeycomb panel of
40 kg/m^3 core density is 2.67/0.035, or 76.29 W/m^2 K.

Honeycomb conductance may be verified by tests performed in cryogenic vacuum compartments (usually a bell jar) in an arrangement similar to that shown in
Fig. 2.3. The sample is made of sufficiently large surface area to dampen edge
effects in the central region, where measurements are being taken. As indicated,
the instrumentation consists of a heater and thermocouples. In this configuration,
accurate monitoring of heater power is not essential because the value of K_t is
obtained only from temperature data and knowledge of the emissivities.

A heat balance in steady state on the small central region (radiation factor $\approx \varepsilon_2$)
shows that the heat rate per unit area q_{12} across the honeycomb must equal the
heat radiated from surface 2 to the walls. On neglecting radiation between the face

sheets,

$$q_{12} = K_t(T_1 - T_2) = \varepsilon_2(\sigma T_2^4 - \sigma T_w^4)$$

where T_w is the average temperature of the surrounding cryogenic compartment as measured by temperature monitors located on the walls. Hence,

$$K_t = \frac{\varepsilon_2(\sigma T_2^4 - \sigma T_w^4)}{T_1 - T_2}$$

which is calculated from measured temperatures and emissivity.

Values of K_t obtained by these methods may be preferred to analytical determinations because they contain all effects other than simple conduction, including radiation and imperfections in construction. The heat balance with radiation exchange between the face sheets is given by

$$q_{12} = K_t(T_1 - T_2) + F(\sigma T_1^4 - \sigma T_2^4) = \varepsilon_2(\sigma T_2^4 - \sigma T_w^4)$$

where F is a radiation exchange factor proportional to the product of emissivities. It is left for an exercise to show that the contribution by radiation is comparatively negligible for aluminum constructions with dissipation densities on the order normally encountered in satellites.

Example: Conductivity of Composites–Lateral Conductance

Orbiting structures undergo dimensional changes induced by temperature variations due to heating and cooling. Limiting the change is important for the precise pointing of instruments and antennas and for maintaining tolerances during on-orbit replacements or repairs. Dimensional stability is also a factor in reducing the dynamic effect imparted on a satellite's main body by time-varying distortions of appended structures (see second example in Chapter 5, Sec. V.B).

Composite structures with optimized lay-ups that give very low coefficients of thermal expansion have occasionally been used to enhance stability. The principle follows from the fact that fibers usually have a considerably lower coefficient of expansion than binders, and hence they restrain the matrix from stretching along their direction (top of Fig. 2.4). Thus, the transverse (across the fibers) coefficient of expansion of the whole composite can be made larger than that of the free matrix because, being prevented from growth in the longitudinal direction, it is forced to expand more in the transverse direction to keep the volume constant. By selecting the appropriate volume fraction of fibers and stacking up the plies in certain predetermined orientations, the net expansion of a laminate in any or all directions can theoretically be made to cancel out.

There could be, however, some serious compatibility ramifications to this application. For instance, a lay-up that can produce the lowest coefficient of expansion in a given direction might be too weak to sustain launch or deployment loads, and its benefits may have to be at least partially compromised by material and structural enhancements. Another consideration is possible long-term matrix outgassing in space and its effects on surface properties.[3,4] Finally, a hybrid design of composites and conventional materials must contend with potentially radical thermal loads in mechanically coupled structures of vastly different coefficients of thermal expansion. In the case of the sixth Application Technology Satellite,[5] an improved

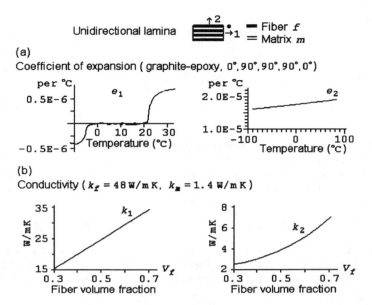

Fig. 2.4 Thermal properties of graphite-epoxy: a) typical values of e from ATS-6 GFRP test samples and b) k_1 from law of mixtures, k_2 from Halpin–Tsai equation.

bonding agent, EPON-934, was developed specifically to withstand the shearing forces at the interface between the antenna support trusses, which were made of graphite fiber reinforced plastic (GFRP), and their titanium fittings at the hub of the antenna.

Various models are found in the literature for representing a composite's thermal conductivity and other transport properties (for a recent contribution, see Ref. 6). But although there is a certain attraction to the mathematics of these models, their practical use has not been conclusive. Much of the discrepency is due to unavoidable nonuniformities in the manufactured product, which are the result of less-than-perfect bonding of fiber to matrix and incomplete (but not necessarily detrimental) mixing and curing. For unidirectional lamina the rule of mixtures for longitudinal conductivity (k_1 in Fig. 2.4) and the Halpin–Tsai equation for transverse values (k_2) are convenient to use. These are given by

$$k_1 = V_f k_f + (1 - V_f) k_m$$

$$k_2 = k_m (1 + \xi \eta V_f)/(1 - \eta V_f)$$

where V_f is fiber volume, $\eta = [(k_f/k_m) - 1]/[(k_f/k_m) + \xi]$, and $\xi = (a/b)^{1/3}$, with a representing the fiber dimension along measurement and b the dimension normal to measurement; that is, $a/b = 1$ for circular or square cross-sectional fibers. In critical structures of several lamina, however, most thermal engineers prefer testing the actual hardware.

In thermal analysis, composite platforms of few lay-ups are often considered thin in the sense that edge effects are mostly local and have minor influence on

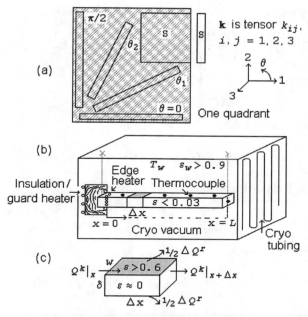

Fig. 2.5   Test determination of a composite lateral conductivity: a) pattern of test specimens (one quadrant), S for k_{33}, others for k_θ; b) arrangement in test chamber, edge heater is uniform over $w \times \delta$; and c) heat transfer in an elemental volume.

the average behavior.[7] Heat transfer in these structures is generally dominated by radiation from the two main surfaces with a one-dimensional transverse coupling between them. Hence, the driving parameters in predicting the temperature under a given heating are surface properties and the transverse conductance, which can be determined by analysis or a test similar to that described earlier for honeycomb panels.

In cases where lateral conductance must be known, test data may be accumulated from samples sliced out of a single representative panel fabricated specifically for this purpose. A typical pattern is shown in Fig. 2.5a. For adequate representation, the number of slices should be proportional to the degree of anisotropy; that is, the larger the anisotropy (k_{11}/k_{22} in Fig. 2.5 is very different than 1.0) the more samples at different alignments will be needed.

The samples are cut thin (ratio of width to length much less than 1.0), and their edges treated with low-emissivity tape ($\varepsilon < 0.03$) to minimize the effective contribution by cross conductivities. The radiating surfaces have a high-emissivity treatment to overwhelm residual edge radiation.

Referring to Fig. 2.5b, the test specimen is equipped with an end heater powered to generate a fin temperature profile detected by a set of thermocouples. Unlike testing for transverse conductivity, net heater power into the sample must be known accurately in this test. By using an insulation wrap and auxiliary heater (called a guard heater), the energy conducted at the sample's edge will approach the measured value of the electrical power ($EI = Q_{\text{heater}}$) when thermocouples on the auxiliary heater and sample heater are made to give the same reading. A heat balance on the elemental volume shown in Fig. 2.5c gives

$$Q_x^k - Q_{x+\Delta x}^k = -\left[\left(\frac{dQ^k}{dx}\right)_x \Delta x + \left(\frac{1}{2!}\right)\frac{d^2Q^k}{dx^2}\bigg)_x (\Delta x)^2 + \cdots\right] = \Delta Q^r$$

where

$$\Delta Q^r = 2\varepsilon w \Delta x \left(\sigma T^4 - \sigma T_w^4\right)$$

and, from Fourier's formula,

$$Q_x^k = -k_\theta w \delta \frac{dT}{dx}\bigg|_x$$

Hence, in the limit as $\Delta x \to 0$,

$$\frac{d^2T}{dx^2} = \frac{2\varepsilon}{k_\theta \delta}\left(\sigma T^4 - \sigma T_w^4\right)$$

with boundary conditions

$$-k_\theta w \delta \frac{dT}{dx}\bigg|_{x=0} = Q_{htr} \quad \text{and} \quad \frac{dT}{dx}\bigg|_{x=L} = 0$$

The radiation exchange factor is replaced here by ε based on the assumption that the region represented by a thermocouple is small in comparison to the enclosure, and the factor 2 in the differential equation implies radiation from both sides. The second boundary condition, $dT/dx|_{x=L} = 0$, indicates no heat loss at the end of the sample, which is very nearly the case when the cross-sectional area is small and the edge is treated with low-emissivity tape.

By multiplying both sides of the differential equation by dT/dx (see Application in Sec. V of this chapter), a first integral is found as

$$\left(\frac{dT}{dx}\right)^2 - \left(\frac{Q_{htr}}{k_\theta w \delta}\right)^2 = \frac{4\varepsilon\sigma}{k_\theta \delta}\left[\frac{1}{5}(T^5 - T_0^5) - T_w^4(T - T_0)\right]$$

Applying the second condition of no heat loss at the end where the measured temperature is T_L,

$$k_\theta = \frac{(Q_{htr})^2}{4\varepsilon\sigma w^2\delta\left[\frac{1}{5}(T_0^5 - T_L^5) - T_w^4(T_0 - T_L)\right]}$$

which is lateral conductivity at angle θ off the established planer reference. Hence, with known dimensions and surface emissivity, the conductivity can be calculated from measured temperatures and heater power.

Example: Conduction versus Radiation

Normally, heat exchange among satellite components is more efficient when conduction, rather than radiation, is the main mode of transfer. The lower resistance to heat flow with conduction paths dampens spacial variations in temperature and moderates potentially hot and cold regions.

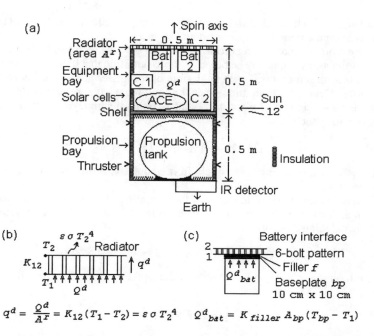

$$q^d = \frac{Q^d}{A^r} = K_{12}(T_1 - T_2) = \varepsilon \sigma T_2^4 \qquad Q^d_{bat} = K_{filler} A_{bp}(T_{bp} - T_1)$$

Fig. 2.6 Radiator for small satellite: a) batteries mounted to satellite radiator, b) all equipment dissipation is transferred to the radiator, and c) battery thermal interface.

Consider the sun-oriented (+0-deg, −12-deg Earth-oriented twilight orbit) spinning small satellite represented by the 0.5-m-diam and 1.0-m-high cylindrical canister shown in Fig. 2.6. The outer skin is covered by panels of solar cells and is thermally isolated from the rest of the satellite. Half the canister (the propulsion bay) is occupied by a propellant tank and accessories and is isolated from the equipment bay, which houses two batteries (equal activity) and other dissipators. There is an isolated and separately controlled infrared detector at the propulsion end, and the other end is reserved for a radiator. Thermal control makes an initial assessment and summarizes the input data as follows. 1) mounting platforms temperature requirements (specifications) include batteries, 0 to 15°C; attitude control electronics (ACE), 0 to 40°C; propulsion bay (isolated), 10 to 50°C; and all other equipment, 0 to 45°C. 2) Orbital average dissipation consists of 20 ± 2 W for batteries (shared activity), 15 ± 1.5 W for the ACE, and 15 ± 1.5 W for all other components, for a total of 50 ± 5 W in the equipment bay.

Faced with a problem in packaging, but recognizing the urgent need to keep the batteries cool, the designer proposes installing an interior mounting shelf and offers for review various arrangements for placing the boxes, always keeping at least the batteries mounted to the radiator. The radiator is tentatively selected as a 2-cm aluminum honeycomb panel of 40 kg/m³ core density and 0.3-mm aluminum face sheets, with an estimated transverse conductance $K_{12} = 133.5$ W/m² K. From Fig. 1.3, the conductance across a 10×10 cm², 6-bolt mounting pattern on RTV filler is K_{filler} (100% contact) $= 467$ W/m² K. The interior surfaces and the radiator are treated with a high-emissivity coating (Chemglaze black, $\varepsilon \approx 0.85$) to enhance heat exchange among the equipment and to space.

There is no external heating on the radiator for the given orbit and orientation, and a first-order assessment may be based on the assumption that both shelf and radiator are at uniform tempratures with negligible heat transfer across the interfaces and insulation. Hence, all dissipation Q^d in the equipment bay is rejected at the radiator. Thus, as shown in Fig. 2.6b, with T_1 and T_2 being the honeycomb temperature at battery and space sides,

$$Q^d/A^r = K_{12}(T_1 - T_2) = \varepsilon \sigma T_2^4$$

where A^r is the radiating area which, from geometry, is the same as the heat conduction area.

To ensure that the batteries' side of the mounting platform does not exceed the specified $T_1 = 15°C$ during maximum heating ($Q^d = 55$ W), the radiating area must be calculated according to

$$55/A^r = 133.5[(15 + 273) - T_2] = 0.85 \sigma T_2^4$$

Iteration on the middle and right side of the equation (trial and error, hand calculator, MATHCAD, etc.) gives

$$T_2 = 287.5 \, \text{K} \, (12.6°C)$$

and, therefore,

$$A^r = 55/[(133.5)(15 - 12.6)]$$
$$= 55/[(0.85)(5.67E-8)(12.6 + 273)^4] = 0.1717 \, \text{m}^2$$

The available area in the radiator region is

$$A_{\text{available}} = \pi(0.5)^2/4 = 0.1964 \, \text{m}^2$$

and, hence, $0.1717 \, \text{m}^2$ can be accommodated.

A suitable design would have insulation covering just enough of the radiator's exterior rim to eliminate (or at least reduce to a great degree) radiation by excess area. The needed width of the insulation can be found from

$$0.1964 - 0.1717 = 0.0247 = \pi[(50)^2 - (50 - 2w)^2]/[(4)(100)^2]$$

or w just over 1.6 cm.

The temperatures during least total equipment dissipation (45 W) are

$$T_2 \, (\text{lowest}) = \{45/[(0.85)(0.1717)(5.67E-8)]\}^{0.25} = 271.56 \, \text{K} \, (-1.4°C)$$

and

$$T_1 \, (\text{lowest}) = Q^d/[A^r K_{12}] + T_2$$
$$= 45/[(0.1717)(133.5)] + 271.56 = 273.5 \, \text{K} \, (0.5°C) \, .$$

Hence, the requirement to keep the batteries' side of the mounting platform between 0 and 15°C is satisfied.

Consider next the temperature of the battery baseplate as described in Fig. 2.6c. Assuming that all battery heating is to be transferred by conduction to the baseplate, a steady-state heat balance gives

$$Q_{battery}^d = K_{filler} A_{baseplate}(T_{baseplate} - T_1)$$

The baseplate area $A_{baseplate}$ is given as $0.01 \, m^2$ and the interface filler conductance can be represented by

$$K_{filler} = fk/\delta$$

where k and δ are the conductivity and thickness of the filler and f is a factor that accounts for deviations from perfect surface contact. Thus, f is 1.0 (and $K_{filler} = 467$ $W/m^2 \, K$) when there are no gaps, and the value decreases (or becomes related to radiation) as contact diminishes. Using the given data, a calculation with maximum heating ($Q_{battery}^d = 11 \, W$, $T_1 = 15°C$) gives

$$T_{baseplate} \text{ (highest)} = T_{bp} = 15 + 11/[f(467)(0.01)]$$

leading to the following:

f:	1.0	0.5	0.3
$T_{bp}(°C)$:	17.4	19.7	22.9

It can be seen that unless there is a decidedly inadequate mount, the temperature of the battery baseplate will remain below $20°C$, which is often cited as the uppermost limit for nickel cadmium or nickle hydrogen batteries.

Radiation heat exchange between shelf and radiator (same high emissivity) can be conveniently described by

$$Q_{shelf}^d = F\varepsilon^2 A_{shelf}\left(\sigma T_{shelf}^4 - \sigma T_1^4\right)$$

where $\mathbf{F}$ is a blockage factor accounting for the presence of covers, harnessing, separators, etc. A full view of the radiator is defined by $\mathbf{F} = 1.0$, and no view at all by $\mathbf{F} = 0$. With a shelf of the same area as the full radiator,

$$\sigma T_{shelf}^4 \text{ (maximum heating)} = \sigma T_{15C}^4 + Q_{shelf}^d/[\mathbf{F}(0.85)^2 (0.1964)]$$

For $\mathbf{F} = 1.0$,

Q_{shelf}^d (W):	10	20	25
T_{shelf} (highest)(°C):	27.2	38.1	43.1

Therefore, under the best circumstances ($\mathbf{F} = 1.0$) the total dissipation on the shelf must not exceed some 22 W if the ACE is to be mounted there and its platform maintained below $40°C$. Because the maximum power in the ACE is 16.5 W, other components selected for mounting on the shelf must, at best, not exceed about 5.5 W in dissipation. For 20-W, total dissipation on the shelf, the influence of $\mathbf{F}$ is as follows:

F:	1.0	0.90	0.80	0.75
T_{shelf} (hot bias)(°C):	38.1	40.4	43.1	44.8

Comparison with previous results demonstrates the superiority of mounting dissipators directly to main radiators. Not only does the approach provide an intimate response of the equipment to a temperature under complete management by the engineer (working with areas and surface coatings), but there is more room for tolerating uncertainties. Comparing $\Delta T / \Delta f$ with $\Delta T / \Delta \mathbf{F}$, it is noted that the temperature is considerably more sensitive to variations in exchange factors when radiation is the dominant mode of heat transfer. In a true situation of equipment in a satellite canister, the presence of cables, connectors, partitions, and other blockages makes estimates of $\mathbf{F}$ doubtful even with the most detailed mathematical modeling. Uncertainties can be reduced by adding minor conduction paths, such as conductive straps, or including in the calculations heat conducted along the honeycomb face sheets and across brackets and structural angles. At times there may be required heat pipes and conductive doublers. In the design of an actual satellite similar in concept to the one in this example, the problem in temperature control was managed by repackaging the ACE into two smaller boxes, which were then mounted (with another, lower dissipator) on side brackets with their baseplates viewing space through cutouts made in two solar panels. The effect of sun impingement through the cutouts was reduced by treating the baseplates with white paint (Z93), which has a low solar absorptivity but high emissivity.

IV. Convection

Convection is heat transfer from a surface to a fluid flowing over it either as a result of a pressure difference (forced convection) or due to variations in gravity and density (natural or free convection). In its simplest description, convection follows Newton's law of cooling, which states that

$$q^c \equiv Q^c / A^c = h(T_s - T_\infty)$$

where $q^c (\text{W/m}^2)$ or $Q^c (\text{W})$ is the heat convected over surface area A^c at temperature T_s, T_∞ is the fluid temperature far from the surface, and $h(\text{W/m}^2 \text{ K})$ is the convection coefficient, which is the primary quantity sought in the physics of convection. Generally, h depends on temperature and, in the case of free convection, the temperature difference between surface and fluid.

Many formulas (empirical and otherwise) comprising a wide variety of geometries and flow conditions have been given for determining h. The data are usually reported in dimensionless combinations of properties, fluid speed u (m/sec) and a characteristic length L, with the dependence of h announced in a dimensionless Nusselt number expressed as a function

$$Nu = \phi(Re, Gr, Ec, Pr)$$

where Nusselt number $Nu \equiv hL/k$, Reynolds number $Re \equiv uL\rho/\mu$, Grashoff number $Gr \equiv L^3 g\beta\rho^2(T_s - T_f)/\mu^2$, Eckert number $Ec \equiv u^2/[c_p(T_s - T_f)]$, Prandtl number $Pr \equiv \mu c_p/k$, and Rayleigh number $Ra \equiv Gr \, Pr$. Here, T_f is a characteristic temperature of the fluid and k, ρ, c_p and μ are fluid conductivity, density, specific heat at constant pressure, and viscosity (N s/m^2). Gravitational acceleration is g (9.8 m/s^2 on Earth and ≈ 0 m/s^2 in space) and β is the volumetric coefficient of expansion (per K) given by

$$\beta = -\frac{1}{\rho}\left(\frac{\partial \rho}{\partial T}\right)_P$$

Table 2.3   Values of h (W/m^{2}K) for flows over surfaces

Free convection
Air: 2–20
Water: 30–300
Forced convection
Air: 20–200
Water: 150–15,000
Condensing steam
5,000–100,000
Boiling water
3,000–50,000

where P (N/m^2) is pressure. The magnitude of h (W/m^2 K) may be appreciated from Table 2.3.

The Mach number, $M = u/u_s$, where u_s is the local speed of sound, enters the calculations when fluids flow at high speed and compressibility effects become important. The influence is usually included when $M > 0.3$. The Knudsen number, $Kn = \lambda/L$, where λ is molecular mean free path (average distance traveled by a molecule between collisions), is a measure of fluid rarefaction and is important in the analysis of convection heat transfer at high altitudes where pressure and density are low. The Knudsen number can be calculated from

$$Kn = 1.26(\gamma M/Re)^{\frac{1}{2}}$$

with $\gamma = c_p/c_v$ being the ratio of the fluid specific heats at constant pressure and at constant volume.

The friction factor f is a useful parameter in formulating flow problems in terms of geometric factors and a mean or average rate of fluid motion. It represents the ratio of fluid driving force to kinetic energy. For flow (along x) in a circular tube in the absence of body forces, the Fanning friction factor is defined in the equation

$$\frac{f}{D} \equiv -\frac{1}{4}\frac{\partial P/\partial x}{\frac{1}{2}\rho u^2}$$

where D is tube diameter and $\partial P/\partial x$ the pressure gradient. Because the gradient is usually uniform along the length of a pipe, the equation is often written

$$f = \frac{D\Delta P}{2L\rho u^2}$$

where ΔP is pressure drop in distance L. In a conduit of general cross section, the diameter is replaced by the hydraulic diameter defined as

$$D_h \equiv 4(\text{cross-sectional area/wetted perimeter})$$

so that

$$f = \frac{D_h\Delta P}{2L\rho u^2}$$

Characteristic speed in a circular tube may be taken as a weighted, integrated average of the radial r distribution $v_x(r)$ around the tube perimeter (direction θ); that is,

$$u = v_{av} \equiv \int_{\theta=0}^{2\pi} \int_{r=0}^{R} v_x(r) r \, dr \, d\theta \bigg/ \int_{\theta=0}^{2\pi} \int_{r=0}^{R} r \, dr \, d\theta$$

where R is tube radius. In a Hagen–Poiseuille flow[8]

$$v_x(r) = \frac{\Delta P}{4\mu L}(R^2 - r^2)$$

and, therefore,

$$u = \frac{\Delta P}{8\mu L} R^2$$

Because the rate of mass flow $\dot{m}$ (kg/s) is

$$\dot{m} = \rho A_{flow} u$$

where A_{flow} is the area normal to flow, the characteristic speed in a tube is also

$$u = \dot{m}/\rho \pi R^2$$

Convection and friction factor formulas for all kinds of applications are listed and discussed in most heat transfer and fluid dynamics books and handbooks.[9-11] Some expressions considered relevant in satellite thermal analysis are given in Table 2.4. However, although these and other similar expressions are helpful in correlating and justifying trends, they should not be considered final statements in predicting convection. It is not unusual, for instance, that slight disturbances or variations in surface roughness lead to differences of as much as 30% between actual data and those obtained by using formulas. Fortunately, convection analyses in satellite thermal control seldom require exact formulations. The problem occurs mainly in dealing with ground cooling and the study of fluid flow in heat pipes and pumped fluid loops, and in these applications thermal engineers rely on test data and established operational procedures. In ground cooling, proven air-conditioning equipment exists, and satellite thermal control there lies in the freedom to adjust the temperature of the cooling medium (usually air) or its flow rate or both to achieve the desired temperatures as monitored by sensors. In the analysis of closed fluid systems, there seems to be a general preference to using actual data obtained from tests.

Example: Launchpad Cooling

Practically all satellites require air conditioning while inside the fairing on the launchpad. A typical configuration is schematically shown in Fig. 2.7. The inlet temperature and rate of flow of cooling air (or drier nitrogen) can generally be adjusted within an adequate range so that the monitored temperatures are maintained manually or automatically within specified limits. Potential problems arising from a component's insufficient exposure to the cooling medium may be solved by introducing special ducts and air deflectors or, preferably, by operating the component intermittently to avoid overheating.

Table 2.4   Formulas for the average Nusselt number and friction factor

Exposed radiators in forced airflow (launch pad cooling inside a shroud):
Flat plate (L = length along flow) with the following usually prevailing conditions:
Laminar flow, $Re_L < 5.0E5$, based on local T_∞ and freestream speed $u = u_\infty$

$$Pr > 0.6, \text{ based on } T_\infty$$

$$Nu = 0.664 Re_L^{\frac{1}{2}} Pr^{\frac{1}{3}}$$

Mixed flow, $5.5E5 < Re < 1.0E8$, based on average $T_f \equiv 0.5(T_s + T_\infty)$ and $u = u_\infty$

$$0.6 < Pr < 60, \text{ based on average } T_f$$

$$Nu = \left(0.037 Re_L^{\frac{4}{5}} - 871\right) Pr^{\frac{1}{3}}$$

Exposed radiators in still air (controlled bay environment without cooling fans):
Vertical flat plate (L = vertical dimension)
No restrictions on Ra_L, $T_f \equiv 0.5(T_s + T_\infty)$

$$Nu = \left\{0.825 + \frac{0.387 Ra_L^{1/6}}{\left[1 + (0.492/Pr)^{9/16}\right]^{8/27}}\right\}^2$$

Upper surface of flat plate at angle θ (up to 60 deg or so) from vertical (L = vertical dimension)
No restrictions on Ra_L, $T_f \equiv 0.5(T_s + T_\infty)$

Equation same as vertical flat plate with g in Ra_L replaced by $g \cos \theta$

Horizontal flat plate facing upward (L = square root of area)
$10^6 < Ra_L < \approx 10^7$, $T_f \equiv 0.5(T_s + T_\infty)$

$$Nu = 0.54 Ra_L^{1/4}$$

$\approx 10^7 < Ra_L < 10^{10}$, $T_f = 0.5(T_s + T_\infty)$

$$Nu = 0.15 Ra_L^{1/3}$$

Horizontal flat plate facing downward (L = square root of area)
$10^5 < Ra_L < 10^{10}$, $T_f \equiv 0.5(T_s + T_\infty)$

$$Nu = 0.27 Ra_L^{1/4}$$

Internal Flow (fluid headers, heat pipes, and capillary pumped loops):
Fully developed flow in circular tubes ($L \equiv$ tube diameter D), characteristic speed is based on rate of mass flow per unit area of cross section

$$T_f \equiv 0.5(T_s + T_\infty)$$

$Re_D < \approx 2.3E3$; $M < 0.3$

$$f Re_D = 16 \text{ (Moody diagram, Fanning friction factor)}$$

$$Nu_D = 4.36 \text{ for uniform heat flux on the wall}$$

$$Nu_D = 3.66 \text{ for constant wall temperature}$$

$Re_D < \approx 2.3E3$; $M > 0.3$

$$f Re_D = 16[1 + M^2(\gamma - 1)/2]^{-1/2}$$

$\approx 2.3E3 < Re_D < 1.0E5$; $M < 0.3$

$$f Re_D^{1/4} = 0.079 \text{ (Moody diagram, Fanning friction factor)}$$

$\approx 2.3E3 < Re_D < 1.0E5$; $M > 0.3$

$$f Re_D^{1/4} = 0.079[1 + M^2(\gamma - 1)/2]^{-3/4}$$

(cont.)

Table 2.4   (Continued)

$Re_D > 1.0E5; M < 0.3$

$$fRe_D^{1/5} = 0.046$$

$Re_D > 1.0E5; M > 0.3$

$$fRe_D^{1/5} = 0.046[1 + M^2(\gamma - 1)/2]^{-3/4}$$

$Re_D > \approx 1.0E4, 0.6 < Pr < 16,700$ at average T_f

$$Nu_D = 0.027Re_D^{4/5}Pr^{1/3}(\mu/\mu_s)^{0.14}$$

$1.0E4 < Re_D < 1.0E6, 0.1 < Pr < 1.0E4$ at average T_f, pipe length/diameter > 25

$$Nu_D = 5 + 0.016Re_D^c Pr^d$$

$$c = 0.88 - 0.24/(4 + Pr)$$

$$d = 0.33 + 0.5E^{-0.6Pr}$$

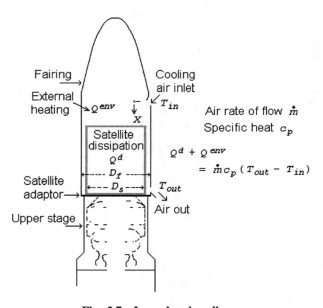

Air rate of flow $\dot{m}$

Specific heat c_p

$$Q^d + Q^{env} = \dot{m}c_p(T_{out} - T_{in})$$

Fig. 2.7   Launchpad cooling.

Energy conservation relates the rise in air temperature from inlet to outlet $(T_{out} - T_{in})$ to collected heat Q by

$$Q = \dot{m}c_p(T_{out} - T_{in})$$

where $\dot{m}$(kg/s) is the rate of mass flow and c_p the average specific heat of air at constant pressure. The heat being removed consists of satellite dissipation Q^d and environment heating Q^{env} transferred across the fairing wall.

For a spacecraft of average diameter D_s centered within a fairing of inside diameter D_f, the air flows across a section whose area can be approximated by

$$A_{\text{flow}} = (\pi/4)\left(D_f^2 - D_s^2\right)$$

Table 2.5 Properties of air

T, K	c_p, J/kg K	ρ, kg/m^3	$\mu E7$, N s/m^2	$kE3$, W/m K	Pr, $\mu c_p/k$
270	1006	1.301	169.6	23.9	0.714
275	1006	1.278	172.1	24.3	0.712
280	1006	1.255	174.6	24.7	0.711
285	1007	1.231	177.1	25.1	0.711
290	1007	1.208	179.6	25.5	0.709
295	1007	1.185	182.1	25.9	0.708
300	1007	1.161	184.6	26.3	0.707
305	1008	1.138	187.1	26.7	0.706
310	1008	1.115	189.6	27.1	0.705
312	1008	1.091	192.1	27.5	0.704
287 (14°C)	1007	1.222	178.1	25.3	0.709

Hence, average air speed over the satellite is

$$u = \dot{m}/\rho A_{\text{flow}} = 4\dot{m}/\left[\pi\rho\left(D_f^2 - D_s^2\right)\right]$$

and the Reynolds number at distance X meters from the inlet is

$$Re = \rho u X/\mu = \dot{m}X/\mu A_{\text{flow}}$$

Typical data are as follows: $\dot{m}_{\max} \approx 0.70$ kg/s (ARIANE IV or Titan IV facility); $0 < T_{\text{in}} < 40°C$ (ARIANE IV or Titan IV facility with air properties as given in Table 2.5); and $D_s = 2.0$ m, $D_f = 3.5$ m, and $X = 4.5$ m (approximately NASA's MMS in ARIANE IV).

Now, the dissipation in an average satellite during launchpad check-out modes may be somewhere between 50 and 150 W, whereas environmental heating during a hot, sunny day at Cape Kennedy or Kourou in French Guiana may be in the range of 1–2 kW. If, for example, $Q = 2000$ W, $\dot{m} = 0.50$ kg/s, and $T_{\text{in}} = 10°C$, then the average outlet temperature is

$$T_{\text{out}} = 10 + 2000/[0.5)(1007)] = 14°C$$

Air properties at that temperature may be extrapolated from standard data as given in the last line of Table 2.5, and at the outlet region

$$Re = (0.50)(4.5)/\{(178.1E{-}7)(\pi/4)[(3.5)^2 - (2.0)^2]\} = 19{,}497.27$$

which is an order magnitude less than $5.0\,E5$, indicating laminar flow. The Prandtl number for the expected range of temperature is greater than 0.6. Hence, the convection equation to use from Table 2.4 is

$$Nu = 0.664 Re^{\frac{1}{2}} Pr^{\frac{1}{3}}$$

$$Nu = 0.664(19{,}497.27)^{\frac{1}{2}}(0.709)^{\frac{1}{3}} = 82.67$$

Suppose now that there are dissipating electronics mounted on an exposed satellite panel measuring 0.5×0.5 m^2, which is located near the outlet. If it is assumed that the dissipated heat on the panel is small compared to the total heating involved, the effect on air outlet temperature may be neglected, and heat transfer from the panel Q_{panel} to local air is given by

$$Q_{panel} = h A_{panel}(T_{panel} - T_{out})$$

with $T_{out} = 14°C$. Using $L = 0.5$ m, $k = 25.3E{-}3$ W/m K, and $A_{panel} = 0.25$ m^2, h is found from

$$h = Nuk/L = (82.67)(25.3E{-}3)/0.5 = 4.18 \text{ W/m}^2 \text{ K}$$

which is not very large and is comparable to free convection in open atmosphere. This indicates that increasing the rate of flow on the launchpad may not be as effective as maintaining the air at a cool temperature.

Calculating panel temperature as function of dissipated heat,

Q_{panel} (W):	10.0	20.0	30.0	40.0
T_{panel} (°C):	23.6	33.1	42.7	52.3

If the specifications limit is 40°C for the panel, then no more than about 27 W may be dissipated on a steady-state basis. The restriction can be relieved either by decreasing the inlet temperature or, less effectively, by supplying a higher rate of airflow, or by doing both.

Obviously, the actual results will depend on exposure and weather conditions that can only be presaged. Other uncertainties relate to flow distortions as the coolant passes over wavy insulated surfaces (see Fig. 6.1), stowed mechanisms, edges, wires, and other obstructions within the fairing. Effects near the interface with the launch vehicle can also be difficult to predict or ascertain. In the case of TOPEX/Poseidon, for example, the satellite was secured to ARIANE's adaptor (about −60°C) at fittings that were integral to the satellite's hydrazine system, which had to maintained at 10°C or greater as a margin against freezing (hydrazine freezes at about 1.5°C) and failure to fire after orbit insertion. This led to an unplanned and bizarre activity whereby the heaters in the propulsion module were being operated at maximum override power while, at the same time, the air conditioner was running at maximum capability to keep batteries in a nearby module relatively cool. Inasmuch as this problem was not revealed in a detailed launchpad computer model adds substance to the argument that one should view calculations on satellite convective cooling only as ballpark estimations. To ensure that temperatures remain within specifications, spacecraft operation and air conditioning must be controlled and manipulated in response to real time telemetry data.

V. Combined Modes of Heat Transfer

Digital routines in computer modeling require that heat transfer between two isothermal nodes (1 and 2, with no other intervening nodes) be given by the linear form

$$Q = U(T)[T_1 - T_2]$$

U is a total conductance combining all heat transfer paths between the two nodes. It contains geometric factors and material and surface properties and usually enters the program following selected rules that relate it to the values of T_1 and T_2.

The linear proportionality is explicit in conduction and convection. When conductivity depends on temperature, the input conduction conductance may be calculated based on mean geometric parameters and the arithmetic average of T_1 and T_2. Other options may be preferred but, in any case, they must all coincide as the nodes become smaller and their centers approach one another. Similarly, for convection heat transfer, C (the product of convection coefficient and area) is calculated based on mean areas and fluid properties at an average temperature or other selected reference. A radiation conductance R can also be introduced in the same form by noting that

$$T_1^4 - T_2^4 = [(T_1^2 + T_2^2)(T_1 + T_2)][T_1 - T_2]$$

Hence, the conductances connecting two nodes, 1 and 2, may be calculated from

$$K_{12} \text{ (conduction conductance, W/K)} = k_{(T_1+T_2)/2}(A^k/L)_{\text{mean}}$$

$$C_{12} \text{ (convection conductance, W/K)} = (hA^c)_{\text{avT}}$$

$$R_{12} \text{ (radiation conductance, W/K)} = \Im_{12}A_1^r\sigma(T_1^2 + T_2^2)(T_1 + T_2)$$

and the total conductance is

$$U_{12} = K_{12} + C_{12} + R_{12}$$

The concept of total conductance is perhaps most useful when knowledge of the temperature of the medium or media connecting two nodes is not essential to drawing conclusions. Consider the steady-state flow of heat $Q_{1,n}$ from node 1 to node n through intervening nodes 2, 3, ..., $n-1$. Conservation of energy requires

$$Q_{1,n} = U_{12}(T_1 - T_2) = U_{23}(T_2 - T_3) = \cdots = U_{n-1,n}(T_{n-1} - T_n)$$

Hence

$$T_j - T_{j+1} = Q_{1,n}/U_{j,(j+1)}, \qquad j = 1, 2, \ldots, n$$

and by writing out the equation for all j and adding,

$$Q_{1,n} = U_{1,n}(T_1 - T_n)$$

where $U_{1,n}$, the net conductance from node 1 to node n, is calculated from the equation

$$1/U_{1,n} = 1/U_{12} + 1/U_{23} + \cdots + 1/U_{(n-1),n}$$

Application: Convection Along a Conducting Fin and the Radiation Problem

Structures protruding from the main body of a satellite are often represented by fin models when evaluating heat leaks into and out of the satellite. Figure 2.8 gives a description of the convection version in one dimension.

A heat balance in steady state, on a cross-sectional element, at distance x from the base, and of temperature $T(x)$ gives

$$Q_x^k - Q_{x+\Delta x}^k = -\left[\left(\frac{dQ^k}{dx}\right)_x \Delta x + \left(\frac{1}{2!}\right)\frac{d^2Q^k}{dx^2}\Big)_x (\Delta x)^2 + \cdots\right] = \Delta Q^c$$

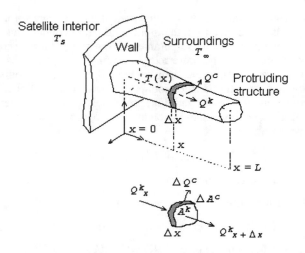

Fig. 2.8   One-dimensional conducting–convecting fin.

with

$$Q_x^k = -kA^k(x)\frac{\mathrm{d}T}{\mathrm{d}x}\bigg|_x$$

$$\Delta Q^c = h\Delta A^c(x)(T - T_\infty)$$

Assuming constant conductivity k, in the limit as $\Delta x \to 0$,

$$\frac{\mathrm{d}^2 T}{\mathrm{d}x^2} + \left(\frac{1}{A^k}\frac{\mathrm{d}A^k}{\mathrm{d}x}\right)\frac{\mathrm{d}T}{\mathrm{d}x} - \left(\frac{1}{A^k}\frac{h}{k}\frac{\mathrm{d}A^c}{\mathrm{d}x}\right)(T - T_\infty) = 0$$

For uniform cross section (A^k const), $A^c(x) = Px$, where P is perimeter, and

$$\frac{\mathrm{d}^2 T}{\mathrm{d}x^2} = \frac{hP}{kA^k}(T - T_\infty)$$

When the convection coefficient h is constant, the equation is linear and has the solution

$$T - T_\infty = C_1 \cosh(hP/kA^k)^{1/2}x + C_2 \sinh(hP/kA^k)^{1/2}x$$

in which the constants C_1 and C_2 are evaluated by substituting at least two known boundary conditions. One of these is invariably the satellite temperature T_s at the fin base ($x = 0$), which can be used as a parameter in the analysis. Substituting $T(0) = T_s$,

$$C_1 = T_s - T_\infty$$

C_2 must now be determined to find the heat transferred from the satellite, which is given by

$$Q^k(0) = -kA^k \frac{\mathrm{d}T}{\mathrm{d}x}\bigg|_{x=0} = -C_2(hPkA^k)^{1/2}$$

The condition for finding C_2 is often defined at the fin's end ($x = L$) where an interaction with the environment or another structure is taking place. It is frequently one of the following: 1) adiabatic interface, signifying a free end, which is thin-walled or insulated,

$$-kA^k\frac{\mathrm{d}T}{\mathrm{d}x}\bigg|_{x=L} \approx \frac{\mathrm{d}T}{\mathrm{d}x}\bigg|_{x=L} \approx 0$$

2) fixed end temperature, indicating a low resistance coupling to a massive body at known temperature T_b,

$$T|_{x=L} \approx T_b$$

3) end heat loss by convection,

$$-kA^k\frac{\mathrm{d}T}{\mathrm{d}x}\bigg|_{x=L} \approx hA^k(T_{x=L} - T_\infty)$$

4) end temperature equals approximately the environment's, indicating a long or poorly conductive fin,

$$T|_{x=L} \approx T_\infty \qquad \text{and} \qquad \frac{\mathrm{d}T}{\mathrm{d}x}\bigg|_{x=L} \approx 0$$

and 5) slope jump to conserve temperature and heat transfer continuity from fin end $(L-)$ to coupled body $(L+)$,

$$T|_{x=L-} \approx T|_{x=L+} \qquad \text{and} \qquad -kA^k\frac{\mathrm{d}T}{\mathrm{d}x}\bigg|_{x=L-} \approx -kA^k\frac{\mathrm{d}T}{\mathrm{d}x}\bigg|_{x=L+}$$

The solutions for the first four cases in dimensionless form are given in Table 2.6. The results for the fifth case and their implications are left for an exercise.

Curiously, this classic problem in convection heat transfer does not appear with any particular regularity in satellite thermal control. But there is interest in its analogy to a conducting fin in a radiation environment. The difference, of course, is that the differential equation for radiation is nonlinear due to the presence of T^4.

The corresponding equation with heat radiation is obtained by replacing the heat convection term (ΔQ_x^c in Fig. 2.8) by

$$\Delta Q_x^r = \Im_x \Delta A^r (\sigma T^4 - \sigma T_\infty^4)$$

A^r, as before, is equal to Px and $\Im_x$ is the radiation exchange factor with surroundings at temperature T_∞. In deep space ($T_\infty = 0$ K), no blockage, and emissivity ε as exchange factor, the differential equation reads

$$\frac{\mathrm{d}^2T}{\mathrm{d}x^2} = \frac{\varepsilon P}{kA^k}\sigma T^4$$

A general solution in elementary functions is not known for this equation when the boundary conditions are of the same nature as those just listed. However, a first

**Table 2.6   Temperature distribution in a uniform
one-dimensional conducting–convecting fin**

End condition	Temperature distribution
Adiabatic $a = 0, c = 0$	$\dfrac{\cosh \beta (1 - \xi)}{\cosh \beta}$
Temperature $a = 1, b = 0,$ $c = \tau(1)$	$\dfrac{\tau(1)\sinh \beta \xi + \sinh \beta (1 - \xi)}{\sinh \beta}$
Convection $a = hL, b = k,$ $c = 0$	$\dfrac{\cosh \beta (1 - \xi) + (hL/\beta k)\sinh \beta (1 - \xi)}{\cosh \beta + (hL/\beta k)\sinh \beta}$
Very long fin $L \to \infty$	$\exp(-\beta \xi)$

$$\frac{d^2 \tau}{d\xi^2} - \beta^2 \tau = 0$$

$$\tau(0) = 1, \left(a\tau + b\frac{d\tau}{d\xi} \right)_{\xi=1} = c$$

$$\tau \equiv \frac{T - T_\infty}{T_s - T_\infty}, \qquad \xi \equiv \frac{x}{L}, \qquad \beta^2 \equiv \frac{hPL^2}{kA^k}$$

integral can be obtained from multiplying both sides by dT/dx. Thus,

$$\frac{dT}{dx}\frac{d^2 T}{dx^2} = \frac{\varepsilon P}{kA^k}\sigma T^4 \frac{dT}{dx}$$

$$\frac{1}{2}\frac{d}{dx}\left(\frac{dT}{dx} \right)^2 = \frac{\varepsilon P}{kA^k}\sigma T^4 \frac{dT}{dx}$$

$$d\left(\frac{dT}{dx} \right)^2 = \frac{\varepsilon P}{kA^k}\sigma T^4 dT$$

Integrating from $x = 0$ (where $T = T_s$ and $dT/dx = -Q^k(0)/kA^k$) to general position x,

$$\left(\frac{dT}{dx} \right)^2 - \left(\frac{Q^k(0)}{kA^k} \right)^2 = \frac{\varepsilon P}{5kA^k}\sigma \left(T^5 - T_s^5 \right)$$

or

$$dx = \frac{dT}{\left[(Q^k(0)/kA^k)^2 + \frac{1}{5}(\varepsilon \sigma P/kA^k)\left(T^5 - T_s^5 \right) \right]^{\frac{1}{2}}}$$

The second integration, utilizing end conditions, can now be performed numerically. For example, if the end condition is zero slope, the relation between end

temperature T_L and heat lost by the fin [which is the same as $Q^k(0)$] is

$$-\left(\frac{Q^k(0)}{kA^k}\right)^2 = \frac{\varepsilon P}{5kA^k}\sigma\left(T_L^5 - T_s^5\right)$$

and integration (numerical) from $x = 0$ to $x = L$ gives T_L [or $Q^k(0)$] when $T(0) = T_s$ is known.

Approximate analytical solutions for the radiation–conduction problem are obtained by first replacing T^4 with a linear expression of the form $MT + N$, then seeking an optimization procedure for M and N that will minimize the resulting errors. The linearized differential equation will then have the same form as that for convection with solutions readily amenable to the application of the boundary conditions. A detailed discussion on this procedure is given in Chapter 5, Sec. IV.B.

References

[1] *CRC Handbook of Chemistry and Physics*, CRC Press, Boca Raton, FL, 1995.

[2] Lucks, C. F., and Sauer, H. J., "International Thermal Conductivity Conferences, 1961–1963: A Historical Profile," *Proceedings of the Twenty-Second International Conference on Thermal Conductivity*, edited by T. W. Tong, Vol. 22, Technomic, Lancaster, PA, 1994, pp. 3–29.

[3] Karam, R. D., "Outgassing of Spacecraft Composites," *Proceedings of the Fifth International Conference of Composite Structures*, edited by I. H. Marshall, Elsevier Applied Sciences, London, 1989, pp. 547–559.

[4] Mallon, J. J., Uht, J. C., and Hemminger, C. S., "Surface Analyses of Composites Exposed to the Space Environment on the Long Duration Exposure Facility Satellite," *Journal of Spacecraft and Rockets*, Vol. 30, No. 4, 1994, pp. 495–501.

[5] Wales, R. O. (ed.), *ATS-6 Final Engineering Performance Report, Vol. I—Program and Systems Summaries; Mechanical and Thermal Details*, NASA RP-1080, 1981.

[6] Klemens, P. G., "Effective Thermal Conductivity of a Matrix with Two Kinds of Inclusion," *International Journal of Thermophysics*, Vol. 17, No. 4, 1996, pp. 979–981.

[7] Karam, R. D., "Thermal Engineering of Spacecraft Composite Structures," *Proceedings of the Third International Conference of Composite Structures*, edited by I. H. Marshall, Elsevier Applied Sciences, London, 1985, pp. 100–117.

[8] Bird, R. B., Stewart, W. E., and Lightfoot, E. N., *Transport Phenomena*, Wiley, New York, 1964, p. 47.

[9] Incropera, F. P., and DeWitt, D. P., *Fundamentals of Heat Transfer*, Wiley, New York, 1981, pp. 263, 417.

[10] Rohsenow, W. M., and Hartnett, J. P., *Handbook of Heat Transfer*, McGraw–Hill, New York, 1973, Chap. 7.

[11] Schlichting, H., *Boundary Layer Theory*, 4th ed. McGraw–Hill, New York, 1960, pp. 502–533.

Chapter 3

Radiation in Thermal Control

I. Introduction

R ADIATION is the only means for rejecting satellite waste heat in space and is often the dominant mode of heat transfer among large structures. This has naturally led to considerable involvement by thermal engineers who, in the process, have advanced some of the most practical methods for dealing with radiation problems. Much of the development has been in computer software (known as thermal programs) with the versatility to treat practically any situation. The speed and numerical accuracy of most of these programs are presently at a very satisfactory level, with more being done to simplify utility and improve graphics.

Radiation mathematics involves integral equations that cast the energy exchange at a surface as function of direction, wavelength, temperature, and surface properties. But because only average values of properties are usually known and radiation distribution data are limited and often inconsistent, it is found both practical and useful in thermal control to replace some formally defined relations with equations and quantities that represent averages over the regimes of applicability. It is in this context that radiation exchange factors and optical properties (emissivity, absorptivity, and reflectivity) are quoted and used. The approach simplifies an otherwise extremely complicated problem and establishes a basis for a tractable analysis and for finding properties by measurements. By way of justification, it may be pointed out that the methods adopted have not been seriously refuted by evidence from test or flight data.

Foremost among the relations to formal radiation physics are the spectral and directional characteristics of surfaces. The spectral property refers to the fact that the energy of radiation varies with wavelength and that emission from a surface is a spectrum of energies at different wavelengths. The directional character relates to observed preferential emission and absorption in certain directions, creating a radiation distribution that is a function of angles of emission and incidence. A perspective is found by introducing generalized definitions of radiation intensity.

II.   Radiation Intensity and Exchange Factors

The theory of radiation exchange among surfaces is given in intermediate and advanced books on heat transfer. The following borrows from Incropera and Dewitt[1] and Siegel and Howell.[2]

A general treatment utilizes the concept of spectral directional emission intensity $I_{\lambda,\theta}^e$ (W/m^2 per steradian per micrometer), which has the lengthy definition: the rate of emitted energy ($d^3 Q_{\lambda,\theta}^e$) at wavelength λ from an elemental surface area dA, in the $\theta-\varphi$ direction, per unit of normal surface area $(dA \cos \theta)$ to this direction, per unit solid angle $d\omega$, about this direction, per wavelength interval $d\lambda$ about λ. The compact translation is

$$I_{\lambda,\theta}^e(\lambda, \theta, \varphi) \equiv \frac{d^3 Q_{\lambda,\theta}^e/dA \cos \theta}{d\omega \, d\lambda}$$

or, for emitted heat flux from surface dA,

$$d^3 q_{\lambda,\theta}^e(\lambda, \theta, \varphi) \equiv \frac{d^3 Q_{\lambda,\theta}^e}{dA} = I_{\lambda,\theta}^e(\lambda, \theta, \varphi) \cos \theta \, d\omega \, d\lambda$$

The subscript λ, θ is for spectral directional, with φ conventionally not included. Thus, $Q_{\lambda,\theta}^e$ or $q_{\lambda,\theta}^e$ is called spectral directional emission. A visual description is given in Fig. 3.1a but, basically, the intensity is a definition with the third derivative in $d^3 Q_{\lambda,\theta}^e$ implying that three elemental quantities are involved: dA, $d\omega$, and $d\lambda$.

In spherical propagation, $d\omega \, (= dA_n/s^2)$ can be replaced by $\sin \theta \, d\theta \, d\varphi$ and two more definitions emerge: the spectral hemispherical emission (also referred to as spectral emission)

$$q_{\lambda}^e(\text{W/m}^2 \mu\text{m}) = \int_{\varphi=0}^{2\pi} \int_{\theta=0}^{\pi/2} I_{\lambda,\theta}^e(\lambda, \theta, \varphi) \cos \theta \sin \theta \, d\theta \, d\varphi$$

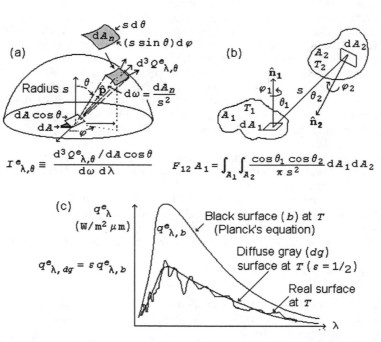

Fig. 3.1 Hemispherical radiation: a) solid angle $d\omega = dA_n/s^2 = \sin \theta \, d\theta \, d\varphi$; b) definition of shape factor, $F_{ij} A_i = F_{ji} A_j$; and c) spectral emission as function of wavelength for black and diffuse gray surfaces, $q_{\lambda,dg}^e = \varepsilon q_{\lambda,b}^e$.

and the total hemispherical emission (usually called emissive power and sometimes simply emission) q^e encompassing all wavelengths (total) and all directions (hemispherical)

$$q^e (\text{W/m}^2) = \int_\lambda q^e_\lambda(\lambda)\, d\lambda = \int_{\lambda=0}^\infty \int_{\varphi=0}^{2\pi} \int_{\theta=0}^{\pi/2} I^e_{\lambda,\theta}(\lambda, \theta, \varphi) \cos\theta \sin\theta\, d\theta\, d\varphi\, d\lambda$$

Spectral emission q^e_λ is used to characterize radiation as a function of wavelength as shown in Fig. 3.1c, whereas q^e(or Q^e) is the quantity used when performing overall energy balances in heat transfer.

When reference is made to the portion of radiation emitted by dA_1 that is intercepted by dA_2, the nomenclature changes slightly, and the transfer equation is written

$$d^3 Q^r_{dA_1 \to dA_2} = I^e_1(\lambda, \theta_1, \varphi_1) \cos\theta_1\, dA_1\, d\omega_{12}\, d\lambda$$

with the superscript r implying radiation exchange (vs emission e) between the surfaces and $d\omega_{12}$ the solid angle subtended by dA_2 on dA_1; that is,

$$d\omega_{12} = dA_2 \cos\theta_2/s^2$$

Therefore

$$d^3 Q^r_{dA_1 \to dA_2} = I^e_1(\lambda, \theta_1, \varphi_1) \cos\theta_1 \cos\theta_2\, dA_1\, dA_2\, d\lambda/s^2$$

with the angles and s as described in Fig. 3.1b. Similarly for dA_2,

$$d^3 Q^r_{dA_2 \to dA_1} = I^e_2(\lambda, \theta_2, \varphi_2) \cos\theta_2 \cos\theta_1\, dA_2\, dA_1\, d\lambda/s^2$$

Hence, the net radiation at dA_1 from the exchange between dA_1 and dA_2 (written $d^3 Q^r_{dA_1 - dA_2}$, with a dash in the subscript instead of an arrow) is

$$d^3 Q^r_{dA_1 - dA_2} = \frac{I^e_1(\lambda, \theta_1, \varphi_1) - I^e_2(\lambda, \theta_2, \varphi_2)}{s^2} \cos\theta_1 \cos\theta_2\, dA_1\, dA_2\, d\lambda$$

The object now is to find the intensities in terms of temperatures so that a relation can be obtained between temperature and surface radiation exchange.

A. Black Surface Exchange and Shape Factor

Radiation exchange is relatively easy to formulate when surfaces are idealized as black. A black surface is diffuse (which means that emission intensity is independent of the angles of emission) and its spectral intensity $I^e_{\lambda,b}$ as function of wavelength and temperature is stated exactly by Planck's equation

$$I^e_{\lambda,b} = \frac{2hc_0^2}{\lambda^5[\exp(hc_0/\lambda\kappa T) - 1]}$$

or

$$q^e_{\lambda,b} = \int_0^{2\pi} \int_0^{\pi/2} I^e_{\lambda,b} \cos\theta \sin\theta\, d\theta\, d\varphi = \frac{2\pi hc_0^2}{\lambda^5[\exp(hc_0/\lambda\kappa T) - 1]}$$

where the subscript b refers to black, and h, c_0, and κ are, respectively, Planck's constant (6.6256×10^{-34} J s), the speed of light in vacuum (2.998×10^8 m/s), and

Boltzmann's constant (1.3805×10^{-23} J/K). Thus, the total intensity (total implies all wavelengths) is found by integration with respect to λ from 0 to ∞ (the steps are given in most heat transfer textbooks) and is found to be

$$I_b^e = (1/\pi)\sigma T^4$$

where σ is the Stefan–Boltzmann constant ($= 5.670 \times 10^{-8}$ W/m^2 K^4) calculated from the other constants. The total hemispherical emission of a black surface q_b^e is obtained by integrating over the hemisphere,

$$q_b^e = \int_{\varphi=0}^{2\pi} \int_{\theta=0}^{\pi/2} \frac{\sigma T^4}{\pi} \cos\theta \sin\theta \, d\theta \, d\varphi = \sigma T^4$$

This is the Stefan–Boltzmann law for blackbody radiation.

As functional dependence on λ can be separated in the equation for net exchange between two black elemental surfaces, the integration with respect to λ (limits 0 to ∞) yields

$$d^2 Q_{dA_1-dA_2,\text{black elements}}^r = \left(\sigma T_{dA_1}^4 - \sigma T_{dA_2}^4\right)\frac{\cos\theta_1 \cos\theta_2}{\pi s^2} dA_1 dA_2$$

Notice that the order of the derivative is reduced by one, and, therefore, the net radiation over the whole of black surface 1 interacting with the whole of black surface 2 can be obtained from the double integral

$$Q_{12,\text{black surfaces}}^r = \int_{A_1} \int_{A_2} \left(\sigma T_{dA_1}^4 - \sigma T_{dA_2}^4\right)\frac{\cos\theta_1 \cos\theta_2}{\pi s^2} dA_1 dA_2$$

When each of the interacting surfaces is at a constant temperature,

$$Q_{12,\text{black surfaces}}^r = \left(\sigma T_1^4 - \sigma T_2^4\right) \int_{A_1} \int_{A_2} \frac{\cos\theta_1 \cos\theta_2}{\pi s^2} dA_1 dA_2$$

which is usually stated as

$$Q_{12,\text{black surfaces}}^r = F_{12} A_1 \left(\sigma T_1^4 - \sigma T_2^4\right)$$

where the product $F_{12} A_1 (= F_{21} A_2)$, with F_{12} standing for the shape factor from surface 1 to surface 2, is the result of the integration.

When a surface i views N others, the additive property of energy permits writing

$$Q_{i-\text{all }N}^r \equiv Q_i^r = \sum_j F_{ij} A_i \left(\sigma T_i^4 - \sigma T_j^4\right)$$

with the summation taken through N. And because (in steady state) $Q_{ij}^r = -Q_{ji}^r$,

$$F_{ij} A_i = F_{ji} A_j$$

Also, from physical considerations, in an enclosure,

$$\sum_j F_{ij} = 1$$

with the summation to include i if i sees itself.

Thus defined, the shape factor is a function of geometry only and remains unchanged when dealing with transients or situations involving other modes of heat transfer. It represents the fraction of the energy emitted by a surface that is intercepted by another. Therefore, in the presence of a hemispherically emitting surface 1 of emission Q_1^e, the radiation intercepted by surface 2, Q_2^i, is given by

$$Q_2^i = F_{12} Q_1^e$$

where the superscript i indicates normal incidence. From the reciprocity $F_{12} A_1 = F_{21} A_2$,

$$q_2^i \equiv Q_2^i / A_2 = F_{12} Q_1^e / A_2 = [(F_{21} A_2 / A_1) Q_1^e] / A_2 = F_{21} Q_1^e / A_1 \equiv F_{21} q_1^e$$

which states that the normal incident energy per unit area of the receiving surface is the emitted energy per unit area of the source adjusted by the shape factor from surface to source. This conclusion is very useful when dealing with the recurring exercise of determining planetary radiation on an orbiting satellite.

B. Diffuse Gray Surface and Script F

The underlying simplicity of black surface radiation exchange lies in the fact that black surfaces are perfect absorbers. Hence, there are no residual radiation interactions beyond what is initially emitted or absorbed. In contrast, a nonblack surface must reflect part of the incident radiation, which must then be accounted for as it interacts with other surfaces and itself.

Being nonblack, real surfaces are practically immune to an exact analysis. However, many satellite surfaces that require radiation computations (especially those interior to canisters) behave much like diffuse emitters and reflectors in that their outgoing and incoming radiation are almost hemispherically uniform (at least at angles that are not too shallow) and their optical properties approximately independent of the direction of emission or incidence. In addition, emission at common satellite temperatures may be averaged, as shown in Fig. 3.1c, to give it a gray trend in which the emissivity ε is constant and, in accordance with Kirchhoff's law, equals absorptivity α for the given frequency range. Under this diffuse gray approximation, an extension to black surface radiation exchange is found by replacing the shape factor by the so-called script F. Thus, for two surfaces,

$$Q_{12}^r = \Im_{12} A_1 (\sigma T_1^4 - \sigma T_2^4)$$

and, in the presence of N surfaces,

$$Q_i^r = \sum_j \Im_{ij} A_i (\sigma T_i^4 - \sigma T_j^4)$$

with reciprocity relation

$$\Im_{ij} A_i = \Im_{ji} A_j$$

The characteristics of $\Im$ emerge from an analysis that introduces the concept of radiosity, which is defined as the amount of heat leaving a surface as electromagnetic radiation. Referring to the enclosure shown in Fig. 3.2, the radiosity J_i on surface i is the reflected portion ($q_i^{\text{reflected}}$) of incident energy (irradiation q_i^i) plus the natural emission q_i^e by a surface at temperature T_i above 0 K. Using this definition

Enclosure of diffuse gray surfaces

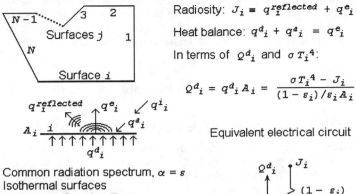

Radiosity: $J_i \equiv q_i^{reflected} + q^e{}_i$

Heat balance: $q^d{}_i + q^a{}_i = q^e{}_i$

In terms of $Q^d{}_i$ and $\sigma T_i{}^4$:

$$Q^d{}_i = q^d{}_i A_i = \frac{\sigma T_i{}^4 - J_i}{(1 - \varepsilon_i)/\varepsilon_i A_i}$$

Equivalent electrical circuit

Common radiation spectrum, $\alpha = \varepsilon$
Isothermal surfaces
Uniform surface heating

$$q_i^{reflected} = \rho q^i{}_i = (1 - \varepsilon_i) q^i{}_i$$

$$q^a{}_i = \alpha_i q^i{}_i = \varepsilon_i q^i{}_i$$

$$q^e{}_i = \varepsilon_i \sigma T_i{}^4$$

Fig. 3.2   Radiosity and the equivalent electric circuit for radiation in a diffuse gray enclosure.

in view of the energy balance, radiosity can be interjected into an equation relating the heat generated (dissipated) Q_i^d in surface i to its temperature. Thus, from

$$J_i \equiv q_i^{\text{reflected}} + q_i^e = (1 - \varepsilon_i)\, q_i^i + \varepsilon_i \sigma T_i^4$$

$$q_i^i = \left(J_i - \varepsilon_i \sigma T_i^4\right)/(1 - \varepsilon_i)$$

and

$$\varepsilon_i \sigma T_i^4 = J_i - (1 - \varepsilon_i) q_i^i$$

Substituting the heat balance

$$q_i^e = q_i^d + q_i^a$$

with

$$q_i^e = \varepsilon_i \sigma T_i^4$$

$$q_i^a = \varepsilon_i q_i^i$$

one obtains

$$q_i^d = \left(\sigma T_i^4 - J_i\right)/[(1 - \varepsilon_i)/\varepsilon_i]$$

or

$$Q_i^d = \left(\sigma T_i^4 - J_i\right)/[(1 - \varepsilon_i)/\varepsilon_i A_i]$$

which indicates that if the surface radiosity is known, then either the temperature can be found corresponding to a given heat dissipation or the net heating (emission less absorption) can be found corresponding to a given temperature. Because the basic analytical problem in satellite thermal control is to find temperatures (and sometimes net heat input), a complete solution to an exclusively radiation exchange system can be obtained by finding the radiosities.

The last equation is reminiscent of the one for an electric current that equals Q_i^d, which is flowing due to potential difference $(\sigma T_i^4 - J_i)$ across resistance $[(1-\varepsilon_i)/\varepsilon_i A_i]$. The circuit is shown on the right in Fig. 3.2. Following Oppenheim's suggestion,[3] the analogy is exploited to reduce radiation exchange in an enclosure to an equivalent electrical network. The needed result is obtained by recalling that the normal incident heating on i from surfaces j is

$$A_i q_i^i = \sum_j F_{ji} A_j J_j$$

As, from reciprocity,

$$A_i q_i^i = \sum_j F_{ij} A_i J_j$$

then A_i may be canceled from both sides to give

$$q_i^i = \sum_j F_{ij} J_j$$

Now,

$$q_i^d + \varepsilon_i q_i^i = \varepsilon_i \sigma T_i^4 = J_i - (1 - \varepsilon_i) q_i^i$$

Hence,

$$q_i^d = J_i - q_i^i$$

But

$$\sum_j F_{ij} = 1$$

and, hence,

$$q_i^d = \sum_j F_{ij} J_i - q_i^i = \sum_j F_{ij} J_i - \sum_j F_{ij} J_j = \sum_j F_{ij}(J_i - J_j)$$

Therefore,

$$Q_i^d = \sum_{j=1}^{N} \frac{J_i - J_j}{(F_{ij} A_i)^{-1}}$$

and by substituting the radiosity relation between Q_i^d and σT_i^4,

$$\sigma T_i^4 = J_i + \frac{(1 - \varepsilon_i)}{\varepsilon_i A_i} \sum_{j=1}^{N} \frac{J_i - J_j}{(F_{ij} A_i)^{-1}}$$

These are sets of simultaneous linear algebraic equations that can be stated in the matrix form

$$a_{is} J_s = C_i$$

where a_{is} are known coefficients of geometry and surface properties and C_i is a column vector whose elements are Q_i^d when the dissipation is known and σT_i^4 when the temperature is known. Hence, J_s are obtained from the inversion

$$J_s = (a_{is})^{-1} C_i$$

and, with radiosities found, the temperatures can now be calculated as function of dissipation, or vice versa.

Network representation can be extended to encompass every surface in an enclosure by repeated use of the terms $(J_i - J_j)/(F_{ji} A_j)^{-1}$. Application to two simple but important cases is shown in Fig. 3.3. As noted, by placing radiosities as intermediary junctions in a circuit with current Q_i^d and net potential $(\sigma T_i^4 - \sigma T_j^4)$, a total electrical conductance is obtained corresponding to the definition of script F in the heat transfer equation.

An important observation from the two-surface enclosure is that when 1 is small compared to its surrounding 2, then $A_1/A_2 \approx 0$, $F_{12} = 1$, and

$$\Im_{12} = \varepsilon_1$$

Two parallel planes with ratio of size to separation $\gg 1$ may be approximated by

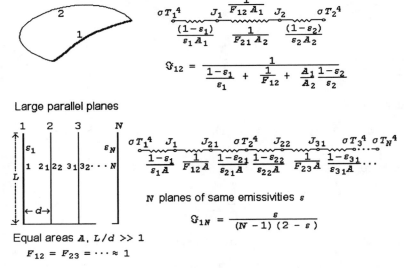

Fig. 3.3   Equivalent circuit and script F for a two-surface diffuse gray enclosure and a series of diffuse gray parallel planes.

an enclosure in which $A_1 = A_2$, $F_{12} = 1$, and

$$\Im_{12} = \frac{1}{(1/\varepsilon_1) + (1/\varepsilon_2) - 1}$$

For very long concentric cylinders of gap $r_1 - r_2$, $A_1/A_2 = r_1/r_2$, $F_{12} = 1$, and

$$\Im_{12} = \frac{1}{(1/\varepsilon_1) + (1/\varepsilon_2 - 1)(r_1/r_2)}$$

For concentric spheres,

$$\Im_{12} = \frac{1}{(1/\varepsilon_1) + (1/\varepsilon_2 - 1)(r_1/r_2)^2}$$

The equation for N infinitely parallel planes given in Fig. 3.3 is sometimes used to assess satellite insulations made from stacked thin sheets of aluminized phenolics (Chapter 6, Sec. II). It is easy to confirm that some 15 separated layers, aluminized on both sides ($\varepsilon < 0.03$), will decrease the radiation effectiveness of a surface by at least an order of magnitude.

Analytical expressions for configurations other than the two shown in Fig. 3.3 might be more difficult to derive; but a numerical value for script F for diffuse gray surfaces can always be found when the shape factors and properties are given. One procedure consists of evaluating a dummy case in which arbitrary values of temperatures are assigned to the surfaces of a given enclosure to obtain corresponding fictional radiosities and heat exchanges. Script F are then obtained as the solution of the resulting set of linear equations. For example, when $N = 3$,

$$Q_{1f}^d = \Im_{12} A_1 \left(\sigma T_{1f}^4 - \sigma T_{2f}^4 \right) + \Im_{13} A_1 \left(\sigma T_{1f}^4 - \sigma T_{3f}^4 \right)$$

$$Q_{2f}^d = \Im_{21} A_2 \left(\sigma T_{2f}^4 - \sigma T_{1f}^4 \right) + \Im_{23} A_2 \left(\sigma T_{2f}^4 - \sigma T_{3f}^4 \right)$$

$$Q_{3f}^d = \Im_{31} A_3 \left(\sigma T_{3f}^4 - \sigma T_{1f}^4 \right) + \Im_{32} A_3 \left(\sigma T_{3f}^4 - \sigma T_{2f}^4 \right)$$

where Q_{if}^d are the (fictitious) values of heat generated found from a radiosity analysis with arbitrarily assigned temperatures T_{if} (see the inversion matrix). Because $\Im_{ij} = (A_j/A_i)\Im_{ji}$, there are only three unknown $\Im_{ij}$ in the preceding three equations.

Other, equivalent techniques (including Hottel's, Eckert's, and Gebhart's[4]) can also be used, and any one of them can be the central routine in a thermal program for evaluating exchange factors. It must be remembered, however, that these standard methods are confined to the idealized diffuse gray model, which is only an approximation of actual behavior. In addition, as the number of interacting surfaces increases, inherent numerical errors and uncertainties in the values of properties will propagate into final results that can be very different from those in real life. Constraining procedures and error allocation schemes that optimize accuracy in terms of selected parameters have been devised based on the enclosure requirement that the shape factors must add up to one,[5,6] but the problem is restrained most effectively in moderating the number of surfaces by using combinations and other simplifications.

As noted in Chapter 2, a convenient substitution for $\Im$ is a product generally written $F_E F_A$, where F_A is shape factor and F_E a factor involving emissivities and dimensional ratios. This allows writing

$$d^2 Q^r_{dA_1-dA_2,\text{diffuse gray}} = (F_E)_{dA_1-dA_2} (\sigma T^4_{dA_1} - \sigma T^4_{dA_2}) \frac{\cos\theta_1 \cos\theta_2}{\pi s^2} dA_1\, dA_2$$

and

$$Q^r_{12,\text{diffuse gray}} = \int_{A_1} \int_{A_2} (F_E)_{dA_1-dA_2} (\sigma T^4_{dA_1} - \sigma T^4_{dA_2}) \frac{\cos\theta_1 \cos\theta_2}{\pi s^2} dA_1\, dA_2$$

which is a convenient formula for analytical derivations (see Application in Chapter 2, Sec. II). A list for some basic geometries is given in Table 2.1. It is noted that for high emissivities

$$[1/\varepsilon_1 + 1/\varepsilon_2 - 1]^{-1} \approx \varepsilon_1\varepsilon_2$$

In satellite thermal control the diffuse gray assumption has been found to give good correlation with test and flight data when radiation is in the infrared range. The approximation also seems satisfactory when specular (but gray) surfaces are present in an enclosure of mainly diffuse gray surfaces.[7,8] The model is the dominant one in satellite thermal analysis, particularly in dealing with interior (electronics canisters) radiation, and many thermal engineers prefer using thermal coatings and finishes with characteristics that approach diffuse gray to avoid more complicated analyses with questionable results.

C. Specular Reflectors and Solar Reflection off Specular Surfaces

Radiation among diffuse emitters that reflect in a purely specular manner and among combinations of purely specular and purely diffuse reflectors have been discussed in a number of papers.[9-11] A collective review with references can be found in Chapter 9 of Ref. 2.

Generally, the analysis is performed by ray tracing and the construction of images using the rules of optics geometry. But although the approach is productive in obtaining analytical expressions for a few idealized geometries, it is rather tedious when making simple calculations. Table 3.1 is a list for some standard geometries for which a script F can be determined.

For complicated geometries, thermal engineers resort to computers. The NEVADA Monte Carlo ray tracing program[12] has the capability to handle partly diffuse and partly specular reflections and is a popular tool in the analysis of solar impingement on radiators that are mostly diffuse in the infrared regime but have specular reflections in the solar range. (Quartz mirrors and unembossed silvered Teflon® exhibit such characteristics.[13]) However, the input parameters to these programs are invariably difficult to quantify due to a lack of data, especially in applications to multiple surfaces. It is therefore recommended that limiting analyses be performed in these cases to ensure that no reflected radiation interferes with an instrument's field of view or aggravates heating at other locations in the satellite. Conservative estimates, confined to specific areas of concern, can be made by assuming maximum specularity at solar incident angles that result in maximum impingement on the region under investigation. Such an approach may seem draconian, but conclusions based on it could at times be the only dependent guideline for finding the optimum orientation of a radiator or for selecting a surface finish.

**Table 3.1 Script F of diffuse gray emitters that are specular
or diffuse reflectors**

From infinite, purely specular, or diffuse surface 1 to infinite, parallel, purely specular
surface 2:

$$\Im_{12} = [1/\varepsilon_1 + 1/\varepsilon_2 - 1]^{-1}$$

From purely specular or diffuse inner to diffuse outer concentric long cylinders, $r_1 < r_2$:

$$\Im_{12} = [1/\varepsilon_1 + (r_1/r_2)(1/\varepsilon_2 - 1)]^{-1}$$

From purely specular or diffuse inner to purely specular, outer, concentric, long cylinders
1 and 2:

$$\Im_{12} = [1/\varepsilon_1 + 1/\varepsilon_2 - 1]^{-1}$$

From purely specular or diffuse inner to diffuse outer concentric spheres, $r_1 < r_2$:

$$\Im_{12} = [1/\varepsilon_1 + (r_1/r_2)^2(1/\varepsilon_2 - 1)]^{-1}$$

From purely specular or diffuse inner to purely specular outer spheres 1 and 2:

$$\Im_{12} = [1/\varepsilon_1 + 1/\varepsilon_2 - 1]^{-1}$$

Across N infinitely long, purely specular, parallel sheets with all emissivities $= \varepsilon$:

$$\Im_{12} = \varepsilon[(N - 1)(2 - \varepsilon)]^{-1}$$

D. Nondiffuse, Nongray Surfaces

Generalized mathematical formulations for nondiffuse and nongray surfaces
usually forgo the script F approach and retain the form of multiple integrals.
As noted earlier, the key quantities in these analyses are generalized radiation
intensities that are defined as functions of radiation and geometric parameters.

As an example, when dependence on wavelength and emission angles must be
included, the emissivity ε_i of elemental surface i of area dA_i in a specific direction
$(\theta_i^e, \varphi_i^e)$ at given wavelength λ may be related to the spectral directional emission
intensity $I_i^e(\lambda, \theta_i^e, \varphi_i^e)$ by the defining equation

$$I_i^e\left(\lambda, \theta_i^e, \varphi_i^e\right) = \varepsilon_i\left(\lambda, \theta_i^e, \varphi_i^e\right)I_{i,b}^e(\lambda)$$

where $I_{i,b}^e(\lambda)$ is the corresponding intensity for a black surface. Similarly, the portion of impinging intensities through solid angles $d\omega_i$ onto dA_i, from all elemental
surfaces j forming the one surface A_j, which end up reflecting in the same direction $(\theta_i^e, \varphi_i^e)$ and at the same wavelength as emission intensity, can be defined in
context of a spectral directional reflectivity $\rho_i(\lambda, \theta_i^e, \varphi_i^e)$ by the equation

$$I_i^{\text{reflected}}\left(\lambda, \theta_i^e, \varphi_i^e\right) = \int_{Aj} \rho_i\left(\lambda, \theta_i^e, \varphi_i^e, \theta_i, \varphi_i\right)I_i^i(\lambda, \theta_i, \varphi_i)\cos\theta_i \, d\omega_i$$

where I_i^i are incident intensities on dA_i from elemental areas dA_j and θ_i, φ_i are
zenith and azimuth angles made at dA_i by line s connecting dA_i with dA_j (see
Fig. 3.1b). Noting that the incident intensities on dA_i are the same as the outgoing
intensities from dA_j, the total outgoing spectral intensity from dA_i, I_i^o, in the

direction $(\theta_i^e, \varphi_i^e)$ is

$$I_i^o = \varepsilon_i(\lambda, \theta_i^e, \varphi_i^e) I_{i,b}^e + \int_{A_j} \rho_i(\lambda, \theta_i^e, \varphi_i^e, \theta_i, \varphi_i) I_j^o(\lambda, \theta_j, \varphi_j) \cos \theta_i \, d\omega_i$$

Similar considerations for elemental surface dA_j give

$$I_j^o = \varepsilon_j(\lambda, \theta_j^e, \varphi_j^e) I_{j,b}^e + \int_{A_i} \rho_j(\lambda, \theta_j^e, \varphi_j^e, \theta_j, \varphi_j) I_i^o(\lambda, \theta_i, \varphi_i) \cos \theta_j \, d\omega_j$$

As in a radiosity analysis for two diffuse gray surfaces, there are here two simultaneous (integral) equations in the two unknowns I_i^o and I_j^o, which, theoretically, can be solved when the functional relationships $\varepsilon(\lambda, \theta^e, \varphi^e)$ and $\rho(\lambda, \theta^e, \varphi^e)$ are given. With outgoing intensities found, the differential energy balance relating heat dissipated dQ_i^d to heat emitted dQ_i^e and heat absorbed dQ_i^a,

$$dQ_i^d = dQ_i^e - dQ_i^a$$

can now be written

$$dQ_i^d = dA_i \int_\lambda \int_{\text{hemisphere}} \varepsilon_i(\lambda, \theta_i^e, \varphi_i^e) I_{i,b}^e(\lambda) \cos \theta_i \, d\omega_i \, d\lambda$$

$$- dA_i \int_\lambda \int_{A_j} \alpha_i(\lambda, \theta_i^e, \varphi_i^e) I_i^e(\lambda) \cos \theta_i \, d\omega_i \, d\lambda$$

or

$$dQ_i^d = \varepsilon \sigma T_i^4 \, dA_i - dA_i \int_\lambda \int_{A_j} \alpha_i(\lambda, \theta_i^e, \varphi_i^e) I_j^o(\lambda, \theta_j^e, \varphi_j^e) [\cos \theta_i \cos \theta_j / s^2] \, dA_j \, d\lambda$$

where ε is emissivity averaged over all wavelengths and directions. Total heat exchange from the whole of surface i to N surfaces becomes

$$Q_i^d = \varepsilon A_i \sigma T_i^4 - A_i \sum_j \int_\lambda \int_{A_j} \alpha_i(\lambda, \theta_i^e, \varphi_i^e) I_j^o(\lambda, \theta_j^e, \varphi_j^e) \left[\frac{\cos \theta_i \cos \theta_j}{s^2} \right] dA_j \, d\lambda$$

Notwithstanding the doubtful validity of such equations to predict actual performance, thermal analysts continue to investigate various techniques for solving them. Numerical solutions of the quadratures are the conventional approach with direct simulation Monte Carlo techniques lately gaining the greatest preference. A discussion on Monte Carlo in radiation heat transfer is given in Ref. 14. Chapter 10 in Ref. 2 contains a sample problem. But in any case, one must be cautious not to draw critical design conclusions from these methods without support from an analysis of behavior under limiting conditions.

III.   Satellite Enclosures

Interacting satellite surfaces may be fashioned into enclosures of individual walls that can include space as a black closeout at 0 K. As illustrated in Fig. 3.4, one enclosure may be thermally coupled to others to form an overall control volume for thermal analysis. For diffuse gray and other simple exchanges, the methods

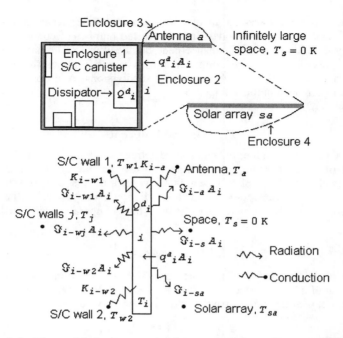

Fig. 3.4   Thermal balance on an isothermal surface within an enclosure.

discussed earlier can be used for calculating (usually with a computer) the script F from given geometries and surface properties.

Script F completes the conductance listing comprising radiation, conduction, and, when necessary, convection elements. These couplings must be known if one is to compute the temperatures for given heating. Referring to Fig. 3.4, if the individual surfaces are assumed isothermal, then a heat balance on each would result in a set of equations whose coefficients are the conductances connecting the various temperatures. For example, considering surface i in steady state yields

$$Q_i^d + q_i^a A_i = K_{i-w1}(T_i - T_{w1}) + K_{i-w2}(T_i - T_{w2}) + K_{i-a}(T_i - T_a)$$

$$+ \sum_j \Im_{i-wj} A_i \left(\sigma T_i^4 - \sigma T_{wj}^4\right) + \Im_{i-a} A_i \left(\sigma T_i^4 - \sigma T_a^4\right)$$

$$+ \Im_{i-sa} A_i \left(\sigma T_i^4 - \sigma T_{sa}^4\right) + \Im_{i-s} A_i \sigma T_i^4$$

In principle, the total number of equations for the coupled enclosures will be sufficient to determine all unknown temperatures.

It should be clarified at this point that an isothermal surface is indeed a rarity in an actual satellite. Large, uniformly heated structures, such as solar arrays and panels with heat pipes, might have regions approaching isothermal conditions, but variations in localized heating and jumps in thermal couplings more often cause a noticeable temperature difference from one location to another on the same structure. Thus, although techniques such as the one outlined earlier are found useful in doing preliminary or systems level assessments, particularly when only one or two interfaces are involved, they must not be viewed as a substitute for a detailed analysis. A modest increase in the number of connected sections (nodes) will, of course, be an enhancement, but the ultimate alternative in thermal control analysis is prepared computer software with routines that can handle quickly and accurately a large number of subdivisions. Additional discussions on this subject are found in Chapter 5.

IV. Optical Properties

Optical properties in thermal control are entrenched in notions relating the temperature of an isothermal surface in space to its heating environment. Thus, a solitary isothermal body of surface area A^r, generating heat at a constant rate Q^d and free of all heat transfer except surface radiation to deep space, will emit all its dissipation in accordance with a steady-state, heat-temperature relationship given by

$$Q^d = Q^e = \varepsilon A^r \sigma T^4$$

where ε is the emissivity of the surface, which is a fraction that can only approach that of a black surface for which $\varepsilon = \varepsilon_b = 1$. Moreover, measurements of temperature, heating, and area will yield a numerical value for ε,

$$\varepsilon = (Q^e/A^r)/\sigma T^4 = q^e/\sigma T^4$$

Similarly, solar absorptivity α^s is perceived such that the same body when free of all heating except solar radiation will have an equilibrium temperature in accordance with the equation

$$\varepsilon A^r \sigma T^4 = \alpha^s A^i q^S$$

where A^i is the area exposed to normal incident solar radiation q^S. Knowledge of the areas, temperature, solar heating, and emissivity (from the first set of measurements) will give a value for α^s.

Reflectivity ρ is thought of in the same manner. Here, however, complications might arise because the amount of reflected energy can depend on the spectrum of incident radiation. But in any event, the values of ε, α^s, and ρ derived from these notions are considered material (surface) properties that may be used in general energy balance equations that can include transients, edge effects, and heat exchanges with neighboring sources such as planets and surrounding structures. How these bulk properties relate to their formal counterparts in radiation physics is the subject of the following paragraphs.

A. Emissivity

Being associated with radiation emission, emissivity is defined in terms of emission intensity, which, as described in Fig. 3.5, generally has spectral as well as directional dependence. A definition that highlights this characteristic is the spectral directional emissivity given by

$$\varepsilon_{\lambda,\theta} \equiv I^e_{\lambda,\theta}(\lambda, \theta, \varphi; T) \big/ I^e_{\lambda,b}$$

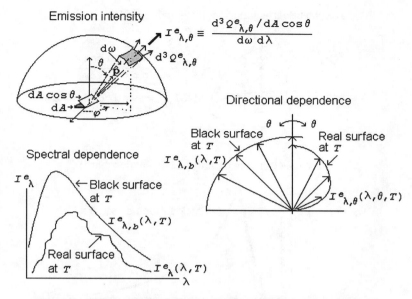

Fig. 3.5 Emission intensity, spectral and directional distribution.

where T is deliberately inserted to acknowledge that emissivity can depend on temperature. In terms of this definition, the spectral hemispherical emission given earlier (Sec. II of this chapter) by

$$q_\lambda^e(\lambda; T) = \int_{\varphi=0}^{2\pi} \int_{\theta=0}^{\pi/2} I_{\lambda,\theta}^e(\lambda, \theta, \varphi; T) \cos\theta \sin\theta \, d\theta \, d\varphi$$

can be restated as

$$q_\lambda^e(\lambda; T) = \int_{\varphi=0}^{2\pi} \int_{\theta=0}^{\pi/2} \varepsilon_{\lambda,\theta} I_{\lambda,b}^e \cos\theta \sin\theta \, d\theta \, d\varphi$$

This gives a meaning to a spectral hemispherical emissivity defined by

$$\varepsilon_\lambda(\lambda; T) \equiv q_\lambda^e(\lambda; T) / q_{\lambda,b}^e(\lambda; T)$$

from which the total hemispherical emission (emissive power) is found as

$$q^e(T) = \int_0^\infty \varepsilon_\lambda(\lambda; T) q_{\lambda,b}^e(\lambda; T) \, d\lambda$$

Comparing to the emissivity used in thermal control,

$$\varepsilon = \frac{q^e(T)}{\sigma T^4} = \int_0^\infty \varepsilon_\lambda(\lambda; T) q_{\lambda,b}^e(\lambda; T) \, d\lambda / \sigma T^4$$

This is also the emissivity used in most other engineering work. It is measured (Sec. IV. C, this chapter) in view of the preceding integral and is an average value for all directions and wavelengths.

(a) Spectral directional incident intensity

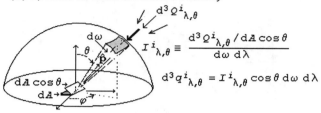

(b) Solar spectrum near Earth (outside atmosphere)

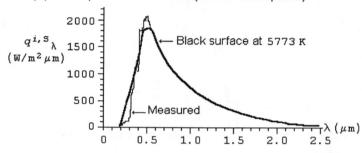

Fig. 3.6   Incident radiation intensity and the solar spectrum outside the atmosphere in the vicinity of Earth; solar radiation is approximately that from a black disk at 5500°C with view factor $F_{E-S} = 2.143E - 5$ (Chapter 4, Sec. V.A).

B.   Absorptivity and Solar Absorptivity

A formal definition of absorptivity utilizes the concept of spectral directional incident intensity $I^i_{\lambda,\theta}$ (W/m^2 per steradian per micrometer), defined as the rate $d^3 Q^i_{\lambda,\theta}$ of incident radiation (irradiation) at wavelength λ in the θ–φ direction, per unit area of the intercepting surface normal to this direction, per unit solid angle about this direction, per wavelength interval $d\lambda$ about λ. A schematical description is given in Fig. 3.6a. Thus,

$$I^i_{\lambda,\theta}(\lambda, \theta, \varphi) \equiv \left[\frac{d^3 Q^i_{\lambda,\theta}}{(dA \cos \theta)} \right] \bigg/ d\omega \, d\lambda$$

or, more to the point,

$$\frac{d^3 Q^i_{\lambda,\theta}}{dA} = d^3 q^i_{\lambda,\theta}(\lambda, \theta, \varphi) = I^i_{\lambda,\theta}(\lambda, \theta, \varphi) \cos \theta \sin \theta \, d\theta \, d\varphi \, d\lambda$$

$Q^i_{\lambda,\theta}(\lambda, \theta, \varphi)$ and $q^i_{\lambda,\theta}(\lambda, \theta, \varphi)$ are called spectral directional irradiation. The spectral hemispherical irradiance is obtained from

$$q^i_\lambda(\lambda)(\text{W/m}^2\mu m) = \int_{\varphi=0}^{2\pi} \int_{\theta=0}^{\pi/2} I^i_{\lambda,\theta}(\lambda, \theta, \varphi) \cos \theta \sin \theta \, d\theta \, d\varphi$$

and total irradiance is

$$q^i \, (\text{W/m}^2) = \int_0^\infty q_\lambda^i(\lambda) \, d\lambda = \int_{\lambda=0}^\infty \int_{\varphi=0}^{2\pi} \int_{\theta=0}^{\pi/2} I_{\lambda,\theta}^i(\lambda, \theta, \varphi) \cos \theta \sin \theta \, d\theta \, d\varphi \, d\lambda$$

A spectral directional absorptivity can be defined by

$$\alpha_{\lambda,\theta} \equiv I_{\lambda,\theta}^{i,a}(\lambda, \theta, \varphi) / I_{\lambda,\theta}^i(\lambda, \theta, \varphi)$$

where $I_{\lambda,\theta}^{i,a}(\lambda, \theta, \varphi)$ is the absorbed portion of the spectral directional incident intensity. Denoting the corresponding spectral hemispherical absorbed irradiance by $q_\lambda^{i,a}$,

$$q_\lambda^{i,a}(\lambda) = \int_{\varphi=0}^{2\pi} \int_{\theta=0}^{\pi/2} \alpha_{\lambda,\theta} I_{\lambda,\theta}^i(\lambda, \theta, \varphi) \cos \theta \sin \theta \, d\theta \, d\varphi$$

and a spectral hemispherical absorptivity is obtained from the ratio

$$\alpha_\lambda \equiv q_\lambda^{i,a}(\lambda) \big/ q_\lambda^i(\lambda)$$

Hence, the total absorbed irradiance is

$$q^{i,a} = \int_0^\infty q_\lambda^{i,a}(\lambda) \, d\lambda = \int_0^\infty \alpha_\lambda q_\lambda^i(\lambda) \, d\lambda$$

and the total hemispherical absorptivity, averaged over all wavelengths and directions is

$$\alpha = \frac{q^{i,a}}{q^i} = \int_0^\infty \alpha_\lambda q_\lambda^i(\lambda) \, d\lambda \bigg/ \int_0^\infty q_\lambda^i(\lambda) \, d\lambda$$

With the limits on λ extending the whole radiation spectrum, α in the preceding equation becomes the absorptivity that must equal the emissivity under Kirchhoff's law for radiative exchange in isothermal enclosures with all surfaces absorbing and emitting uniformly in all directions.

The measured spectral distribution of the sun in the vicinity of Earth, beyond atmospheric interference, is approximately as shown in Fig. 3.6b. It is concentrated in the 0.2–3 μm range and compares well with a spectral distribution based on considering the solar disk as a black surface at 5500°C (see Example in Chapter 4, Sec. V.A). The absorptivity of a surface in this spectrum is the solar absorptivity α^s used to characterize thermal coatings and other treatments. Using the last equation as a basis for measurement, solar absorptivity is calculated from

$$\alpha^s \approx \int_0^\infty \alpha_\lambda q_{\lambda,b}^e(\lambda; 5773 \text{ K}) \, d\lambda \bigg/ \int_0^\infty q_{\lambda,b}^e(\lambda; 5773 \text{ K}) \, d\lambda$$

where $q_{\lambda,b}^e(\lambda; 5773 \text{ K})$ is spectral emission (W/m²μm) from a black surface at 5773 K.

Some surfaces exhibit a noticeable directional variation in reflected solar radiation at shallow incident angles.[15] It is not certain how representative integrated averaging techniques become under such conditions. Most thermal engineers, however, seem satisfied to bypass the problem by performing limiting case analyses in which maximum local variations are assumed on the value of α^s.

C. Reflectivity and Reflectometers: Properties Measurement

From the conservation of energy, unabsorbed incident radiation on an opaque surface must be reflected. Here, however, definitions become mired by the possibility that the reflection may be scattered at different intensities and may be either diffuse, specular, or a combination of both. Although there have been in this connection some generalized formulations and tests,[13] thermal engineers, by and large, tend to consider the spectral, directional character of a surface in reference only to the nature of incident radiation. They also follow the recommendation of the International Commission on Illumination and express reflectivity as the ratio of the energy actually reflected to what would be reflected off a theoretically perfect reflector that is also perfectly diffuse.[16] Instruments for measuring reflectivity are designed to operate on this basis.

Reflectivity can be measured by means of a reflectometer. Basically, the device contains a source of radiation, a monochromator for selecting the incident spectral band, and a detector that intercepts the radiation reflected off the surface being tested and determines its ratio to that reflected off the calibrated standard. (Opaque layers of sodium chloride and barium sulfate make an excellent standard as the stack is very nearly perfectly reflective, as well as perfectly diffuse, up to 80 deg off normal direction.) Ideally, the instrument should have no limitations to the selection of spectral bands or the angles of incidence, and information on the spectral directional reflectivity, denoted by $\rho_{\lambda,\theta}(\lambda, \theta, \varphi)$, where θ, φ is direction of incidence, may be obtained for any combination of wavelengths and incident direction. Sophisticated instruments (called goniospectrophotometers) that approach ideal operation are available but they are not commonly used in thermal control.

Practical, easy-to-operate reflectometers have somewhat limited capabilities. But various instruments have been built with enhancements for accuracy in specific wavelength bands and incident angles. The Cary 140 integrating sphere reflectometer, for example, is a laboratory-fixed instrument that gives accurate data in the wavelengths range of 0.3–1.6 μm (covering about 95% of the solar spectrum) and permits incidence and detection in practically any direction. It can also be used to determine the portions of diffuse and specular reflection. The Gier Dunkle MS-251 portable reflectometer is convenient in the range 0.3–2.5 μm, and the Gier Dunkle DB-100 portable infrared (IR) reflector is widely used at normal incidence ($\theta = 0$ deg) in the range 0.5–35 μm, which covers over 90% of the radiation spectrum of a black surface at room temperature. The Gier Dunkles are often operated directly on a standing spacecraft for periodic preflight monitoring of coatings' properties.

A major benefit derived from measuring reflectivity is that the value of $\rho_{\lambda,\theta}(\lambda, \theta, \varphi)$ can be averaged over a given range of λ to obtain the average absorptivity in that spectral band. Thus, at angle of incidence θ,

$$\alpha(\theta) = 1 - \rho(\theta)$$

and, using weighted integration for averaging in the range λ_1 to λ_2,

$$\alpha_{\lambda_1,\lambda_2}(\theta) = 1 - \left[\int_{\lambda_1}^{\lambda_2} \rho_{\lambda,\theta}(\lambda, \theta, \varphi)\, q_\lambda^{\text{reflected}}(\theta)\, d\lambda \bigg/ \int_{\lambda_1}^{\lambda_2} q_\lambda^{\text{reflected}}(\theta)\, d\lambda \right]$$

$q_\lambda^{\text{reflected}}(\theta)$ is the magnitude of reflected energy detected off the standard sample, which should be very nearly equal to the controlled incident radiation.

Reflectometers are more appropriately called emissometers when the radiation source is a heated cavity (approximating a blackbody) emitting in the IR spectrum. With a controllable range from 0.5 to 35 μm, the average emissivity at incident angle θ, which is equal to the average absorptivity at that angle, is determined from

$$\varepsilon(\theta; T) = 1 - \left[\int_{0.5\,\mu m}^{35\,\mu m} \rho_{\lambda,\theta}(\lambda, \theta)\, q_{\lambda,b}^e(\lambda; T)\, d\lambda \Big/ \int_{0.5\,\mu m}^{35\,\mu m} q_{\lambda,b}^e(\lambda; T)\, d\lambda \right]$$

For measurements made at normal incidence ($\theta = 0$),

$$\varepsilon_n = 1 - \left[\int_{0.5\,\mu m}^{35\,\mu m} \rho_n(\lambda)\, q_{\lambda,b}^e(\lambda; T)\, d\lambda \Big/ \int_{0.5\,\mu m}^{35\,\mu m} q_{\lambda,b}^e(\lambda; T)\, d\lambda \right]$$

In the solar spectrum between 0.3 and 2.5 μm (1.6-μm upper value is usually considered sufficient), average solar absorptivity for incident angle θ is obtained from

$$\alpha^s(\theta) = 1 - \left[\int_{0.3\,\mu m}^{2.5\,\mu m} \rho_{\lambda,\theta}(\lambda, \theta)\, q_\lambda^S(\lambda)\, d\lambda \Big/ \int_{0.3\,\mu m}^{2.5\,\mu m} q_\lambda^S(\lambda)\, d\lambda \right]$$

in which $q_\lambda^S(\lambda)$ is energy in the solar spectrum (equal in magnitude to incident radiation at angle θ), which is diffusely reflected off the standard sample.

Representative as-applied data from various measurements made by or for Fairchild Space Company between 1975 and 1990 are shown in Fig. 3.7a. The

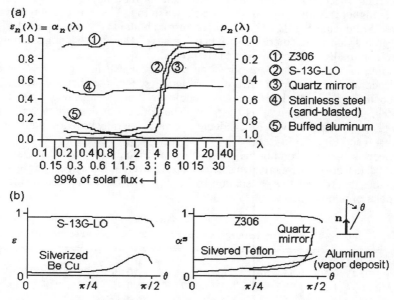

Fig. 3.7 As-applied optical properties of thermal control surfaces: a) normal ($\theta = 0$) properties, variation with wavelength; b) total wavelength average properties, variation with direction.

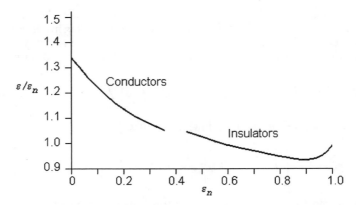

Fig. 3.8   Ratio of total hemispherical emissivity to the normal value.

spectral data (variations with λ) were taken at normal incidence on various occasions, sometimes using different instruments. They are connected here on the semilog scale. The normal solar absorptivity α_n^s is found by integration from $\lambda_1 \approx 0.2$ to $\lambda_2 \approx 3$, whereas total normal emissivity ε_n is obtained by integration over the whole spectrum shown in Fig. 3.7a.

It is noted that variations with λ of Z306 black paint and sand-blasted stainless steel are small, and the ratio α_n^s/ε_n for both finishes remains about one throughout the spectrum, even though the two coatings have considerably different optical properties. In contrast, buffed aluminum yields a ratio α_n^s/ε_n of the integrated values higher than one, and for the case shown in Fig. 3.7a, it is possibly three or more. White paint S-13G-LO and quartz mirrors have low as-applied α_n^s/ε_n (integrated values), but there is a shift in the relative values that becomes important when marginally higher emissivities are needed.

Variations in α^s and ε with direction are represented in Fig. 3.7b. There is evidence of deviation from diffuse characteristics. Finishes that are electrically nonconductive (mostly high ε) show a small variation in emissivity up to about 70 deg, where a sharp drop is observed. Theoretical considerations verify this trend and opposite behavior for conductors, which generally have low emissivities (Sec. 4.6 and Fig. 4.8 in Ref. 2.)

Spectroscopic data on normal emissivities are much more abundant than integrated data from hemispherical measurements. Conversions may be made by using equations that relate ε to ε_n as summarized in Fig. 3.8. The curves are the locus of the pairs of points[17] shown in Table 3.2.

Table 3.2   Relation between normal and hemispherical emissivity

Insulators											
ε_n: 0.45	0.50	0.55	0.60	0.65	0.70	0.75	0.80	0.85	0.90	0.95	1.0
$\varepsilon/\varepsilon_n$: 1.04	1.025	1.005	0.995	0.98	0.97	0.96	0.95	0.94	0.935	0.935	1.0

Conductors							
ε_n: 0.00	0.05	0.10	0.15	0.20	0.25	0.30	0.35
$\varepsilon/\varepsilon_n$: 1.33	1.27	1.225	1.185	1.145	1.105	1.075	1.055

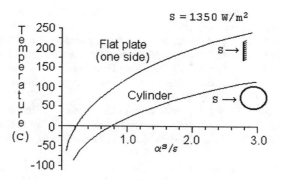

Fig. 3.9   Average temperature vs α^s/ε: for one-sided flat plate, $\sigma T_{av}^4 = (\alpha^s/\varepsilon)q^S$; for cylinder, $\sigma T_{av}^4 = (\alpha^s/\pi\varepsilon)q^S$.

Figure 3.9 shows the effect of α^s/ε on the average temperature of a surface in the sun. The curves are constructed using the relation

$$\varepsilon A^r \sigma T_{av}^4 = \alpha^s A^i q^S$$

where q^S is taken as unobstructed normal solar radiation, $S = 1350\,\text{W/m}^2$ (nominal value), with $A^i = A^r$ for a flat plate (adiabatic shaded side), and $A^i = A^r/\pi$ for a cylinder.

References

[1]Incropera, F. P., and DeWitt, D. P., *Fundamentals of Heat Transfer*, Wiley, New York, 1981, pp. 542–685.

[2]Siegel, R., and Howell, J. R., *Thermal Radiation Heat Transfer*, 2nd ed., Hemisphere, Washington, DC, 1981.

[3]Oppenheim, A. K., "Radiation Analysis by the Network Method," *Transactions of the American Society of Mechanical Engineers*, Vol. 78, 1956, pp. 725–735.

[4]Sparrow, E. M., "On the Calculation of Radiant Interchange between Surfaces," *Modern Developments in Heat Transfer*, edited by W. Ibele, Academic, New York, 1963, pp. 181–211.

[5]Taylor, R. P., and Luck, R., "Closure Enforcement Methods for Radiation View Factors," *Journal of Thermophysics and Heat Transfer*, Vol. 9, No. 4, 1995, pp. 660–666.

[6]Taylor, R. P., Luck, R., Hodge, B. K., and Steele, W. G., "Uncertainty Analysis of Diffuse-Gray Radiation Enclosure Problems," *Journal of Thermophysics and Heat Transfer*, Vol. 9, No. 1, 1995, pp. 63–69.

[7]Toor, J. S., and Viskanta, R., "A Critical Examination of the Validity of Simplified Models for Radiant Heat Transfer Analysis," *International Journal of Heat Transfer*, Vol. 15, 1972, pp. 1553–1567.

[8]Schornhorst, J. R., and Viskanta, R., "An Experimental Examination of the Validity of the Commonly Used Methods of Radiant Heat Transfer Analysis," *Journal of Heat Transfer*, Vol. 90, No. 4, 1968, pp. 429–436.

[9]Eckert, E. R. G., and Sparrow, E. M., "Radiative Heat Exchange between Surfaces with Specular Reflection," *International Journal of Heat Transfer*, Vol. 3, No. 1, 1961, pp. 42–54.

[10]Sparrow, E. M., Eckert, E. R. G., and Jonsson, V. K., "An Enclosure Theory for Radiative Exchange between Specularly and Diffusely Reflecting Surfaces," *Transactions of the American Society of Mechanical Engineers: Journal of Heat Transfer*, Nov. 1962, pp. 294–300; also American Society of Mechanical Engineers, Paper 61-WA-167, 1961.

[11]Sparrow, E. M., and Lin, S. H., "Radiation Heat Transfer at a Surface Having Both Specular and Diffuse Reflectance Components," *International Journal of Heat and Mass Transfer*, Vol. 8, 1965, pp. 769–779.

[12]*NEVADA User's Manual*, Turner Associates Consultants, Incline Village, NV.

[13]Drolen, B. L., "Bidirectional Reflectance and Specularity of Twelve Spacecraft Thermal Control Materials," *Journal of Thermophysics and Heat Transfer*, Vol. 6, No. 4, 1992, pp. 672–679.

[14]Howell, J. R., "Application of Monte Carlo to Heat Transfer Problems," *Advances in Heat Transfer*, edited by T. F. Irvine, Jr., and J. P. Hartnett, Vol. 5, Academic, New York, 1968, pp. 1–54.

[15]Stultz, J. W., "Solar Absorptance of Second Surface Mirrors for High Angles of Incidence," AIAA Paper 74-670, 1974.

[16]MacAdam, D. L., "Spectrophotometry," *Encyclopedia of Physics*, edited by R. G. Lerner and G. L. Trigg, VCH, New York, 1991, pp. 1160–1162.

[17]Jakob, M., *Heat Transfer*, Vol. II, Wiley, New York, 1957, pp. 41–52.

Chapter 4

Heating Fluxes

I. Introduction

HEATING sources are present throughout the life of a satellite, but because operational time in orbit is by far the longest, most of thermal engineering is expended on ensuring long-term temperature stability in space. Of course, this is not to say that the evaluation of other phases in a satellite's history are less important.

The nature of heating and the methods for dealing with it are defined in terms of the source and the surrounding environment. These can be divided broadly into ground, vacuum testing, launch, and space. On the ground, heating is due mostly to equipment dissipation and the conditions of surrounding air. Direct or indirect solar effects must also be considered when the satellite is not sheltered. Vacuum testing eliminates the interaction with air but must contend with the presence of gravity and the limitations of vacuum chambers to simulate the true thermal environment in space. During launch, the effects include exposure to sunlight and air molecular friction. Finally, heating in space is considered due mainly to satellite equipment dissipation and radiation from the sun and planets. Parasitic sources affecting cryogenics, or nuclear power, superhigh rf generation, and other specialized systems are treated by following essentially the same principles but with correspondingly applicable conditions.

II. Ground Environment

While on the ground in air, the desired temperatures are created by air conditioning, a process that continues through launch operations until shortly before liftoff. Heat removal takes place by exchange with either still or flowing cool air (sometimes dry nitrogen) in accordance with the principles of convection. In this respect, it must be noted that although a satellite's thermal design is directed toward operation in space, accommodations must be made to ensure that the necessary coefficients of convection are not frustrated by the air's inability to flow when and where it is needed. When the solution to the problem of freedom of flow appears to entail costly work on the satellite or launch facility, the option of intermittent operation of dissipating components should be considered. Furthermore, in all operations in air, humidity limits must be preserved to prevent moisture condensation, which occurs when the vapor temperature falls below the saturation value.

Temperature and humidity control are also required during storage and transportation, but it is not expected to operate the satellite during these periods.

Shipping containers often feature white coatings on the outside to reflect solar radiation; however, main control is by air conditioning on the inside.

Ground testing in vacuum chambers is a separate and unique aspect of thermal control and a dedicated discussion is presented in Chapter 7.

III. Ascent Heating

Significant spacecraft heating due to air friction (aerodynamic heating) occurs within the span of approximately 2–5 min after liftoff. This is too short a period to affect a heavy satellite as a whole, but unsheltered individual light components could rise to temperatures beyond acceptable levels and may even be damaged from exposure. The solution since the beginning of the space program has been a protective fairing designed to bear the brunt of heating. The fairing itself may rise momentarily (about a minute) to temperatures as high as 180°C, but its effect on most exposed elements is usually minor, especially because it carries on its interior surface an insulation and low-emissivity treatment.

Launch weight considerations require that the fairing be jettisoned as early as possible, an event that usually takes place at an altitude of about 115 km. At this elevation there is free molecular heating q^{fmh}(W/m^2), that can be conservatively estimated from the equation

$$q^{fmh} = (1/2)\rho u^3$$

where ρ(kg/m^3) is local atmospheric density and u(m/sec) the magnitude of velocity parallel to the heated surface. Flight monitors indicate a large variation in the heating value along the vehicle's skin with an average of about 250 W/m^2 for some 20 min. Usually this can be tolerated by most components.

In cases where ascent heating may be a hazard, critical components should be designed with improved thermal mass Mc (product of mass and specific heat) or constructed from materials that can withstand excessive temperatures for relatively short periods of time. Also, it may sometimes be necessary to increase launchpad air cooling to launch at a sufficiently low temperature to ride out the period causing high temperatures.

Example: Estimating Temperatures During Ascent

Estimates of the maximum temperature that a light, nondissipating component reaches during ascent are often based on the conservative assumption that the component has no mass at all and will, therefore, respond instantly to heating. The temperature is then obtained from a steady-state energy balance that equates energy absorbed to that radiated; that is,

$$Q^a = \varepsilon A^r \sigma T^4$$

where Q^a may include free molecular heating (FMH) in addition to radiation from the sun and Earth. (Some NASA specifications require a total incident heating flux equivalent to three solar constants.)

Components of special concern include exposed layers of multilayered insulation, solar arrays, antenna dishes, deployment mechanisms, wires, and extended elements. In the case of a stowed solar array or antenna dish, the side hidden from view could have an interface with the satellite's main canister that renders it nearly adiabatic, and a maximum temperature analysis should then consider radiation only from the exposed surface.

Using a steady, free molecular flux of 250 W/m^2 and undegraded (beginning of life) surface properties, the preceding equation gives the data in Table 4.1. A nominal value of solar flux equal to 1350 W/m^2, incident perpendicular to the surface, is assumed here when solar energy is added.

The fact that combined heating fluxes on the cell side of a solar array can potentially drive the temperature in excess of the recommended short-time limit suggests exposing instead the other side, because it is usually treated with low solar absorptivity. This, however, would erase the possibility of at least partial charging of the satellite batteries by random or deliberate solar impingement on the cells in the event of a delayed array deployment. An alternative is often exercised in a slow rotation (rotisserie mode) imparted by the launch vehicle during ascent. The result of this maneuver is reduced average heating and a lower temperature.

Cooling during ascent does not normally present serious concerns. Because stowed structures generally have a reduced view to space, it is expected that their cool-down rate dT/dt, where t is time, will not be as steep as that experienced during orbital eclipses, for which satellites are designed and tested to sustain. A conservative, low-temperature analysis assumes perfect cool-down beginning at launchpad conditions and continuing according to the equation that relates loss of internal energy to the heat radiated; that is,

$$Mc\frac{dT}{dt} = -\varepsilon A^r \sigma T^4$$

with initial condition

$$T(0) = T_{\text{launchpad}} = T_0$$

Here, A^r (m^2) is exposed area, M is mass (kg) and c the specific heat in J/kg K when $\sigma = 5.67E{-}8$ W/m^2 K^4. By separating the variables and integrating from time zero, when the temperature is T_0, the temperature after t seconds is found as

$$T_t = \frac{T_0}{\left[1 + \left(3\varepsilon A^r \sigma T_0^3 t / Mc\right)\right]^{\frac{1}{3}}}$$

For example, a solar array of specific thermal mass $Mc/A^r = 0.7$ W-h/m^2 K (typical) with exposed cell side ($\varepsilon = 0.78$) would, in 30 min of perfect cooling, drop from 20°C (typical liftoff temperature) to a temperature

$$T_{30\,\text{min}} = (20 + 273)/[1 + (3)(0.76)(1/0.7)(5.67E{-}8)(20 + 273)^3(30/60)]^{1/3}$$
$$= 196.4\,\text{K}\,(-77°\text{C})$$

This value is in the same range as array temperatures at the end of eclipse in a low Earth orbit and considerably higher than those at the end of maximum geosynchronous eclipse.

Table 4.1   Maximum temperature during ascent

Item	α^s	ε	Impinging flux	q^a, W/m^2	T, °C	Short-time limit (°C)
Kapton® insulation sheet	0.48	0.81	FMH	250	−2	150
(125 μm thick with Al backing)			FMH and solar	250 + (0.48)1350	101	
Solar array	0.80	0.78	FMH	250	1	120
(cell side exposed, adiabatic back)			FMH and solar	250 + (0.80)1350	143	
Solar array	0.20	0.84	FMH	250	−4	120
(back exposed, adiabatic cell side)			FMH and solar	250 + (0.20)1350	50	
Antenna dish	0.20	0.85	FMH	250	−5	70
(coated side exposed, adiabatic back)			FMH and solar	250 + (0.20)1350	49	
Tubular element	0.12	0.08	FMH	250	211	250
(shiny, buffed Al surface)			FMH and solar	250 + [(0.12)1350]/π	143	

IV. Orbit Environment Heating Fluxes

Satellite heating in Earth orbit is considered by most thermal analysts to be caused only by equipment $I^2 R$ dissipation and radiation received from the sun and the Earth. Effects from other heavenly bodies, elementary particle bombardment, friction with low-density matter, and actual space background (≈ 2.7 vs 0 K for deep space) are usually assumed negligible when determining the temperature. Heating from Earth also becomes negligible at geosynchronous orbit and during most of the trajectory in a high-eccentricity orbit (Molniya).

Impinging radiation from the sun on a surface is characterized by the solar flux S (W/m^2), commonly known as the solar constant, and by orientation with respect to the sun. The great distance from the sun justifies the assumption that radiation is in parallel rays, which gives rise to the term solar or sun vector, defining a vector of magnitude S with direction along the rays. Solar impingement is, therefore, reduced by the cosine of incident angle off the surface normal.

Earth radiation is separated into two entities: IR emission (Earth flux E, W/m^2) and reflected sun (albedo A, W/m^2), which is usually stated as a fraction f of the solar constant. Earth flux and albedo are commonly assumed diffuse and, therefore, reduced by the shape factor to the receiving surface. Thus, once environment fluxes (S, A, and E) are specified and the orientation and orbital parameters of a satellite surface defined, calculations of normal incident fluxes (indicated by the symbols q^S, q^A, and q^E, in W/m^2) become solely an exercise in geometry.

NASA's "Space and Planetary Environment Criteria Guidelines for Use in Space Vehicle Development"[1] is sometimes cited as a standard reference on values of environment heating fluxes. The document is occasionally revised to reflect on-going measurements by instruments carried on various satellites. These have included Nimbus-7, Solar Maximum Mission (SMM), Upper Atmosphere Research Satellite (UARS), and others. The Earth Radiation Budget Experiment (ERBE),[2] consisting of Earth Radiation Budget Satellite (ERBS) and two National Oceanic and Atmospheric Administration (NOAA) spacecraft, is a dedicated program for global measurements. Data collection is a continuing effort with many of the findings regularly reported in the *Journal of Geophysical Research*. A summary and comparison of the yearly mean value of solar constant as measured by different experiments is tabulated in Ref. 3.

ERBE confirms that a rather wide range of values for environment fluxes must be used in thermal analysis to encompass all eventualities. It also indicates that albedo is the least determinable effect, with strong dependence on Earth topography, solar elevation, and such variables as cloud formation and atmospheric conditions. Earth flux is found to vary with the planet's local surface temperature and cloud cover, and solar radiation is affected mainly by seasonal variations with changing distance from the sun.

But despite rather consistent trends in the data, there has not been strict unanimity among thermal engineers on selecting specific values and tolerances. In general, the disagreements are due to varying specifications by source contractors who often differ in their requirements for margins. The values given in Table 4.2 appear to be within most requirements encountered in practice.

A. Solar Radiation

Most measurements above the Earth's atmosphere indicate a solar constant between 1365 and 1373 W/m^2 at mean distance from the sun. The spectral distribution

Table 4.2   Commonly used ranges of environment fluxes

Solar flux: typically, $S = 1350 \pm 2$ W/m^2 at 1 a.u.
   summer solstice, 1310 ± 10 W/m^2
   winter solstice, 1390 ± 10 W/m^2
Albedo factor f (fraction of solar flux): 0.33 ± 0.13
Earth flux E: 236 W/m^2 $\pm$ 38 W/m^2

(estimated from tables in Ref. 4) is about 7% uv in the 0.31–0.40 μm range, 46% visible (0.40–0.69 μm), and 47% IR above 0.70 μm. Solar IR has shorter wavelengths than the IR emitted at normal satellite temperatures, and as noted in Sec. II., Chapter 2, one can take advantage of this difference and condition a surface to have simultaneously a high reflectivity in the solar spectrum and high emissivity in IR. (The opposite is attempted in making so-called solar collectors.) As mentioned before, the property connected with this idea is solar absorptivity α^s, which is the fraction of unhindered solar energy that is absorbed by a surface; that is,

$$S^a(\text{unhindered}) = \alpha^s S \cos \theta$$

where S^a (W/m^2) is absorbed solar energy when the solar vector (magnitude S) impinges at angle θ off the surface normal. Values of α^s and associated emissivities for various satellite surface treatments are given in Table 6.1.

B.   Albedo

Albedo (from the Latin *albus*, for whiteness) is heating from sunlight reflected off Earth. It is usually considered to be in same spectrum as solar radiation and often quoted as a fraction of the solar constant; that is,

$$A = f S,$$

with f known as the albedo factor.

Albedo appears more significant at the Earth's polar ice caps and can be estimated in those regions with some accuracy as function of the sun's elevation and the satellite's orbital parameters. However, predictions for overland and above oceans become distorted by the highly variable effects of cloud formations and water distribution in the atmosphere.

Observations by ERBE indicate that albedo may not be strictly diffuse and could have a dependence on the solar zenith angle (angle between line connecting the satellite to Earth center and the line from Earth to sun) that exceeds a simple shape factor adjustment at high-orbit inclinations. Updates for thermal modeling and analysis have been proposed but there remains a reluctance to depart from standards in use since the early 1970s. Perhaps this is because monitored temperatures on orbiting satellites have generally supported traditional data, particularly when applied to solar arrays. Albedo effects on solar arrays are critical near local noon in a high-noon orbit (Sec. V.C, this chapter) and ERBE's data at this position translate into significantly higher temperatures than those monitored in flight.

In dealing with albedo, thermal engineers generally assume the radiation to be fully diffuse with its absorbed fraction equal to α^s. NASA has occasionally required using the values given in Table 4.3. Comparison is made with ERBE's data at three inclinations, with f corrected to account for the effects of zenith elevation.

Table 4.3 Albedo factor as function of orbit inclination

Orbit inclination, deg	f (NASA TM-82478)			f (ERBE, Stephens et al.[2]) corrected average
	minimum	average	maximum	
± 90	0.38	0.42	0.46	0.56
± 80	0.34	0.38	0.42	
± 70	0.30	0.34	0.38	
± 60	0.26	0.30	0.34	0.31
± 50	0.22	0.28	0.32	
± 40	0.19	0.25	0.29	
± 30	0.20	0.24	0.28	0.22
± 20	0.20	0.24	0.28	
± 10	0.20	0.24	0.28	

C. Earth Flux

Emitted radiation from the planet Earth is considered diffuse and equivalent in intensity and wavelength to that from a black surface at about $-20°C$, which gives a nominal value of Earth flux E equals 236 W/m^2. A tolerance of ±38 W/m^2 (16%) is sometimes imposed in thermal analysis. On the average, the data appear valid despite satellite observations that show large and changing localized variations.

Because the spectrum of Earth radiation is in the same band as that normally emitted by satellites, the fraction of impinging Earth flux absorbed by a satellite's radiator is its emissivity ε. Hence, a surface treatment intended to reflect emitted Earth radiation would also reduce by the same proportion the surface's emission ability.

V. Orbit Normal Incident Fluxes

Environment fluxes must be known to calculate the normal incident fluxes that contribute to the heating of an orbiting satellite. Also needed for this calculation are orbit dates, inclination, eccentricity, elevation, and satellite surface orientations with respect to the sun and Earth. Dates relate Earth's distance from the sun, and altitudes define the reduction in the intensities of Earth flux and albedo that, when considered diffuse, are inverse functions of the distance squared. Orbit inclination and eccentricity define the orbital period and the times the satellite spends in sunlight and Earth shadow. Surface orientations are used to find the normal component of incident flux.

A. Incident Flux Determination

As the calculations of incident fluxes are based on geometric concepts, a first-order determination of normal incident fluxes is often possible from examining a spacecraft aspect in a few positions in orbit, including daytime and nighttime. But, more conveniently, there are a number of orbital flux programs,[5-7] many of which are accessible to personal computer users, that can be run with little difficulty to quickly obtain the data both as function of orbit time and as integrated orbital averages on each designated surface. The computed values are usually normal components from which the absorbed fluxes (denoted by superscript a) are obtained according to

$$q^{S,a} = \alpha^s q^S$$

$$q^{A,a} = \alpha^s q^A$$

$$q^{E,a} = \varepsilon q^E$$

Mathematical models for predicting orbital fluxes can be based on actual Earth and celestial geometries (which differ slightly from the spherical and pointing assumptions made when doing hand calculations) and can be modified to include penumbra effects. Most flux programs also work in conjunction with surface-generating routines (PATRAN, SURTRAN, and others) with options to account for shadowing and reflection exchange among the various surfaces. Such information can be useful in evaluating the effects of nonuniform heating on a solar array and other structures that may be prone to thermally induced distortions. It is also possible and, in fact, common to integrate a flux program into a thermal network analyzer[8] for uninterrupted determination of temperatures from a single input file.

Normal incident fluxes are usually supplied by thermal control immediately as the mission and the general external shape of a satellite are defined. Often, initial data are obtained for the surfaces of a uniform polyhedron (generally not more than eight sides, with a top and a bottom) imagined in the same orbit as the satellite. Because of their simplicity, it is not uncommon to continue using these geometries throughout a project when doing parametric and systems level studies.

Example: Estimating Earth Flux and the Solar Constant

By knowing the average surface temperature and emissivity of planet Earth, estimates can be made of its emitted radiation that falls on a satellite surface at given altitude and orientation. Similar calculations can be made of incident solar radiation near Earth by assuming the sun at that distance to be equivalent to a disk of known effective temperature and emissivity. The problem is related to finding the shape factor from a flat surface to a sphere.

The geometry for vertical and horizontal orientations (normals n_{1V} and n_{1H}) is shown in Fig. 4.1. The cosine of the angle θ_{1H} that the normal of a horizontal element dA_1 makes with the line s, connecting to an element dA_2 at zenith φ and azimuth ψ on the sphere's surface, is found by noting that $\sin\theta_{1H}/R = \sin\varphi/s$. Hence, from trigonometric relations,

$$\cos\theta_{1H} = [s^2 - R^2\sin^2\varphi]^{1/2}/s$$

The angle θ_2 that the sphere's normal makes with the connecting line is

$$\theta_2 = \theta_{1H} + \varphi$$

and, hence,

$$\cos\theta_2 = \cos\theta_{1H}\cos\varphi - \sin\theta_{1H}\sin\varphi$$

Also,

$$\cos\theta_{1V} = M/s$$

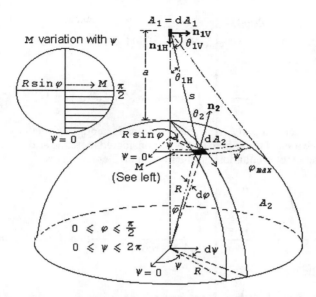

Fig. 4.1   Geometry for calculating the shape factor from an external elemental area to a sphere; angles θ_{1V} are θ_{1H} are θ_1 in the shape factor equation for the vertical and horizontal elements; the maximum view to the sphere is through φ_{max}; half-chord M is parallel to n_{1V}.

where, as shown in Fig. 4.1, M is half the length of the chord parallel to n_{1V} and passing through dA_2. Thus,

$$\cos\theta_{1V} = (R\sin\varphi)\sin\psi/s$$

It is also noted that the maximum value of φ that can be viewed by dA_1 is

$$\varphi_{max} = \cos^{-1}[R/(R+a)]$$

From

$$s^2 = R^2 + (R+a)^2 - 2R(R+a)\cos\varphi$$

$$s^2/R^2 = 1 + (1+v)^2 - 2(1+v)\cos\varphi$$

where v is the ratio of elevation to radius ($v = a/R$). Hence, in terms of v, φ, and ψ

$$\cos\theta_{1H} = \frac{1 + (1+v)^2 - 2(1+v)\cos\varphi - \sin^2\varphi}{[1 + (1+v)^2 - 2(1+v)\cos\varphi]^{1/2}}$$

$$\cos\theta_{1V} = \frac{\sin\varphi\sin\psi}{[1 + (1+v)^2 - 2(1+v)\cos\varphi]^{1/2}}$$

$$\cos\theta_2 = \frac{[1 + (1+v)^2 - 2(1+v)\cos\varphi - \sin^2\varphi]^{1/2}\cos\varphi - \sin^2\varphi}{[1 + (1+v)^2 - 2(1+v)\cos\varphi]^{1/2}}$$

and

$$\varphi_{max} = \cos^{-1}[1/(1+v)]$$

Now,

$$A_1 F_{12} = \int_{A_1} \int_{A_2} (\cos\theta_1 \cos\theta_2/\pi s^2) \, dA_1 \, dA_2$$

where $dA_2 = [(R\sin\varphi)\,d\psi]R\,d\varphi$, and when A_1 is negligibly small by comparison to the surface area of the sphere it may be considered the same as dA_1. Substitution into the shape factor equation for the horizontal element gives

$$F_{12H} = 2\int_0^{\cos^{-1}\frac{1}{(1+v)}}$$

$$\times \frac{[1+(1+v)^2 - 2(1+v)\cos\varphi - \sin^2\varphi]^{\frac{1}{2}}}{[1+(1+v)^2 - 2(1+v)\cos\varphi]^2}(\sin\varphi\cos\varphi - \sin^3\varphi)\,d\varphi$$

For the vertical element,

$$F_{12V} = \frac{2}{\pi}\int_0^{\cos^{-1}1/(1+v)}$$

$$\times \frac{[1+(1+v)^2 - 2(1+v)\cos\varphi - \sin^2\varphi]^{\frac{1}{2}}\sin^2\varphi\cos\varphi - \sin^4\varphi}{[1+(1+v)^2 - 2(1+v)\cos\varphi]^2}\,d\varphi$$

Numerical integration (programmable calculator, MATHCAD, etc.) yields the results plotted in Fig. 4.2.

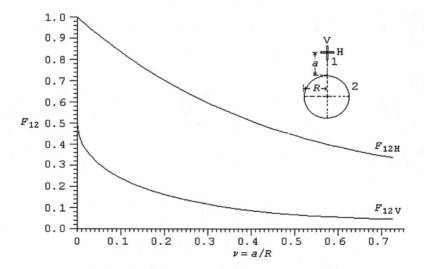

Fig. 4.2 Shape factor from a surface element to a sphere.

Similar integrals for other ranges of tilt angle can also be obtained and all made part of an orbital flux program designed to calculate planetary heating on various satellite surfaces. Irregular or curved surfaces can be approximated by sets of adjoining flat planes.

Emission from the heating source must also be known in determining the incident flux. Earth's hemispherical emission (Earth flux) based on considering Earth as a black radiator at $-20°C$ is found from

$$E = \sigma T^4_{-20°C} = 236 \text{ W/m}^2$$

From the definition of shape factor and using the reciprocity principle (Chapter 3 Sec. II.A.), the normal incident energy from Earth on a unit area of receiving surface is

$$q^E = F_{\text{surface to Earth}} E$$

By comparison to the size and distance from the Earth (mean radius $= 6340$ km) the surfaces of orbiting satellites may be considered elements. Hence, for example, in a 1000-km orbit

$$v = 1000/6340 = 0.158$$

and the shape factor from a flat radiator always parallel to Earth's horizon is found from Fig. 4.2 to be

$$F_{12H}(0.158) = 0.750$$

Therefore, normal incident Earth radiation per unit radiator area is

$$q^E(\text{horizontal flat surface at 1000 km}) = (0.750)(236) = 177.0 \text{ W/m}^2$$

To estimate the solar flux (solar constant) in the vicinity of planet Earth, one considers a surface element above the atmosphere with its normal pointing toward the sun. At this distance, the sun is seen as a disk with $\varphi_{\max} = \pi/2$ and s practically a constant, whereas θ_1 is nearly zero at all projections. Hence, θ_2 is equal to φ, and the shape factor integral equation becomes

$$F_{12}A_1 \approx A_1 \frac{R^2}{\pi s^2} \int_0^{2\pi} d\psi \int_0^{\pi/2} \sin\varphi \cos\varphi \, d\varphi$$

or

$$F_{12} \approx R^2/s^2$$

Here F_{12} is the shape factor from the element to the sun, R the sun's radius, and s the distance to the surface of the sun. Using a value of $R = 6.88E5$ km and mean distance $s = 148.6E6$ km, the equation gives

$$F_{12} = 2.143E-5$$

If the solar disk is considered effectively a black surface at 5773 K (5500°C), then normal incident solar flux near Earth is

$$S = (2.143E-5)\left(\sigma T^4_{5773 \text{ K}}\right) = 1350 \text{ W/m}^2$$

which is frequently given as nominal value for the solar constant at 1 a.u. Another number often quoted is 1352 W/m^2, corresponding to a surface temperature of 5503°C. Data from ERBS give nominal $S = 1371$ W/m^2 and the sun surface at 5489°C. The shape factor corresponding to these values is calculated as $F_{12} = 2.194E-5$, which can be obtained from the shape factor equation by slight variations on s or R or both.

B. Flat Plate in Space: The Sink Temperature

There are two powerful simplifications often invoked in thermal computations. One is the assumption of an isothermal panel, which indicates a radiator whose temperature remains uniform independent of applied localized heating, and the other is the concept of orbital average temperature, which assumes that equipment mounted on an orbiting platform will react to an average platform temperature whose value is based on the heating input averaged over an orbital period. The validity of both assumptions is approached when a satellite is sufficiently massive and its thermal design directed toward reducing the spatial variations in temperature.

Figure 4.3a shows an isothermal flat plate, with one active side, being simultaneously heated by dissipators and solar, albedo, and Earth flux. No other energy sources are present and only radiation heat transfer is involved. The absorbed fraction of both solar and albedo is α^s and, hence, they can be combined as

$$q^{S,a} + q^{A,a} = \alpha^s(q^S + q^A)$$

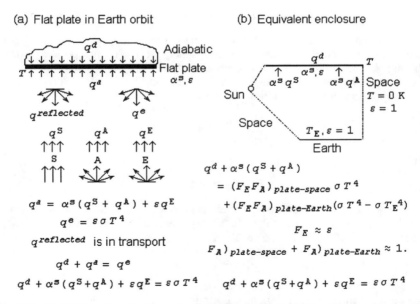

(a) Flat plate in Earth orbit

$$q^d = \alpha^s(q^S + q^A) + \varepsilon q^E$$

$$q^e = \varepsilon \sigma T^4$$

$q^{reflected}$ is in transport

$$q^d + q^d = q^e$$

$$q^d + \alpha^s(q^S + q^A) + \varepsilon q^E = \varepsilon \sigma T^4$$

(b) Equivalent enclosure

$$q^d + \alpha^s(q^S + q^A)$$
$$= (F_E F_A)_{plate\text{-}space}\, \sigma T^4$$
$$+ (F_E F_A)_{plate\text{-}Earth}(\sigma T^4 - \sigma T_E^4)$$

$$F_E \approx \varepsilon$$

$$F_A)_{plate\text{-}space} + F_A)_{plate\text{-}Earth} \approx 1.$$

$$q^d + \alpha^s(q^S + q^A) + \varepsilon q^E = \varepsilon \sigma T^4$$

Fig. 4.3 Orbital average heating of a flat plate in Earth orbit; sink temperature is defined by $\sigma T_\infty^4 \equiv (\alpha^s/\varepsilon)(q^S + q^A) + q^E$ so that $q^d = \varepsilon(\sigma T^4 - \sigma T_\infty^4)$.

Absorbed Earth flux is

$$q^{E,a} = \varepsilon q^E$$

and an energy balance at plate temperature T with radiation to deep space gives

$$q^d + \alpha^s(q^S + q^A) + \varepsilon q^E = \varepsilon \sigma T^4$$

The equivalent enclosure shown in Fig. 4.3b is dominated by the presence of Earth and space with the sun represented by a small surface at a very great distance $[(F_A)_{E-S} \approx 0]$. The equations of enclosure radiation exchange give the same result when noting that the plate is small in comparison to black Earth and infinitely large space (F_E becomes the plate emissivity ε). Enclosure formulations are more useful when other interacting surfaces must be considered (see Chapter 3, Sec. III).

In the analysis of radiating systems it is sometimes convenient to define the influence of environment heating flux in terms of an equivalent sink at temperature T_∞, representing an imaginary surrounding enclosure with which the system exchanges radiation. Thus,

$$q^d = \varepsilon \sigma T^4 - \{\alpha^s(q^S + q^A) + \varepsilon q^E\}$$

is written

$$q^d = \varepsilon(\sigma T^4 - \sigma T_\infty^4)$$

with

$$\sigma T_\infty^4 \equiv (\alpha^s/\varepsilon)(q^S + q^A) + q^E$$

In general,

$$q^d = \Im(\sigma T^4 - \sigma T_\infty^4)$$

with $\Im$ being an exchange factor that depends on the properties and geometry of the radiator and its view of the surroundings.

The usefulness of this concept is evident in the analysis of radiation shields (Chapter 6, Secs. II.A and V.B) and when simulating heat fluxes in test chambers by imposing an equivalent wall temperature (Chapter 7, Sec. II.C).

Example: Thermal Performance of a Flat Plate in Space

The isothermal flat plate equation can be used to make simple and immediate calculations. Consider as an example a communications and data handling system intended for low Earth orbit with its radiator always at the same slant angle of 60 deg to Earth horizon (C&DH module of NASA's Explorer Platform). The maximum orbital averages of normal incident fluxes have been given as

$$q^S = 110.3 \text{ W/m}^2 \qquad q^A = 34.1 \text{ W/m}^2 \qquad q^E = 95.8 \text{ W/m}^2$$

The surface treatment is silverized Teflon, which reduces the impact of solar and albedo heating. Long-term properties of this coating in a low Earth orbit have been specified as

$$\alpha^s = 0.28 \qquad \text{and} \qquad \varepsilon = 0.75$$

Substituting into the flat plate equation, the equipment dissipation emitted per square meter of radiator area can be calculated in terms of the radiator temperature. Thus, for $T = 35$, 40, and 45°C, the corresponding $q^d = 270.4$, 295.9, and 322.6 W/m². Hence, if the expected dissipation is, say, 200 W, and the equipment platform is to be kept below 40°C, then the radiator must have an area of at least

$$A^r = Q^d/q^d = 200/295.9 = 0.676 \, \text{m}^2$$

Example: Orbital Flux from SSPTA and TRASYS: Low Earth Orbit (LEO)

A compilation of values of normal incident fluxes to encompass all practical orbits and radiator orientations would be quite voluminous and completely impractical. Usually, incident fluxes are generated from flux programs specifically for each spacecraft, even when there are similarities to others, and always with sufficient variations on the orbit to bound all possible aspects of the mission. The process is systematic, and with currently available computational tools, it is quick and can be done at a reasonable cost.

Typical outputs from flux programs contain listings of the normal incident fluxes on each modeled surface as a function of selected orbit increments (time). Often, there is also an appended summary line giving orbital averages calculated from such relations as

$$q_{av} = \sum_j q_j(t_{j+1} - t_{j-1})/p$$

where p is orbit period and q_j the value of normal incident flux midway in time increment $(t_{j+1} - t_{j-1})$.

Orbital averages can be used to predict platform temperatures of massive components, whereas time variations become important in the analysis of items whose ratios of mass to radiating area are relatively low. The latter include solar arrays, antenna elements, and secluded light components such as deployment mechanisms (springs, latches, hinges, etc.), printed circuit boards on a solar array, and other external systems that cannot be isolated from environment heating without affecting their operation.

Some common low Earth orbits (LEOs) are schematically shown in Fig. 4.4. The orbital averages of normal incident fluxes in these orbits for various fixed satellite orientations are summarized in Figs. 4.5–4.8. In these figures the value of normal incident solar is essentially the solar constant adjusted by incident angle and ratio of time in sunlight to orbital period. Normal incident albedo has adjustments by the shape factor to Earth and the ratio of suntime to orbit period, and normal incident Earth flux is Earth emission adjusted by the shape factor. The scenarios are patterned after results from SSPTA for Fairchild Leasecraft and TRASYS for NASA/JPL TOPEX/Poseidon on an orbiting, small (2.54-cm) cube. The trends of the curves were established from computer data for two orbit inclinations (23.5 deg and 57 deg for nonpolar orbits) at 400, 1000, and 1400 km altitudes. Clearly, however, profiles presented in this fashion cannot exhibit the numerical accuracy found in the actual printout of a flux program, nor do they include intershadowing effects that can significantly affect the outcome of a thermal analysis. Nonetheless, the curves are convenient for performing preliminary worst-case analyses as illustrated in some of the numerical examples given in this book.

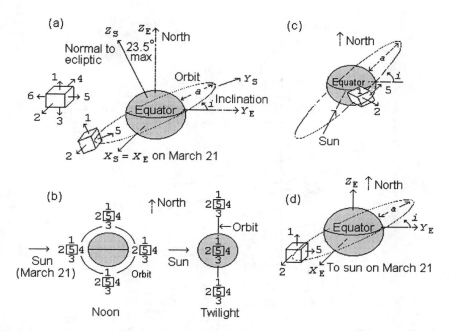

Fig. 4.4 Typical low Earth orbits and surface orientations: a) sun-oriented inclined orbit with vector 2 parallel to solar vector; b) sun-oriented polar orbits with vector 2 parallel to solar vector; c) Earth-oriented, vector 2 passes through Earth center; and d) star pointer, vector 1 points toward celestial north.

It is noted that nominal values of environment fluxes are used in the sun-oriented and Earth-oriented missions, and maximum values in a star (Polaris) pointer. For first-order analysis, orbital averages pertaining to other dates and tolerances may be calculated by ratio adjustments. For example, orbital average normal incident albedo on the antisun side of the cube (surface 4) in an inclined sun-oriented orbit at 1000 km is found from the curves in Fig. 4.5 to be about 80 W/m² for $S = 1350$ W/m² and $f = 0.33$. At 1000 km with $S = 1400$ W/m² and $f = 0.46$, the estimate is $80(1400/1350)(0.46/0.33)$, or 116 W/m².

C. Beta Angle: Effect on Solar Array Temperature

A mental picture of incident fluxes in LEO may be obtained by referring to the beta angle, β, defined as the angle the solar vector makes with the orbit plane. Because of Earth's oblateness and the sun's right ascension from vernal equinox and declination from the equatorial plane, the beta angle varies continuously over a year, passing through zero and reaching a maximum equal to the absolute value of the sum of orbit inclination and the ±23.5-deg greatest solar declination.[9] Hence, incident fluxes calculated with beta angle as a variable will provide the whole range of orbital heating.

As may be surmised by studying the orbit sketches in Fig. 4.9, albedo is greatest when noon at a point on Earth coincides with the satellite being at or nearest the directly overhead position. This is high noon in $\beta = 0$ deg orbit, and it is where

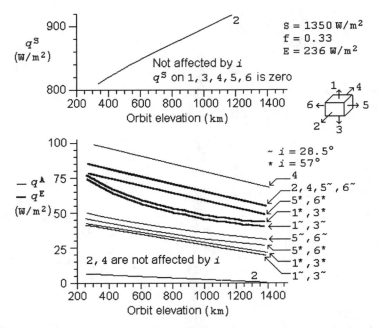

Fig. 4.5   Orbital average normal incident heating flux in sun-oriented inclined orbit, with vector 1 perpendicular to the ecliptic plane and vector 2 parallel to solar vector (March 21).

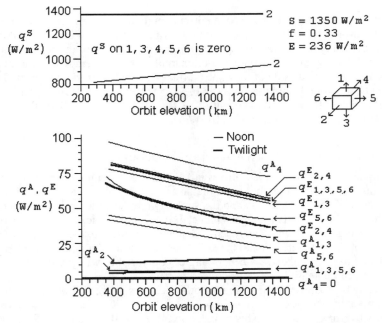

Fig. 4.6   Orbital average normal incident heating flux in sun-oriented polar orbit with vector 1 pointing north and vector 2 parallel to solar vector (March 21).

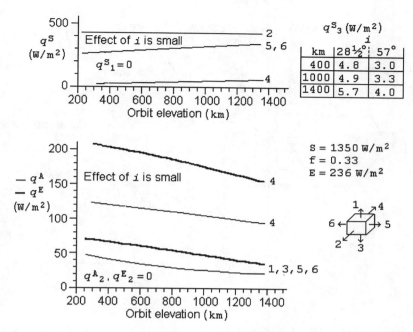

Fig. 4.7 Orbital average normal incident heating flux in Earth-oriented inclined orbit with vector 2 passing through Earth center and vector 5 in the velocity direction (March 21).

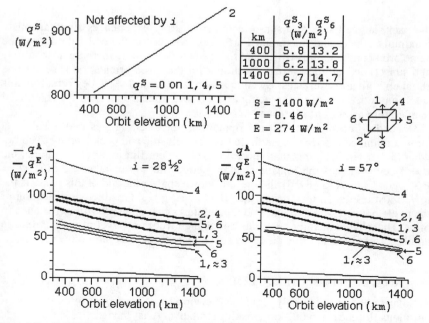

Fig. 4.8 Orbital average normal incident heating flux in star-pointing orbit with vector 1 pointing to Polaris (North Star) and vector 2 toward the sun (December 22).

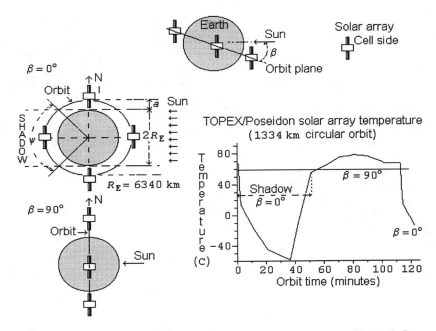

Fig. 4.9   Effect of beta angle on solar array temperature; orbit period $p = 2\pi[(R_E + a)^3/\mu]^{1/2}$ with $\mu = 1.437E9$ km^3/min^2; array temperature based on $S = 1420$ W/m^2, $f = 0.4$, and $E = 261$ W/m^2 with end of life properties of $(\alpha^s/\varepsilon)_{cell} = 0.76/0.81$ and $(\alpha^s/\varepsilon)_{back} = 0.48/0.85$.

the back, anticell side of a sun-oriented solar array experiences simultaneously maximum albedo and maximum Earth flux. Albedo on the cell side during the same orbit can be intuitively judged to be small (with some increase near sunrise and sunset) and orbital average Earth flux for the given orientation is seen to be equal on both array surfaces. By comparison, albedo in $\beta = 90$ deg orbit remains small on both sides of a sun-facing array. These effects appear in the calculated temperature profiles as illustrated in Fig. 4.9.

Depending on inclination and orientation, other beta angles may result in a larger combined albedo and Earth flux on the cell side and back leading to somewhat higher temperatures than those for $\beta = 0$ deg. But at a given altitude, the $\beta = 0$ deg orbit has the longest shadow period (no sun or albedo during $\psi = \pi - 2\cos^{-1}[R_E/(R_E + a)]$) and the cool-down span at the end of this phase is the lowermost orbital temperature. Thus, $\beta = 0$ deg gives a distinctive indication of the cycle between the two extreme temperatures that an array will experience in orbit. These data, including the rates of change from one level of temperature to another, form the basis of solar array specifications and test requirements.

D.   Geosynchronous Orbit (GEO)

Solar radiation is the only significant environment heating at or near geosynchronous altitude ($\approx$36,000 km), making it relatively easy to visualize and calculate

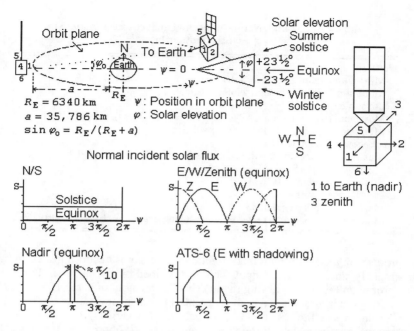

Fig. 4.10   Earth-viewing, three-axes stabilized satellite in geosynchronous orbit, fully sunlit orbit when $\varphi > \varphi_0 \approx 9$ deg; orbit period is 24 h with maximum 72-min eclipse (observed) at $\varphi = 0$ deg.

the incident heating flux as it varies with orbit position and solar incident angle. However, because of the usual complications from intershadowing among satellite structures, thermal engineers often resort to computer flux programs.

A satellite in geosynchronous orbit (GEO) is shown in Fig. 4.10. The representation is typical of most geosynchronous satellites, which are stable along three-body axes with the normal of one plane always pointing toward Earth. Orbit inclination is usually very small (less than 4 deg) and, as in the indicated orientation, the motion with respect to the solar vector is east to west. At equinox, the solar vector remains perpendicular to the north–south direction with no sun impinging on surfaces aligned exactly in those orientations. During and near this time, these surfaces are practically in a steady-state condition of zero solar heating.

With seasonal variations, as the orbit plane tilts above and below the ecliptic, the solar vector begins to cone around the north or south surface, depending on time of year, until it culminates during winter or summer solstice in a maximum solar input at an incident angle of approximately 23.5 deg. Therefore, the greatest normal incident solar heating on a north or south radiator, without specialized pointings or maneuvers, may be assumed to be a steady-state value of

$$q^S_{N/S}(\text{maximum}) = S \sin 23.5 \text{ deg}$$

or 558 W/m^2 on south radiators when $S = 1400$ W/m^2 during winter solstice and 518 W/m^2 on north radiators when $S = 1300$ W/m^2 during summer solstice.

Table 4.4 Orbital average of normal incident solar heating in geosynchronous orbit

North surface
  Maximum (summer solstice; solar elevation = 23.5 deg): 0.399S
  Minimum (equinox through winter solstice): 0
South surface
  Maximum (winter solstice; solar elevation = −23.5 deg): 0.399S
  Minimum (equinox through summer solstice): 0
Earth-viewing (nadir) surface
  Maximum (near equinox with no eclipse; solar elevation beyond ≈
    ±9 deg): 0.314S
  Minimum (equinox): 0.269S
East/west/zenith surfaces
  Maximum (equinox): 0.318S
  Minimum (summer and winter solstice; solar elevation = ± 23.5 deg): 0.269S

Surfaces aligned with the other directions have exposures to a truncated, approximately sinusoidal solar input. This is described by the curves in Fig. 4.10. The profile on the Earth-viewing (nadir) side has a cutout of some 18 deg ($\pi/10$ rad) during equinox about the midway point in Earth shadow.

The time a satellite spends in Earth eclipse also varies. At equinox, it is a continuous, approximately 72-min period in every 24 h of orbit time. When solar elevation goes beyond approximately ±9 deg ($\sin^{-1}[R_E/(R_E + a)]$, with $R_E = 6340$ km and $a = 35,786$ km) the satellite remains in sunlight throughout the orbit.

Orbital averages of incident solar heating are calculated by integrating sinusoidal equations with shifts and truncations. For example, the orbital average without extraneous shadowing on the Earth-viewing side during equinox can be found from

$$q_{\text{av(nadir)}}^{S} = \frac{S}{2\pi}\left[\int_{\pi/2}^{\pi-\pi/20} \sin(\psi - \pi/2)\,\mathrm{d}\psi + \int_{\pi+\pi/20}^{3\pi/2} \sin(\psi - \pi/2)\,\mathrm{d}\psi\right]$$

$$= 0.2685S$$

By considering the change in solar elevation, the maximum and minimum orbital average of the normal incident flux on all surfaces can be similarly found as listed in Table 4.4.

Application: Thermal Design and Analysis of Geosynchronous Satellites

As an integrated thermal system, the north–south surfaces of an Earth-pointing, three-axes stabilized satellite in GEO receive maximum solar heating during summer or winter solstice, with one side in the sun while the other shaded. The fact that these two sides cannot simultaneously be in sunlight for long periods of time has given rise to a popular approach in thermal control, whereby high and moderate dissipators are mounted on platforms in those directions (surfaces 5 and 6 in Fig. 4.10) with enhanced heat paths between them.

Low-power and nondissipaters are usually placed on the east, west, and zenith sides (2, 4, 3) whereas the Earth-viewing radiator mainly carries electronics and accessories associated with experiments, antennas, and various pointing devices. Multilayered insulation generally covers the E–W and zenith sides to suppress their interaction with external heating and, therefore, maintain a benign influence on the north–south radiators. Other standard features include low α^s/ε coatings, heaters, and sometimes louvers on the N–S and Earth-viewing radiators. Upgraded designs may also have a network of heat pipes with a central panel connecting the north and south sides.[10]

In this configuration, and unless there is considerable unsteady influence or significant solar reflection from surrounding structure, the extremes of solar heating are legitimately represented by the two steady-state cases of zero flux during equinox and maximum flux during solstice. Thermal engineers often appraise the performance at these bounds with separate analyses for winter and summer solstice. But the orbital averaging technique may not be valid when radiators in directions other than north and south are exposed. For, compared to near-Earth orbits, the GEO period is quite long, making the validity of averaging fluxes somewhat muted as extended periods are spent under radically different heating. The problem relates to predicting the temperature of a system in which the momentary temperature rate of change (dT/dt) may be small, whereas the overall time span is large. Computer models dealing with these situations are sometimes run more than once at different initial conditions to iterate on a periodically repeating temperature profile. An alternative, particularly in making initial assessments, is to compare orbital average performance with a special hot case, in which the satellite is in a stationary steady-state position with maximum solar incidence, and a special cold case of steady-state eclipse. If the computed temperatures under these assumptions are not substatially different from the results of an orbital average analysis, then either approach could become the basis for a thermal design.

Steady-state calculations with constant incident solar heating could also serve to find an upper bound on the temperature of light structures such as solar arrays, antennas, and other exposed items of relatively low thermal mass. The lowest temperature, normally occurring at the end of longest eclipse, is obtained from a transient analysis. A standard approach consists of calculating the minimum temperature on entering eclipse based on steady-state cold case incident solar flux, then using this value as initial temperature in a perfect cool-down (no heating) for 72 min to simulate maximum shadow time. A transient analysis is also needed to obtain the rates of cool-down on entering eclipse and heat-up as the satellite emerges into sunlight. Chapter 5, Sec. V.A describes how this information is obtained and applied to the thermal qualification of solar array panels.

VI. Charged Particles

Heating caused by charged particles in orbit, particularly in radiation belts, is very low and can be justifiably ignored except in cases of critical cryogenic application. There is, however, a persisting problem associated with the discharge of electric charges accumulated on a satellite surface.[11] Usually, the contention is settled by resorting to conductive exterior surfaces with grounding leads to the spacecraft's unexposed structure.

A whole activity exists to develop conductive thermal coatings and techniques for grounding. (See description in Fig. 6.1 for the grounding of multilayered insulation.) The demands appear most urgent on items with high sensitivity to an

(a) Extremely low heating

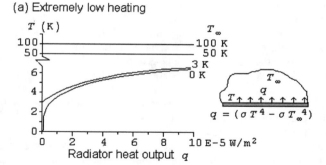

(b) Satellite range heating

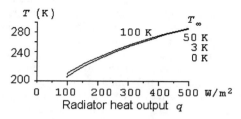

Fig. 4.11   Radiator temperature as function of dissipation, effect of surrounding temperature.

electromagnetic surrounding (for example, a magnetometer boom), but solutions are often not too simple. A particularly complicated interface is that of a moving large structure, such as a rotating solar array, where a large number of wire leads may have to be extended along the surface and through a slip ring assembly for grounding to the relatively stationary main structure.

Example: Sensitivity of Radiator Temperature to Heating–Cryogenic Regimes

The effect of environment heating on a radiator's temperature T can be illustrated by referring to the equivalent sink or environment temperature T_∞ to which radiator heat q is being rejected. In steady state,

$$q = \Im\left(\sigma T^4 - \sigma T_\infty^4\right)$$

where $\Im$ is an exchange factor which, for the purpose of illustration, may be assumed to be equal to 1. Results of calculations at $T_\infty = 0, 3, 50$, and 100 K are shown in Fig. 4.11.

From Fig. 4.11a, one notes that sensitivity to extremely small (parasitic) heating is significant only when operating at very low cryogenic regimes. It is practically nonexistent at temperatures above about 50 K. Hence, whereas energy interaction with actual space background [The Cosmic Background Explorer's (COBE) data indicate $\approx$2.7 K, with $\sigma T_{2.7K}^4$ about $0.3E{-}5$ W/m^2] must be taken into account when dealing with radiator temperatures near absolute zero, the effect is completely

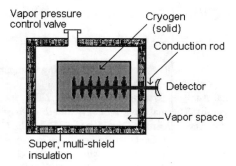

Fig. 4.12   Stored cryogen system.

negligible at common satellite temperatures, and thermal engineers can justifiably assume radiation to deep space at 0 K.

In the other extreme, it is deduced from Fig. 4.11b that at power levels of 100 W/m^2 or greater the feedback from an environment at 100 K or less is not too significant. This makes space simulation in vacuum test chambers by shrouds of liquid or gaseous nitrogen (100 K or somewhat lower) a legitimate approximation for most satellite applications.

The sensitivity of conventional radiators to small heating from charged particles or slight pointing inaccuracies limits their cryogenic application to temperatures not very much below 100 K. Alternatives for producing lower temperatures include active refrigeration systems and stored cryogens.[12] The Infrared Astronomical System (IRAS) and COBE IR detectors were cooled by boiling and venting stored liquid helium in superfluid state (<2.18 K). A schematic is given in Fig. 4.12. Commercial cryogens are available to cover a wide range of operational temperatures, and the device has the advantage that it requires no external electrical power. The limitations are mainly short life (<1year) for a conventional weight and a critical container design.

Acknowledgment

Information on thermal software in NASA's library can be obtained through COSMIC, University of Georgia, Athens, Georgia.

References

[1] "Space and Planetary Environment Criteria Guidelines for Use in Space Vehicle Development, 1982 Revision, Vol. 1," NASA TM-82478, 1983.

[2] Stephens, G. L., Campbell, G. G., and Vonder Haar, T. H., "Earth Radiation Budgets," *Journal of Geophysical Research* (Atmosphere), Vol. 86, No. C10, 1981, pp. 9739–9760.

[3] Mecherikunnel, A. T., "A Comparison of Solar Total Irradiance Observations from Spacecraft: 1985–1992," *Solar Physics*, Vol. 155, No. 2, 1994, pp. 211–221.

[4] Burlov–Vasiliev, K. A., Gurtovenko, E. A., and Matveiev, Yu. B., "New Absolute Measurements of the Solar Spectrum 310-685 nm," *Solar Physics*, Vol. 157, No.1, 1995, pp. 51–73.

[5]"Thermal Radiation Analyzer System (TRASYS) User's Manual," NASA JSC-22964, 1993.

[6]Russell, D. J., "The New Space Shuttle Orbiter/ Payload Thermal Integration Model," AIAA Paper 87-1595, 1987.

[7]Noravian, H., "TRASYS/ SSPTA: A Thermal Modeling Comparison," Society of Automotive Engineers, Paper 840926, 1984.

[8]"Integrated Thermal Analysis System (ITAS)," ANALYTIX, Inc., Lutherville, MD.

[9]Gilmore, D. G., "Satellite Thermal Environments," *Satellite Thermal Control Handbook*, edited by D. G. Gilmore, The Aerospace Corp. Press, El Segundo, CA, 1994, Chap. 2.

[10]Wales, R. O. (ed.), *ATS-6 Final Engineering Performance Report, Vol. I - Program and Systems Summaries; Mechanical and Thermal Details*, NASA RP-1080, 1981.

[11]Anderson, P. C., and Koons, H. C., "Spacecraft Charging Anomaly on a Low-Altitude Satellite in an Aurora," *Journal of Spacecraft and Rockets*, Vol. 33, No. 5, 1996, pp. 734–738.

[12]Donabedian, M., "Cryogenic Systems," *Satellite Thermal Control Handbook*, edited by D. G. Gilmore, The Aerospace Corp. Press, El Segundo, CA, 1994, Chap. 8.

Chapter 5

Satellite Thermal Analysis

I. Introduction

S ATELLITE thermal analysis is concerned with predicting the temperature of a satellite in a known or assumed heating environment. The predictions are made by applying the principle of conservation of energy, generally stated by a second-order partial differential equation with Kelvin temperature T the variable sought as function of position vector $\mathbf{r}$ and time t. In standard notation, the thermal energy equation (or conduction equation) is

$$\nabla \cdot (\mathbf{k} \cdot \nabla T) + q^* = \rho c \left(\frac{\partial T}{\partial t}\right)$$

which is to be solved in view of initial conditions

$$T(r_i, 0) = T_0(r_i)$$

and boundary (surface) conditions that define the heat exchange at location r_i (of unit normal $\mathbf{n}_{ri}$) with surfaces at locations r_j; that is,

$$[-\mathbf{k}(T) \cdot \nabla T)] \cdot \mathbf{n}_{ri} = \sum_j K'_{ij}(T)(T_{ri} - T_{rj}) + \phi(t)$$

Some of the T_{rj} may be functions of time and the summations may be replaced by integrals when dealing with continuously varying temperatures and coefficients.

The nomenclature is as follows:

c = specific heat (at constant pressure for fluids), J/kg K, i.e., W-s/kg K
K'_{ij} = general interface conductances from r_i to r_j, W/m^2 K
$\mathbf{k}$ = thermal conductivity (a symmetric tensor) with components $k_{\mu\nu} (= k_{\nu\mu})$, W/m K
q^* = volumetric heat generation, W/m^3
ρ = density, kg/m^3
ϕ = heating input from extraneous or environment sources, W/m^2 of receiving surface

Finding from these equations exact analytical expressions for the temperature is not generally possible due to the nonlinear dependence of some of the K'_{ij} on temperature. In addition, nonuniformities and jumps in material properties and heating inputs in actual satellites would require an unmanageably large number of equations and boundary conditions if the continuum character of the mathematics is to be preserved. The practical approach is to seek instead answers through simplifications and by resorting to numerical techniques.

The preferred method, commonly referred to as thermal modeling, consists of treating the satellite as a set of regions in which continuum representation is imitated by subdividing each into a number of small isothermal *nodes* connected by the conductances that relate heat transfer to temperature differences. The thermal energy equation is then applied to each node with the differentials replaced by ratios of small but finite differences, while local (nodal) heat dissipation and boundary exchanges are combined into a single heat generation term. In this discrete form, a set of simultaneous algebraic equations is obtained involving nodal temperatures with coefficients that are known functions of thermophysical properties and temperature. Hence, the essence of thermal modeling is first to locate and configure the nodes to give a realistic representation of the actual system, then to devise and apply techniques by which the temperatures are calculated (usually by a computer) as accurately as possible from the resulting algebraic equations.

Modeling must necessarily presuppose an established configuration. Normally this is conceived from simplified forms of the energy equation, formulated into mathematical expressions designed to bound the thermal problem and to display the relative influence on temperature by various controlling parameters. By their very nature, these exercises are an insight into the character of the thermal system, which makes them invaluable in early concepts studies and for systems level assessments. As importantly, simplified analyses are essential in verifying computer results. This point cannot be overemphasized; numerical data from evolving computer models are invariably clouded by modeling complexities and details and, at least initially, are rarely free of errors. A tractable analysis, properly formulated, is the most effective probe for detecting these errors and for revealing subtleties and trends that can be easily overlooked when large models are being used.

II.   Thermal Energy Equation

Transforming the law of energy conservation into a differential equation involving temperature (instead of energy or entropy) is described in all intermediate or advanced books on heat transfer. The following is a brief treatment of the practical case where conductivity is at most a function of temperature and can be represented by components along principal axes ($k_{\mu\nu} = 0$ for $\mu \neq \nu$).

As shown in Fig. 5.1, heat conduction is applied to a small volumetric element residing inside the body (not exposed) and shaped to be compatible with the geometry (coordinates) of the system being considered. For a Cartesian element of dimensions Δx, Δy, Δz (top of Fig. 5.1) the statement, Heat into the element plus internal heat generated equals heat out of element plus heat stored, is written

$$Q_x^k + Q_y^k + Q_z^k + Q^d = Q_{x+\Delta x}^k + Q_{y+\Delta y}^k + Q_{z+\Delta z}^k + Q^{\text{stored}}$$

where Q_x^k, Q_y^k, and Q_z^k are, respectively, the heat rates conducted at x, y, and z along the normals of the yz, xz, and xy planes. The same holds true for $Q_{x+\Delta x}^k$, $Q_{y+\Delta y}^k$, and $Q_{z+\Delta z}^k$ at $x + \Delta x$, $y + \Delta y$, and $z + \Delta z$.

Nodal heat generation (dissipation) Q^d is related to the volumetric heat generated q^* (W/m^3), by

$$Q^d = q^* \Delta x \, \Delta y \, \Delta z$$

Cartesian, $T(x, y, z)$

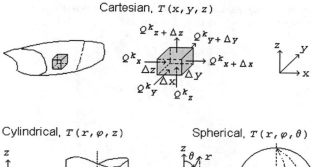

Cylindrical, $T(r, \varphi, z)$ Spherical, $T(r, \varphi, \theta)$

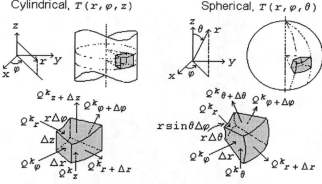

Fig. 5.1 **Heat conduction in a volumetric element in Cartesian, cylindrical, and spherical coordinates.**

and heat stored Q^{stored} is, by definition,

$$Q^{\text{stored}} = \left(\rho c \frac{\partial T}{\partial t} \right) \Delta x \, \Delta y \, \Delta z$$

For conductivity represented by the diagonal matrix

$$\mathbf{k} = \begin{bmatrix} k_x & 0 & 0 \\ 0 & k_y & 0 \\ 0 & 0 & k_z \end{bmatrix}$$

Fourier's law for conduction gives

$$Q_x^k = -k_x(\Delta y \, \Delta z)\frac{\partial T}{\partial x}, \qquad Q_y^k = -k_y(\Delta z \, \Delta x)\frac{\partial T}{\partial y}, \qquad Q_z^k = -k_z(\Delta x \Delta y)\frac{\partial T}{\partial z}$$

with no cross terms. Substituting the Taylor expansion

$$Q_{\xi + \Delta \xi}^k = Q_\xi^k + \left. \frac{\partial Q^k}{\partial \xi} \right)_\xi \Delta \xi + (1/2!)\partial^2 Q^k / \partial \xi^2)_\xi (\Delta \xi)^2 + \cdots$$

into the heat balance equation and dividing by volume, it is found that in the limit, as Δx, Δy, and Δz go to zero,

$$\frac{\partial}{\partial x}\left(k_x \frac{\partial T}{\partial x} \right) + \frac{\partial}{\partial y}\left(k_y \frac{\partial T}{\partial y} \right) + \frac{\partial}{\partial z}\left(k_z \frac{\partial T}{\partial z} \right) + q^* = \rho c \frac{\partial T}{\partial t}$$

The corresponding expression in cylindrical coordinates is obtained following the same procedure for the cylindrical element in Fig. 5.1 while noting that

$$Q_{cyl}^d = q^* r \Delta\varphi \, \Delta r \, \Delta z$$

$$Q_{cyl}^{stored} = \left(\rho c \frac{\partial T}{\partial t}\right) r \, \Delta\varphi \, \Delta r \, \Delta z$$

Thus,

$$\frac{1}{r}\frac{\partial}{\partial r}\left(k_r r \frac{\partial T}{\partial r}\right) + \frac{1}{r^2}\frac{\partial}{\partial \varphi}\left(k_\varphi \frac{\partial T}{\partial \varphi}\right) + \frac{\partial}{\partial z}\left(k_z \frac{\partial T}{\partial z}\right) + q^* = \rho c \frac{\partial T}{\partial t}$$

In spherical coordinates

$$Q_{sph}^d = q^* r^2 \sin\theta \, \Delta r \, \Delta\varphi \, \Delta\theta$$

$$Q_{sph}^{stored} = \left(\rho c \frac{\partial T}{\partial t}\right) r^2 \sin\theta \, \Delta r \, \Delta\varphi \, \Delta\theta$$

and

$$\frac{1}{r^2}\frac{\partial}{\partial r}\left(k_r r^2 \frac{\partial T}{\partial r}\right) + \frac{1}{r^2 \sin^2\theta}\frac{\partial}{\partial \varphi}\left(k_\varphi \frac{\partial T}{\partial \varphi}\right)$$

$$+ \frac{1}{r^2 \sin\theta}\frac{\partial}{\partial \theta}\left(k_\theta \sin\theta \frac{\partial T}{\partial \theta}\right) + q^* = \rho c \frac{\partial T}{\partial t}$$

In accordance with the theory of differential equations, an initial condition and a minimum of two boundary conditions (to account for external influence) in each dimension are required to find the temperature as function of position and time.

When heat generation is absent and the conductivity (now written k) constant and not dependent on direction,

$$\nabla^2 T = \left(\frac{\rho c}{k}\right)\frac{\partial T}{\partial t}$$

which in the steady state reduces to the Laplace equation

$$\nabla^2 T = 0$$

Where ∇^2 in the three coordinate systems is given by

$$\frac{\partial^2 T}{\partial x^2} + \frac{\partial^2 T}{\partial y^2} + \frac{\partial^2 T}{\partial z^2}$$

$$\frac{\partial^2 T}{\partial r^2} + \frac{1}{r}\frac{\partial T}{\partial r} + \frac{1}{r^2}\frac{\partial^2 T}{\partial \varphi^2} + \frac{\partial^2 T}{\partial z^2}$$

$$\frac{\partial^2 T}{\partial r^2} + \frac{2}{r}\frac{\partial T}{\partial r} + \frac{1}{r^2 \sin^2\theta}\frac{\partial^2 T}{\partial \varphi^2} + \frac{1}{r^2}\frac{\partial^2 T}{\partial \theta^2} + \frac{1}{r^2 \tan\theta}\frac{\partial T}{\partial \theta}$$

Although the differential equations may themselves be defined as linear when the thermal properties and heating are not functions of temperature, nonlinearity emerges if the boundary conditions involve T^4. Hence, in general, problems with combined conduction and radiation are difficult to solve without the aid of numerical methods.

Application: Solutions by Finite Differences

Finite differences are used to approximate differential equations by algebraic equations involving discrete values of the variables. The method requires the construction of a grid that fills the geometric domain, together with a scheme that approximates derivatives by actual differences.

Three types of differences of a function in two variables are defined as follows. Forward differencing:

$$\Delta_{+u} f(u, v) \equiv [f(u + \Delta u, v) - f(u, v)]/\Delta u$$

$$\Delta_{+v} f(u, v) \equiv [f(u, v + \Delta v) - f(u, v)]/\Delta v$$

Backward differencing:

$$\Delta_{-u} f(u, v) \equiv [f(u, v) - f(u - \Delta u, v)]/\Delta u$$

$$\Delta_{-v} f(u, v) \equiv [f(u, v) - f(u, v - \Delta v)]/\Delta v$$

Central differencing:

$$\delta_u f(u, v) \equiv [f(u + \Delta u/2, v) - f(u - \Delta u/2, v)]/\Delta u$$

$$\delta_v f(u, v) \equiv [f(u, v + \Delta v/2) - f(u, v - \Delta v/2)]/\Delta v$$

When central differencing is performed twice, the following is obtained:

$$\delta_u^2 f(u, v) = [f(u + \Delta u, v) - 2f(u, v) + f(u - \Delta u, v)]/\Delta u$$

$$\delta_v^2 f(u, v) = [f(u, v + \Delta v) - 2f(u, v) + f(u, v - \Delta v)]/\Delta v$$

A two-dimensional temperature field is used in Fig. 5.2 to illustrate application to the thermal energy equation. For the sake of bookkeeping, it is important to note the convention that relates the spatial coordinate location (in this case x, y) to the indices that define the position (m, n) of the discrete nodes. Also, addition and subtraction of unity in the indices indicate, respectively, a step before and a step after ($\pm 1/2$ refer to half-steps). For example, the index $m - 1$ refers to a backward displacement from location m, whereas $p + 1$ is future time from the present p. The index p is an integer relating time to time interval by $t = p\Delta t$.

Using central differencing, the derivative of T at m, n with respect to x, holding y and t constant (see graph in Fig. 5.2) as approached from left and right, is approximately

$$\left.\frac{\partial T}{\partial x}\right|_{m-1/2,n;p} \approx \frac{T_{m,n;p} - T_{m-1,n;p}}{\Delta x}, \qquad \left.\frac{\partial T}{\partial x}\right|_{m+1/2,n;p} \approx \frac{T_{m+1,n;p} - T_{m,n;p}}{\Delta x}$$

Hence,

$$\left.\frac{\partial^2 T}{\partial x^2}\right|_{m,n;p} \approx \frac{\partial T/\partial x|_{m+1/2,n;p} - \partial T/\partial x|_{m-1/2,n;p}}{\Delta x} \approx \frac{T_{m+1,n;p} + T_{m-1,n;p} - 2T_{m,n;p}}{(\Delta x)^2}$$

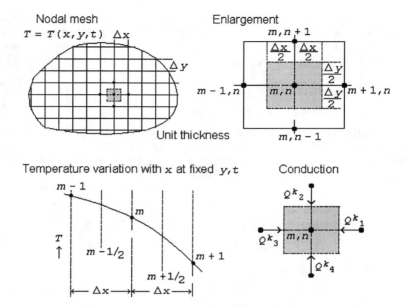

Fig. 5.2   **Two-dimensional nodal mesh designation for finite differences.**

Similarly,

$$\frac{\partial^2 T}{\partial y^2}\bigg|_{m,n;p} \approx \frac{T_{m,n+1;p} + T_{m,n-1;p} - 2T_{m,n;p}}{(\Delta y)^2}$$

The time derivative is approximated by the forward differencing

$$\frac{\partial T}{\partial t}\bigg|_{m,n;p} \approx \frac{T_{m,n;p+1} - T_{m,n;p}}{\Delta t}$$

Another useful result obtained from averaging with forward and backward differencing is

$$\frac{\partial}{\partial x}\left(k_x(T)\frac{\partial T}{\partial x}\right)\bigg|_{m,n;p} \approx k_\mu \frac{T_{m+1,n;p} - T_{m,n;p}}{(\Delta x)^2} - k_\nu \frac{T_{m,n;p} - T_{m-1,n;p}}{(\Delta x)^2}$$

Simpler expressions are normally obtained when square grids ($\Delta x = \Delta y$) are chosen. In this case, the Laplace equation in two dimensions (no time dependence)

$$\frac{\partial^2 T}{\partial x^2} + \frac{\partial^2 T}{\partial y^2} = 0$$

is approximated by (notice that the index p is not necessary)

$$T_{m+1,n} + T_{m-1,n} + T_{m,n+1} + T_{m,n-1} - 4T_{m,n} = 0$$

Similarly, for a square grid and forward differencing in time, the equation

$$\frac{\partial^2 T}{\partial x^2} + \frac{\partial^2 T}{\partial y^2} = \frac{\rho c}{k}\frac{\partial T}{\partial t}$$

results in

$$T_{m,n;p+1} = [(k/\rho c)(\Delta t)/(\Delta x)^2][T_{m+1,n;p} + T_{m-1,n;p} + T_{m,n+1;p} + T_{m,n-1;p}]$$

$$+ [1 - 4(k/\rho c)(\Delta t)/(\Delta x)^2]T_{m,n;p}$$

When uniform heat generation is included [say, $Q^d = q^*(\Delta x \cdot \Delta y \cdot 1)$] and $\Delta x = \Delta y$,

$$\frac{\partial^2 T}{\partial x^2} + \frac{\partial^2 T}{\partial y^2} + \frac{q^*}{k} = \frac{\rho c}{k}\frac{\partial T}{\partial t}$$

becomes

$$T_{m,n;p+1} = [(k/\rho c)(\Delta t)/(\Delta x)^2][T_{m+1,n;p} + T_{m-1,n;p} + T_{m,n+1;p} + T_{m,n-1;p}]$$

$$+ q^*\Delta t/\rho c + [1 - 4(k/\rho c)(\Delta t)/(\Delta x)^2]T_{m,n;p}$$

Because the thermal energy equation is a heat conservation statement, the same results are obtained by using the primitive form of a heat balance. Thus, when conductivity is constant and there is no heat generation, Fourier heat conductions across the faces of the node shown in Fig. 5.2 are

$$Q_1^k = Q_{(m+1,n) \text{ to } (m,n)}^k = k(\Delta y \cdot 1)[T_{m+1,n} - T_{m,n}]/\Delta x$$

$$Q_2^k = Q_{(m,n+1) \text{ to } (m,n)}^k = k(\Delta x \cdot 1)[T_{m,n+1} - T_{m,n}]/\Delta y$$

$$Q_3^k = Q_{(m-1,n) \text{ to } (m,n)}^k = k(\Delta y \cdot 1)[T_{m-1,n} - T_{m,n}]/\Delta x$$

$$Q_4^k = Q_{(m,n-1) \text{ to } (m,n)}^k = k(\Delta x \cdot 1)[T_{m,n-1} - T_{m,n}]/\Delta y$$

With $\Delta x = \Delta y$, the energy balance $\sum Q_i^k = 0$ gives

$$T_{m+1,n} + T_{m-1,n} + T_{m,n+1} + T_{m,n-1} - 4T_{m,n} = 0$$

which is the same as the earlier result for the Laplace equation.

The energy balance method is useful in finding finite difference equations for border nodes, which include heat transfer across boundaries. An illustration with a radiation boundary is given in Fig. 5.3. Performing a heat balance in steady state,

$$Q_1^k + Q_2^k + Q_3^k + Q_4^k = Q_1^r + Q_2^r$$

$$Q_{(m-1,n) \text{ to } (m,n)}^k + Q_{(m,n+1) \text{ to } (m,n)}^k + Q_{(m+1,n) \text{ to } (m,n)}^k$$

$$+ Q_{(m,n-1) \text{ to } (m,n)}^k = Q_{(m,n) \text{ to } \infty}^r$$

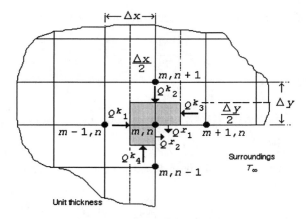

Fig. 5.3   Heat transfer in a boundary node.

$$k\left(\frac{\Delta y \cdot 1}{\Delta x}\right)(T_{m-1,n} - T_{m,n}) + k\left(\frac{\Delta y \cdot 1}{\Delta x}\right)(T_{m,n+1} - T_{m,n})$$

$$+ k\left(\frac{\frac{1}{2}\Delta y \cdot 1}{\Delta x}\right)(T_{m+1,n} - T_{m,n}) + k\left(\frac{\frac{1}{2}\Delta y \cdot 1}{\Delta x}\right)(T_{m,n-1} - T_{m,n})$$

$$= \mathfrak{I}_{(m,n) \text{ to } \infty}\left(\frac{\Delta y}{2} \cdot 1\right)(\sigma T_{m,n}^4 - \sigma T_\infty^4)$$

$$+ \mathfrak{I}_{(m,n) \text{ to } \infty}\left(\frac{\Delta x}{2} \cdot 1\right)(\sigma T_{m,n}^4 - \sigma T_\infty^4)$$

When $\Delta x = \Delta y$,

$$T_{m-1,n} + T_{m,n+1} + \frac{1}{2}(T_{m+1,n} + T_{m,n-1}) + \frac{\mathfrak{I}_{(m,n) \text{ to } \infty}\Delta x}{k}\sigma T_\infty^4$$

$$- \left(3 + \frac{\mathfrak{I}_{(m,n) \text{ to } \infty}\Delta x}{k}\right)\sigma T_{m,n}^4 = 0$$

In the linear convection version, $Q_{ij}^c = hA^c(T_i - T_j)$, and with $\Delta x = \Delta y$

$$T_{m-1,n} + T_{m,n+1} + \frac{1}{2}(T_{m+1,n} + T_{m,n-1}) + \frac{h\Delta x}{k}T_\infty - \left(3 + \frac{h\Delta x}{k}\right)T_{m,n} = 0$$

By repeating the differencing process for all of the nodes in the mesh, a set of algebraic equations, equal to the number of nodes, is obtained. In steady state, the equations are solved simultaneously to obtain the unknown temperatures in terms of propagating boundary conditions. In transients, forward time differencing begins with an initial condition ($p = 0$) defined for all positions, then temperatures are calculated for the next period ($p = 1$) assigned by Δt. The new values are now used as initial condition to calculate the temperatures for the next period, and so on.

Forward differencing in time is explicit because it enables direct computation of a node's future temperature ($T_{m,n;p+1}$) from its value ($T_{m,n;p}$) and the values of neighboring nodes ($T_{m+1,n;p}$, $T_{m-1,n;p}$, $T_{m,n+1;p}$, and $T_{m,n-1;p}$) at a preceding time. But, logical as this may seem, the method is potentially unstable in that the calculated temperature begins to oscillate from one time to the next when the time interval is made larger than a certain lower limit. Stability can be maintained by ensuring that the coefficients of present temperature in the difference equation are nonnegative. In the two-dimensional case with equal spacial increments, this requires

$$\left[1 - 4\left(\frac{k}{\rho c}\right)\frac{(\Delta t)}{(\Delta x)^2}\right] \geq 0$$

The requirement for one dimension is

$$\left[1 - 2\left(\frac{k}{\rho c}\right)\frac{(\Delta t)}{(\Delta x)^2}\right] \geq 0$$

and for three dimensions (equal spacial increments)

$$\left[1 - 6\left(\frac{k}{\rho c}\right)\frac{(\Delta t)}{(\Delta x)^2}\right] \geq 0$$

These restrictions, however, can sometimes lead to having to select extremely small time steps that could make the method impractical even with the use of high-speed digital computers.

A resolution is found in implicit backward differencing of the time derivative, which uses the time step before the time the temperature is being approximated. The simplest illustration is obtained by considering a center point m, between $m - 1$ and $m + 1$, in one-dimensional conduction. Backward differencing in time gives

$$(T_{m-1;p+1} - T_{m;p+1}) - (T_{m;p+1} - T_{m+1;p+1}) = \left[\rho c \frac{(\Delta x)^2}{k\Delta t}\right](T_{m;p+1} - T_{m;p})$$

with two similar equations for nodes $m - 1$ and $m + 1$.

The new form does not give $T_{m;p+1}$ (temperature of x at time $t + \Delta t$) in terms of $T_{m-1;p}$ and $T_{m+1;p}$ (temperatures of the other nodes at time t), which means that for each time step there will be three equations that need to be solved simultaneously for the temperatures. From this, one can imagine the innumerable number of iterations needed to obtain the temperature history of a satellite subdivided into hundreds of nodes. Nevertheless, the method has its power in its being free of instabilities and will permit sizable time steps which, when judiciously selected, will not introduce an undue increase in error.

These springboard techniques, and variations for improving accuracy and speed, are described in books on numerical analysis.[1-3] Heat transfer books present similar treatments using heat transfer language.[4] It should also be noted that in thermal control, as in other fields, available prepared software [such as the System Improved Numerical Differencing Analyzer (SINDA)] offers a variety of computational options to handle even the most complicated nodal breakdowns with only a nominal effort by the user. As a matter of fact, hardly a season goes by without the appearance of a new more user-friendly, or faster, or more accurate computer routine for solving thermal problems. Among the latest developments are routines that have the computer itself make optimum selections and perform many of the calculations for the input data. In addition, many of these programs can work in conjunction with other, simultaneously ongoing programs that generate surfaces and calculate heat transfer couplings and incident orbital fluxes.

III. Thermal Model

The numerical thermal model is the working tool in the development of a satellite thermal control system. It is used to predict temperatures on a large scale, with most structures and other components interacting with one another and with the surrounding environment. Because the environment has a direct bearing on a satellite temperature, it is not unusual that variations on the same basic model are configured to deal with specific situations such as on-orbit, vacuum testing, launchpad operations, and ascent.

Generating the thermal model begins early in a satellite project, with additions and upgrades continuing as notions on design and performance become firmer. Final confirmation follows the thermal balance test (Chapter 7), conducted in a vacuum chamber, when predictions from the model are correlated with test results. The process usually involves correcting modeling errors and updating assumptions related mostly to electronics heat dissipation and tolerances on hardware properties. The correlated model would eventually be used in evaluating the performance in orbit, particularly in the event of an anomaly or when conducting specialized operations and maneuvers.

Advancements in software technology and instruction manuals have greatly simplified the mechanics of thermal modeling. But most significant is an absolute awareness of the relationship between the modeling exercise and the final product, which is made up of physical and not mathematical entities. Ultimately, it is the actual performance of the satellite that is of greatest concern, and the purpose of the model should be to reveal the thermal character of that performance and not to provide an opportunity for constructing extensive networks that give highly detailed data of mostly minor or secondary relevance. Distinguishing and sorting out the overlapping domains of the salient and the less than relevant are essential aspects of the process and can be perfected only after some years of modeling experience seasoned by hands-on involvement in building and testing satellites.

The thermal model is generally considered a deliverable item and must, therefore, be prepared following a standard format intended to make it easy to interpret, run, and modify. An introduction on the SINDA format preferred by NASA Goddard Space Flight Center is given in Ref. 5.

Application: System Improved Numerical Differencing Analyzer (SINDA) Program

SINDA is a general purpose computer program that permits selections and combinations of various finite difference schemes for solving partial differential equations. The code, particularly SINDA 87, referred to as the Gaski version,[6] has been adapted to finding the temperature from the heat transfer equation (parabolic differential equation) by following certain programming rules that reduce the mathematics to thermal nodal grids and an input file. The adaptation is natural because SINDA routines allow diffusion nodes, which in thermal systems can represent normal physical masses; boundary nodes, which can simulate heat reservoirs (sinks) of fixed or time-varying temperature; and arithmetic nodes, which can simulate very light masses such as bond lines, thin sheets of material, and other similar components with a near-instantaneous response to heating. Arithmetic nodes can also be used as imaginary points, lines, or planes that represent interfaces or cavities whose temperature may be set as the average value of the surroundings.

The finite difference form of the thermal energy equation is coded into SINDA by the weighted average (or θ-method) representation, which in the current nomenclature can be described by

$$(Mc)_{ri}\left(\frac{T_{ri;p+1} - T_{ri;p}}{\Delta t}\right)$$

$$= \theta\left[\sum_j K_{ji}(T_{rj;p+1} - T_{ri;p+1}) + \sum_j R_{ji}(T_{rj;p+1} - T_{ri;p+1})\right.$$

$$\left. + Q_i^*(T_{ri;p+1})\right] + (1-\theta)\left[\sum_j K_{ji}(T_{rj;p} - T_{ri;p})\right.$$

$$\left. + \sum_j R_{ji}(T_{rj;p} - T_{ri;p}) + Q_i^*(T_{ri;p})\right]$$

where

$(Mc)_{ri} = \rho c\Delta V_i$, with ΔV_i the element volume, is thermal mass, in J/K, of node centered at r_i.

$K_{ji} = K_{ij} = k\Delta A_i^k/r_{ij}$, with k the average conductivity and ΔA_i^k mean conduction area, are conduction conductances, in W/K, from node centered at r_i to nodes centered at r_j, separated by distances r_{ij}. The factors may be modified to include convection terms.

$R_{ji} = R_{ij} = \sigma\Im_{ij}(\Delta A_i^r)[(T_{ri}^2 + T_{rj}^2)(T_{ri} + T_{rj})]$, with ΔA_i^r the elemental radiating area, are radiation conductances, in W/K, from node centered at r_i to nodes centered at r_j.

$Q_i^* = q_i^*\Delta V_i$ is heat generated, in W, at node centered at r_i. This quantity may be made a function of temperature to simulate thermostatic heater control or modified to incorporate boundary heat exchanges, including environment fluxes and convection.

θ is a dimentionless factor which can be adjusted, together with mesh size and time interval, to optimize computational speed and accuracy. Note that $\theta = 0$ yields fully explicit forward differencing equations, while $\theta = 1$ leads to fully implicit backward schemes. The case of $\theta = 1/2$ is known as the Crank-Nicholson scheme.

The steady-state form of the preceding equation (left side equals 0 and the p index absent) is perfectly suited for finding steady-state temperatures by any one of several routines available in SINDA for solving linear algebraic equations (Cholesky matrix decomposition is good for moderately sized models). Convergence is generally said to occur when a specified maximum temperature change between consecutive iterations is concurrently met with a tolerance on the energy balance of the whole network and a tolerance on nodal heat balance.

For nonsteady-state problems, explicit routines are simple to perform and require only successive arithmetic calculations proceeding in time from a given initial temperature map. But, as noted in Sec. II of this chapter, the process can suffer from instability. In the present formulation, this depends on the size of time interval

Δt in relation to $(Mc)_{ri}$, R_{ij}, and K_{ij}, which contain the nodes' spatial increments, the thermal properties, and temperature. The indication in SINDA is CSGMIN, which signifies the upper limit on Δt that will avert instability. It is calculated at each time step for each node from

$$\text{CSGMIN}_{ri} = (Mc)_{ri} \bigg/ \sum_{j} (R_{ij} + K_{ij})$$

In each iteration SINDA selects the least value of CSGMIN found among all of the nodes in the network and proceeds with calculations at smaller time intervals (usually 95% of CSGMIN). The smaller the value of CSGMIN, the more iterations are needed and the longer the execution time will be.

Orders of magnitude for CSGMIN can be cited by considering representative regions of large platforms, including honeycomb panels, modeled as two-node (1 and 2) transverse conduction systems of uniform lateral temperature. In this case, $(Mc)_{ri}$, R_{ij}, and K_{ij} all contain the same elemental area $(\Delta x\,\Delta y)$, which is canceled from the numerator and denominator in the preceding equation. Assuming unobstructed radiation to space $(T_{rj} = 0\,\text{K})$ from, say, node 2, $\Im$ may be taken as emissivity ε_2, and

$$\text{CSGMIN}_2 = (Mc/A)_2 \big/ \big[\varepsilon_2 \sigma T_2^3 + K_t\big]$$

where Mc/A is thermal mass per unit area, which is now independent of the node size, and K_t is unit transverse conductance in W/m^2 K from node 1 to node 2. For solid materials, $K_t = k/\delta$, where k is conductivity and δ thickness; for honeycomb panels, K_t is defined and found by the methods of Chapter 2, Sec. III.

Typical data for high emissivity are listed in Table 5.1. It is seen that in contrast to structures of nominal mass, the number of time iterations can become extremely large for effectively light systems when total durations are on the order of minutes or hours. The problem is particularly vexing in multinode models made up of combinations of light and massive nodes. It is not unusual in these situations that convergence is delayed or permanently stalled because the least value of CSGMIN happens to pertain to one or a few nodes whose temperature response is radically sensitive by comparison to the remainder of the system. To avoid the difficulty and save on computation time, thermal analysts often convert light diffusion nodes into

Table 5.1   Order of magnitude of transient systems parameters

System [$c \sim \mathcal{O}(10^3)$, J/kg K]	M/A, kg/m^2	T, °C	$\varepsilon_2 \sigma T_2^3 + K_t$, W/K	CSGMIN s
Honeycomb platform with electronics	$\mathcal{O}(10\text{–}10^2)$	-10 to 50	$\mathcal{O}(10)$	$\mathcal{O}(10^3\text{–}10^4)$
Honeycomb solar array or antenna	$\mathcal{O}(1)$	-100 to 100	$\mathcal{O}(10)$	$\mathcal{O}(10^2)$
Phenolic sheets (0.25 mm)	$\mathcal{O}(10^{-1})$	-100 to 100	$\mathcal{O}(10^2)$	$\mathcal{O}(1)$
Thin phenolic sheets (125 μm)	$\mathcal{O}(10^{-1})$	-100 to 100	$\mathcal{O}(10^3)$	$\mathcal{O}(10^{-1})$
Thin phenolic sheets (25 μm)	$\mathcal{O}(10^{-2})$	-100 to 100	$\mathcal{O}(10^3)$	$\mathcal{O}(10^{-2})$
Painted aluminum radiators/ shields (1.5 mm)	$\mathcal{O}(1)$	-50 to 100	$\mathcal{O}(10^5)$	$\mathcal{O}(10^{-2})$

arithmetic nodes (no thermal mass) to exempt them from the CSGMIN restriction. For handling this situation, SINDA requires specifying two constants (DRLXCA and ARLXCA) as separate convergence criteria for diffusion and arithmetic nodes.

Fully implicit schemes are absolutely stable and will permit larger time steps to reduce computer time. However, their accuracy is more difficult to evaluate and they invariably require the use of a matrix inversion routine which, for large models, can frustrate the intended savings in time. In using these methods, the analyst often refers to the calculated CSGMIN as a signpost for selecting time intervals that are not too large (say, selected Δt less than five times CSGMIN) so that accuracy will hopefully remain within reasonable bounds. In this regard, an advantage offered by SINDA when implicit routines are used is that Δt can be made to fit various phases in the calculations. It can be made small when the boundary conditions, including heating inputs, are changing rapidly, and then enlarged during phases of moderate temperature variation.

The Crank–Nicholson scheme (FWDBCK in SINDA) is essentially the average of the fully explicit and fully implicit routines. It combines stability with good accuracy (second order) for sizable time steps. It is a preferred approach that is sometimes supplemented by implicit calculations with finer time steps to accommodate rapidly changing boundary conditions.

In modeling for SINDA, nodal construction and selection of time steps and differencing routines to optimize accuracy and speed, when an exact solution is not available for comparison, are mostly in the realm of intuition and experience and a great deal of interactive experimentation. In many cases the main body of a satellite is sufficiently massive for the temperature to be nearly stationary in time, and modeling for orbital average analysis (steady state) is considered adequate. Transient calculations on solar arrays and antennas are usually made separately from the massive electronics canister and, therefore, need not contend with out-of-proportion stability criteria. Specific regions in the satellite with sudden changes in power are also often modeled as separate systems of normally few uniform nodes. For critical transients, a standard confirmation technique consists of repeating the calculations at smaller time steps and observing trends of convergence to a possible asymptote. Although time consuming, the approach may be unavoidable when utmost accuracy is crucial.

A. Systems Evaluation of the Thermal Model

From the systems' standpoint, a judgement on the effectiveness and viability of the thermal model must take the following into consideration:

1) Designs that are difficult to model are invariably also difficult to check for errors or trace for causes of anomalies in test and flight. Such designs should be questioned and scrutinized. Along the same lines, results from models of uncomplicated designs can generally be confirmed by simplified analytical formulations.

2) Frequently, the degree of complexity in a thermal model is proportional to the multiplicity of internal radiation couplings. Because there is also considerable uncertainty associated with the assumptions made in calculating radiation exchange factors, especially in the presence of harnesses, connectors, protective covers, seperating walls, and other obstructions, the analysis is more reliable for designs that are not too dependent on internal (inside satellite canisters) radiation exchange.

3) Input data for the power profiles must be continuously tracked and verified by subsystem (component) engineering. Footprint power distribution is of particular importance in selecting the shape, size, and location of the nodes.

4) Detailed models should have associated with them reduced versions constructed specifically for parametric, tradeoffs, and interface studies. Reduced models must not be encumbered by particulars so that they can be appreciated by experts in other fields who might not have a prevalent knowledge of thermal analysis. In this connection, the computer program listing pages should be generous in comments and explanations.

5) Flexibility in moving modular array blocks can significantly reduce computer time when analyzing behavior under specific conditions. One example is removing the array block for radiation connections and adding a convection block when analyzing ground cooling. Another example is making removable subblocks for internal radiation connections to check vacuum test results for the relative influences of conduction, which generally can be modeled accurately, and radiation, which is difficult to predict.

6) Support graphics should highlight the pertinent rather than display the capabilities of the softwear. A single curve showing the effect of a radiator's α^s/ε on the maximum temperature of a battery cell is more demonstrative to the informed viewer than a whole colorful map cluttered up with numbers.

B. Tolerances and Margins on the Input Data

As in all mathematical representations, the results from the thermal model are as valid as the values of the input data. These must include the margins and tolerances observed in direct measurements or inferred from other sources. The pertinent parameters are 1) thermal–physical properties: conductivity, density, and specific heat; 2) geometry: shapes, dimensions, and orientations; 3) surface properties: radiation characteristics (α^s, ε, ρ, τ, and specularity); 4) orbital heating fluxes: mainly solar, albedo, and Earth radiation; and 5) components dissipation.

Heat transfer conductances (K, C, and $\Im$ or $F_E F_A$) and other heating factors are calculated in terms of these data or may have predetermined values based on other measurements. There are also margins pertaining to heat transfer parameters defined for specialized or combined systems. These are usually modified by the word "effective". Thus, there is an effective conductance associated with joints, mounting interfaces, and heat transfer by combined modes or phase change. There are also effective emissivity and effective solar absorptivity of radiation shields and self-adjusting radiating systems (Chapter 6, Secs. II and V).

With respect to tolerances and margins, the following should be noted:

1) To be legitimate, tolerances must be based on thorough engineering practice. Data from flawed systems (outgassing adhesives, bad insulation, poorly mounted electronics, leaking fluid components, etc.) are inconsistent and unpredictable and, therefore, cannot be considered parameters in the development of a design.

2) Experience and the use of heritage equipment (off-the-shelf items) increase the chances of tolerances being within expected ranges. Among the many examples are the fabrication and installation of multilayered insulation (Chapter 6, Sec. II), the application of coatings on a radiator, and the laying of a filler to enhance surface contact. Experience is essential in these operations because there are no immediate

means for verifying the performance and repairing a radical departure not revealed until testing could have serious repercussions.

3) Margins on environment heating fluxes should be consistent with available data on actual thermal performance in orbit.

4) The heat dissipated by satellite electronics is obviously a decisive factor in interpreting the results of the thermal model and determining how the design should proceed. Unfortunately, in the majority of cases the dissipation cannot be fully verified until the first integrated satellite functional test is completed, at which time the thermal design should have nearly reached final status. On the positive side, many components in today's satellites have heritage baselines with documented power profiles, and developing components continually undergo breadboard performance testing with power estimates regularly updated and delineated in a satellite resources and margins management report. Generally, the SRMMR serves as the official source for the power values used by thermal control.

An essential element of this activity is systems' coordination of the effort to ensure that the thermal design is not proceeding on unrealistic power profiles. But it is also recognized that power (same as weight) is often different (usually higher) in the final product than initially anticipated. There is really no foolproof answer to the dilemma within the scope of reasonable engineering and cost, but comparisons of analyses with test and flight data in a number of successful programs indicate that most unforseen and contingency deviations fall within the margins on component power dissipation. Typically, a margin of 10% over maximum estimated power and 10% below the minimum is imposed when predicting temperatures. The criterion is especially pursuasive when new components are being developed.

C. Hot Case and Cold Case

To contend with errors, tolerances, and spacecraft power uncertainties, thermal engineers almost universally adopt hot case and cold case analyses to define upper and lower bounds on predicted temperatures. The ground rules may vary, depending on experience and the nature of the mission, but the central guideline, at least in applications to NASA programs, is to build confidence by designing to meet the temperature specifications under accumulated biases, both at upper and lower levels. Thus, the power profile for a hot case analysis may correspond to an operation in which components' activity results in high dissipation, while the orbit is such that the radiators are exposed to considerable combined solar, albedo, and Earth heating. Biased margins and tolerances are then imposed on power, environment heating fluxes, and thermal properties in a direction that makes the analysis produce the maximum possible temperature. Similarly, the input data for the cold case are selected to result in a calculated lowest temperature. It is important, however, that the definitions and assumptions in both cases remain realistic and applicable within the scope of the intended mission so as to restrain the impact on dependent systems. It would be a failing, for instance, not to take advantage of the possibility that sufficient electrical power may be available for heater temperature control during orbits with low electronics dissipation and least heating fluxes on the radiators. Similarly, it would be an excess to assume simultaneous solar impingement on a solar array and on a radiator whose surface is perpendicular to it.

In general, unexpected occurrences, including loss of power, radiator misalignment, and a malfunction in the thermal hardware do not set the basis for a thermal design or performance requirements. These are considered failure modes that are assessed as deviations from normally expected conditions.

IV. Approximate Analytical Solutions

A fundamental feature of a simplified thermal analysis is the representation of a particular region in a satellite by just a few compact equations that are independent or connected by linearized coefficients. The ultimate simplicity in this regard is representation by a single isothermal body i of thermal mass $(Mc)_i$ and surface radiating area A_i^r, whose temperature T_i is indicative of the range expected for a whole region.

To do this, the energy equation is reformulated by discarding the spatial derivatives and substituting for the term of heat generated an expression that includes the collective heat dissipated in the region (Q_i^d) less its net heat transfer to the surroundings. Exchanges with the surroundings are included as a sum of products of temperature differences and modulating conductances. That is, if the heat transferred by radiation to j (at temperatures T_{jr}) is with exchange factors $\Im_{ij}$, and that by conduction (and convection) to j (at temperatures T_{jk}) is with conductances K_{ij} (W/K), then the energy equation reads

$$(Mc)_i \frac{dT_i}{dt} = Q_i^d - \sum_j \Im_{ij} A_i^r \left(\sigma T_i^4 - \sigma T_{jr}^4 \right) - \sum_j K_{ij}(T_i - T_{jk})$$

with initial condition $T_i(0)$. The interpretation recognizes that the radiation terms must encompass exchanges with the whole surrounding, as if it were an enclosure. This could include an exchange with space by the corresponding exchange factor.

When external heating includes exchanges with Earth and the sun (some of the j), the equation may be recast (Chapter 4, Sec. V.B.) with explicit reference to absorbed environment fluxes,

$$(Mc)_i \frac{dT_i}{dt} = Q_i^d + (\alpha^s A^S q^S + \alpha^s A^A q^A + \varepsilon A^E q^E)_i - \sum_j \Im_{ij} A_i^r \left(\sigma T_i^4 - \sigma T_{jr}^4 \right)$$

$$- \sum_j K_{ij}(T_i - T_{jk})$$

with known or assumed $T_i(0)$. Here A^S, A^A, and A^E correspond to the receiving areas for the indicated normal incident fluxes.

In steady state and with all areas equal to the same value A,

$$q_i^d + [\alpha^s(q^S + q^A) + \varepsilon q^E)]_i - \sum_j \Im_{ij} \left(\sigma T_i^4 - \sigma T_{jr}^4 \right) - \sum_j K_{ij} \frac{(T_i - T_{jk})}{A} = 0$$

Hence, knowledge of the dissipation and heating input from the surroundings allows the temperature to be determined as function of thermal couplings, area, solar absorptivity, and emissivity. All of these parameters can be controlled, and combinations may be selected to produce the desired temperature for a given dissipation.

When there are no interactions with other surfaces and there is radiation to deep space with a uniform emissivity ($\sum_j \Im_{ij} = \varepsilon$), the indices may be dropped and the equation written

$$q^d + \alpha^s(q^S + q^A) + \varepsilon q^E = \varepsilon \sigma T^4$$

This last equation (seen before in Chapter 4) is decidedly the one used most in group discussions and when making preliminary thermal assessments. It is especially useful when applied to massive radiators whose spatial temperature variations are not too great, a situation resembling the performance of components-carrying platforms in orbit. In this case, the dissipation and incident heating fluxes may be averaged over the orbit period, and a first-order analysis made to determine whether temperature control can be achieved solely through the use of a surface treatment or if specialized techniques and hardware might be needed.

A. Preliminary Sizing of Radiators and Heaters

Some of the very first calculations made by the thermal analyst pertain to estimating the size of satellite canister radiators and the requirements for heater power. This can be immediately done if there are reasonably valid definitions of the orbits and components' activity and if there is at least a rudimentary image of the shape and dimensions of the satellite. The analysis is usually based on orbital average considerations under hot case and cold case conditions. An outline referring to the simple case of an isothermal, space-facing, radiating panel with dissipating components mounted to its back follows:

1) Coordinate a systems effort to obtain best estimates of expected satellite power and the maximum (T_{max}) and minimum (T_{min}) temperatures considered acceptable for the panel in orbit. Classify the cases for thermal analysis in terms of orbits and corresponding spacecraft powers. Include in the classification possible extended satellite pointings or fail-safe maneuvers.

2) Study available drawings for information on the satellite aspects in orbit. Using data from charts or orbital heat flux programs, distinguish the directions along which the impinging fluxes are lowest and do not vary by large amounts during the course of mission. Regions that are mostly perpendicular to, or shaded from, the solar vector are candidate locations for placing the panel. These locations are also preferred from the standpoint of improved predictability. Other surfaces may have to be insulated to reduce the effects of exposure to space environment.

3) Select a thermal treatment for the radiator surface. A low solar absorptivity will reduce solar and albedo effects; a high emissivity provides the desired potential for keeping the panel cool.

4) Apply tolerances and margins that will categorize the input to the analysis as hot case and cold case data. As a minimum, these must include components dissipations, environment heating fluxes, and solar absorptivity and emissivity.

5) Assuming an isothermal radiator, perform an orbital average (steady-state) hot case energy balance to find the area A^r (one side) that will keep the temperature $T_{hot} \leq T_{max}$,

$$A^r = \frac{Q^d - Q_{loss}}{\varepsilon \sigma T^4 - \alpha^s(q^S + q^A) + \varepsilon q^E}\bigg|_{\text{hot case data}}$$

Q_{loss} is heat lost or gained by the attached and surrounding structure. In a hot case analysis, this quantity is frequently assumed zero when the interacting structure can have the same or lower temperature, and negative when this temperature is higher.

6) Use the calculated A^r in an orbital average cold case energy balance to determine the minimum temperature:

$$\sigma T_{\text{cold}}^4 = \frac{Q^d - Q_{\text{loss}}}{\varepsilon A^r} + \frac{\alpha^s}{\varepsilon}(q^S + q^A) + q^E \bigg|_{\text{cold case data}}$$

A realistic estimate of Q_{loss} (see example in Chapter 6, Sec. V) must be included in this calculation.

7) If the computed $T_{\text{cold}} \geq T_{\text{min}}$, then, depending on the desired margins, a design may be feasible without extensive use of specialized hardware. If $T_{\text{cold}} < T_{\text{min}}$, then heaters or other devices need to be considered. The heater power that can bring the temperature up to T_{min} is found from the cold case energy balance in the form

$$Q_{\text{heater}} = Q_{\text{loss}} + A^r \left\{ \varepsilon \sigma T_{\text{min}}^4 - [\alpha^s(q^S + q^A) + \varepsilon q^E] \right\} - Q^d \big|_{\text{cold case data}}$$

with all quantities known except heater power.

Besides providing an early insight into the nature and magnitude of the power problem, procedures such as the one just described often serve as the starting point to establishing the necessary protocol between thermal control and power. It must be noted, however, that these exercises are only a prelude to eventual detailed modeling and analysis, where spacial temperature distributions and refined values of heat interactions are obtained and used in making final design decisions.

Example: Passive versus Active Thermal Control

As used here, passive thermal control implies achieving the required radiator temperature solely by surface treatment, including the use of insulation. A parametric evaluation of the orbital average temperature of a flat plate gives an indication of the capability of a passively controlled equipment platform to reject dissipated heat at an acceptable temperature.

The analysis must consider performance both at the beginning of the mission [beginning of life (BOL)], when the radiator is new and its properties near their initial values, and at end of life (EOL), when it may be in a degraded state after long-term exposure to space environment (particularly solar and albedo radiation). Similarly, the range of possible orbits and orientations must be scanned for the two cases that signify the maximum and minimum incident solar, albedo, and Earth fluxes.

Consider, for example, a radiator aligned with side 1 in the Earth-oriented orbit cited in Fig. 4.7. Direct solar impingement is negligible and at, say, 600-km elevation, orbital average Earth flux and albedo are read off the curves and found to be about 60 W/m^2 and 32 W/m^2, respectively. The values to use in hot case and cold case analysis may be estimated by applying the ratio of the tolerances on f

and E given in Table 4.2. Thus,

$$q_{600\,\text{km}}^E \text{ (hot case)} = 60(274/236) = 69.7\,\text{W/m}^2$$

$$q_{600\,\text{km}}^E \text{ (cold case)} = 60(198/236) = 50.3\,\text{W/m}^2$$

$$q_{600\,\text{km}}^A \text{ (hot case)} = 32(0.46/0.33) = 44.6\,\text{W/m}^2$$

$$q_{600\,\text{km}}^A \text{ (cold case)} = 32(0.20/0.33) = 19.4\,\text{W/m}^2$$

Assuming a silvered Teflon treatment, the range of the values of properties are given in Table 6.1 as:

$$\alpha^s/\varepsilon \text{ (hot case; 5 years)} = 0.28/0.75$$

$$\alpha^s/\varepsilon \text{ (cold case)} = 0.17/0.79$$

Following the procedure for bounding the problem, extreme values of the parameters are combined to obtain the range of temperatures from the equation

$$Q^d/A + \alpha^s(q^S + q^A) + \varepsilon q^E = \varepsilon\sigma T^4$$

The temperature as a function of power dissipation is shown in Fig. 5.4.

It is found that in a LEO a temperature between 0 and 40°C (allowable flight specifications for many electronics platforms) can be achieved when the ratio of maximum to minimum dissipated power is less than 2 to 1 (350/200 or 1.75). This is possibly within the range of the operational power of a weather satellite or one that carries an experiment with limited requirements. But in a more versatile operation, the power ratio may be nearer to 4, in which case active thermal control using specialized hardware may be necessary.

The same radiator aligned with the south direction in GEO will receive maximum solar input during winter solstice. This is calculated as

$$q_{\text{Geo}}^S(\text{hot case, winter solstice}) = 1400\sin 23.5 \text{ deg} = 558\,\text{W/m}^2$$

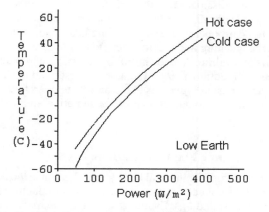

Fig. 5.4   Orbital average temperature of a silvered Teflon radiator in a 600-km, Earth-oriented mission (side 1 in Fig. 4.7); hot case $\alpha^s/\varepsilon = 0.28/0.75$; cold case $\alpha^s/\varepsilon = 0.17/0.79$.

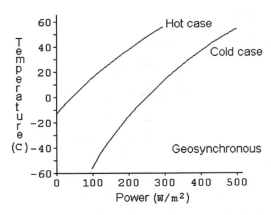

Fig. 5.5 Temperature of a silvered Teflon south-side radiator in GEO; hot case
$\alpha^s/\varepsilon = 0.35/0.75$; cold case $\alpha^s/\varepsilon = 0.17/0.79$.

There is no solar impingement during equinox and geosynchronous cold case is
defined by

$$q^S_{\text{Geo}} \text{ (cold case)} = 0$$

From Table 6.1 for, say, a seven-year mission,

$$\alpha^s/\varepsilon \text{ (hot case; 7 years)} = 0.35/0.75$$

$$\alpha^s/\varepsilon \text{ (cold case)} = 0.17/0.79$$

and the temperature vs power dissipation is as shown in Fig. 5.5.

In this case passive temperature control between 0 and 40°C is not possible for
the given conditions and margins. To obtain a cold case temperature of 0°C requires
250 W/m² of heating, whereas hot case calculations show that the temperature will
exceed 40°C when the dissipation is above 215 W/m². For some communications
satellites, the power is usually nearly a constant, and a solution is found in sizing
the radiator for hot case operation and providing heater power to compensate for
the lack of absorbed heating in the cold case. This, of course, entails added power
requirements on the satellite. Active control by louvers could reduce or eliminate
heater power, but will require a larger radiator. As described in Chapter 6, Sec. V,
the great advantage in using louvers is that a narrow temperature variation can be
maintained by the automatic adjustment of the radiating area to compensate for
heat variations.

B. Conducting–Radiating Fin: Linearization

An upgrade on the single isothermal body simplification is the representation of
a system or region by a collection of one-dimensional conducting fins exchanging
radiation with their surroundings. Estimates can be made from this model of both
temperature levels and spatial variations.

The formulation requires that the term for heat generation in the thermal energy
equation be replaced by an expression combining dissipation and environment

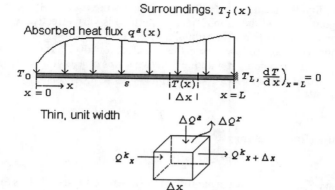

Fig. 5.6 One-dimensional conducting–radiating fin.

heating with radiation losses. But the combination could involve separate quantities for the areas associated with conduction, absobed heating, and radiation. For uniform cross section and surface geometry, A^k is constant and

$$A^r_{\Delta x} = A^r \Delta x/L$$

$$A^a_{\Delta x} = A^a \Delta x/L$$

where A^r and A^a are the total areas of radiation and absorption and Δx is a section along the length L. Referring to the elemental volume in Fig. 5.6, the energy balance

$$Q^k_x - Q^k_{x+\Delta x} = \Delta Q^r - \Delta Q^a$$

becomes

$$\left(-kA^k\frac{dT}{dx}\right)_x - \left(-kA^k\frac{dT}{dx}\right)_{x+\Delta x} = \Im(x)[\sigma T^4 - \sigma T_j^4(x)]A^r\frac{\Delta x}{L} - q^a(x)A^a\frac{\Delta x}{L}$$

where $q^a(x)$ is absorbed energy, which can include dissipation, and $\Im(x)$ is the radiation exchange factor with an environment of temperature $T_j(x)$. In the limit, as $\Delta x \to 0$,

$$\frac{d^2T}{dx^2} + \frac{A^a q^a(x)}{kA^k L} - \frac{\Im(x)A^r}{kA^k L}[\sigma T^4 - \sigma T_j^4(x)] = 0$$

When the coefficients and surrounding temperature are constant,

$$\frac{d^2T}{dx^2} - \Gamma^2 T^4 = \Lambda - Bq^a(x)$$

with Γ, Λ, and B constant.

With appropriate boundary conditions this last equation can be made to apply to a number of actual situations, but it is nonlinear and difficult to resolve into quadratures. However, approximate expressions can be found featuring parameters that may be regulated to improve the accuracy. This is not much different from obtaining solutions by numerical routines that use adjustable weighted factors. But the availability of a formula that relates the temperature to controllable parameters (such as geometry and thermophysical properties) is very effective as a compact description of a problem and its solution.

The approximation follows from the linearization of T^4 (now considered a function of T) by replacing it with the first two terms of a Taylor expansion about a mean temperature T_m; that is, T^4 is replaced by $T^4(T_m) + 4T^3(T_m)[T - T_m]$, which is written

$$T^4 \sim 4T_m^3 T - 3T_m^4$$

But because (from algebra)

$$T^4 - \left(4T_m^3 T - 3T_m^4\right) > 0$$

for arbitrary T_m, the radiative capability of the linearized system is less than that of the exact one, which makes the approximate temperatures always higher than the actual temperatures. Because the actual temperature is unrelated to the selection of T_m, however, the least difference between exact and approximate profiles at general locations X_i and fin parameters P_i is constrained by

$$\left(\frac{\partial T_{approx}}{\partial T_m}\right)_{X_i, P_i} = 0$$

which is the equation from which the best T_m is obtained.

Numerical results derived from using the optimization criterion are in most cases nearly identical to those found by finite differences.[7] It is also discovered that under moderate heating, data obtained by choosing T_m based on physical considerations are not significantly different from those calculated by the tedious and sometimes unstable formal procedure.

The method produces relevant results in a variety of practical problems. Consider, for example, the uniform fin shown in Fig. 5.6 with the indicated boundary conditions that are typical of relatively thin (no end losses) structures protruding from the body of a satellite. For uniform radiation to deep space with constant emissivity ε,

$$\frac{d^2 T}{dx^2} + \frac{A^a q^a(x)}{k A^k L} - \frac{\varepsilon A^r \sigma T^4}{k A^k L} = 0$$

and

$$T(0) = T_0, \qquad \left.\frac{dT}{dx}\right|_{x=L} = 0$$

As was just noted, $q^a(x)$ may include internal heat generation and estimated heating from extraneous sources. Substituting the linearized form for T^4 and introducing the parameters

$$\tau = T/T_0$$

$$\xi = x/L$$

$$\beta = A^a L / k A^k T_0$$

$$\gamma^2 = 4\varepsilon A^r L \sigma T_0^3 \tau_m^3 / k A^k$$

the equation and boundary conditions take the following dimensionless form:

$$\frac{d^2\tau}{d\xi^2} + \beta q^a(\xi) - \gamma^2(\tau - 0.75\tau_m) = 0$$

$$\tau(0) = 1 \quad \text{and} \quad \left.\frac{d\tau}{d\xi}\right|_{\xi=1} = 0$$

with $\tau_m = T_m/T_0$. The solution of this linear system of equations is given by

$$\tau(\xi) = 0.75\tau_m + \frac{(1 - 0.75\tau_m)[\cosh\gamma(1 - \xi)]}{\cosh\gamma}$$

$$- \frac{\beta\sinh\gamma}{\gamma\cosh\gamma}\int_0^1 [\cosh\gamma(1-\chi)]q^a(\chi)\,d\chi - \frac{\beta}{\gamma}\int_0^\xi [\sinh\gamma(\xi-\chi)]q^a(\chi)\,d\chi$$

The integral representation is especially useful when the heat input can be expressed in terms of rational functions or elementary transcendentals.

For constant flux $q^a(x) = U$ (which can be zero) the expression becomes

$$\tau(\xi) = \frac{T}{T_0} = 0.75\tau_m + \left(\frac{\beta U}{\gamma^2}\right) + \left(1 - 0.75\tau_m - \frac{\beta U}{\gamma^2}\right)\frac{\cosh\gamma(1-\xi)}{\cosh\gamma}$$

and, evaluating $\tau(0) - \tau(1)$,

$$1 - \tau(1) = \frac{T_0 - T_L}{T_0} = \left(1 - 0.75\tau_m - \beta\frac{U}{\gamma^2}\right)\frac{\cosh\gamma - 1}{\cosh\gamma}$$

which is the equation for finding the temperature at the adiabatic end.

When $q^a(x)$ is a constant, the temperature distribution is monotonic in x with $|T_0 - T_L|$ forming the largest temperature difference along the fin. Because T_0 is known, the best estimate for $T_0 - T_L$ is found from an optimization at the fin's end; that is, by calculating τ_m according to

$$\left(\frac{\partial\tau}{\partial\tau_m}\right)_{\xi=1} = 0$$

By performing the differentiation, a relation is found between the optimum dimensionless mean temperature τ_{mo} and the corresponding $\gamma = \gamma_o$:

$$\tau_{mo} = \frac{2\gamma_o\sinh\gamma_o + 4(\beta U/\gamma_o^2)(\cosh^2\gamma_o - \cosh\gamma_o - 0.5\gamma_o\sinh\gamma_o)}{\cosh^2\gamma_o - \cosh\gamma_o + 1.5\gamma_o\sinh\gamma_o}$$

A second equation is the definition

$$\gamma_o^2 = 4\varepsilon A^r L\sigma T_0^3\tau_{mo}^3/kA^k$$

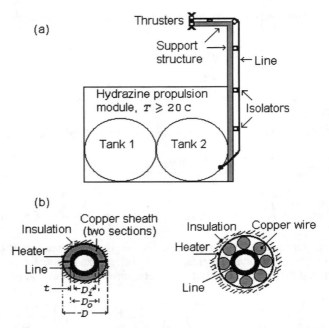

Fig. 5.7 Enhancing heat conduction along hydrazine propellant lines; a) hydrazine line extending outside a propulsion module and b) two methods for placing copper on propellant line.

which is solved simultaneously with the first to give the best selection of mean temperature as function of known parameters.

Example: Thermal Control of an External Hydrazine Propellant Line

Consider a 1-m-long hydrazine propellant line made of circular stainless steel ($k = 16$ W/m K) tubing, with a 1-cm outside diameter D_o, and a 0.15-cm wall thickness t. The line protrudes from a spacecraft propulsion module, which is thermally controlled to remain between 20 and 50°C, and runs along a support structure as shown in Fig. 5.7a to supply fuel to thrusters used for orbit adjustment. The critical problem is that hydrazine freezes at about 1.5°C. A standard specification that includes margins requires hydrazine propellant lines to remain between 10 and 50°C.

Because of combinations of shadowing, orbit orientation, and maneuverings, the line can sometimes be in the sun for long periods, and at other times be completely shaded. The question is whether all locations on the line will remain above 10°C in a cold case, total shadow situation by virtue of drawing heat by conduction from the main module whose temperature is at least 20°C. The possibility improves by wrapping the line with some 10 sheets of thin alumizied Kapton with Dacron mesh separators. (Details on multilayered insulation are found in Chapter 6, Sec. II).

The cold case includes least insulation effectiveness (usually specified by $\varepsilon_{ins} = 0.02$) with the module at its lowest temperature (20°C). The value of γ, the parameter associated with conducting radiating fins, is then obtained as

$$\gamma = \left[4\varepsilon A^r L \sigma T_0^3 \tau_m^3 / k A^k\right]^{1/2}$$

$$= \left\{\left[(4)(0.02)\pi(0.01)(1)(1)(5.67E-8)(20+273)^3 \tau_m^3\right] / [(16)\right.$$

$$\left.\times (\pi/4)(0.01^2 - 0.007^2)]\right\}^{1/2}$$

$$= 2.365 \tau_m^{3/2}$$

With $U = 0$, τ_m is found by iterating on the equations

$$\gamma_o = 2.365 \tau_{mo}^{3/2}$$

and

$$\tau_{mo} = 2\gamma_o \sinh \gamma_o / (\cosh^2 \gamma_o - \cosh \gamma_o + 1.5\gamma_o \sinh \gamma_o)$$

MATHCAD or a programmable hand calculator (HP 28S) gives $\gamma_o = 1.6138$ and $\tau_{mo} = 0.775$.

The lowest temperature along the line is at the farthest end and is calculated from

$$(T_0 - T_L)/T_0 = (1 - 0.75\tau_{mo})(\cosh \gamma_o - 1)/\cosh \gamma_o$$

which gives

$$T_L = 293 - 293\{[1 - (0.75)(0.775)][(\cosh 1.6138 - 1)/\cosh 1.6138]\}$$

$$= 217.3 \, \text{K} \, (-56°\text{C})$$

Clearly, this is unacceptable.

A possible remedy is to supply direct heating by a thin, approximately $\frac{1}{2}$-cm-wide flexible strip heater that may be wrapped helically (no space between pitches) along the entire length of the line. Then, in shadow, the farthest end can certainly be maintained at 10°C or higher if heater power equals or exceeds

$$Q_{\text{heater}} = \varepsilon A^r \sigma (10 + 273)^4$$

Assuming that the heater strip adds about 1 mm to the line diameter and that its surface emissivity is 0.7 (thin Kapton tightly wrapped on stainless steel, $\alpha^s \approx 0.6$), the required heater power without insulation is obtained as

$$Q_{\text{heater}} = (0.7)(\pi)(0.01 + 0.001)(1)(5.67E-8)(283)^4 = 8.8 \, \text{W}$$

A hot case calculation of the average temperature in the sun (assuming no Earth flux or albedo) with the heater on is found from

$$\sigma T^4 = \frac{Q_{\text{heater}}}{\varepsilon A^r} + \left(\frac{\alpha^s}{\varepsilon}\right)\left(\frac{A^s}{A^r}\right)q^s$$

For a cylinder, $A^s/A^r = 1/\pi$ and, based on a maximum incident flux $q^s = 1400$ W/m^2 and negligible conduction to the module, the average hot case temperature is

$$T(\text{in the sun}) = \{(1/\sigma)[8.8/(0.7)(\pi)(0.011)(1) + (0.6/0.7)(1/\pi)(1400)]\}^{1/4}$$

$$= 338.65 \text{ K} (65.7°\text{C})$$

Although 65.7°C is higher than the specified upper limit of 50°C, it might be acceptable because the flashpoint of hydrazine exceeds 160°C (normal boiling point is 114°C). However, limitations on available electrical power must be considered, especially if there are other, additional lines requiring similar thermal control.

Heater power will be reduced considerably if the heater–line system is protected by insulation. Doing this, however, entails a serious risk if the insulation effectiveness turns out to be better than expected. Thus, for an assumed insulation effectiveness $\varepsilon_{\text{ins}} = 0.02$, only about $\frac{1}{4}$ W would be needed to maintain the farthest end at 10°C; but if it happens (which is not impossible following long-term venting of trapped gasses in space) that the effectiveness improves to, say, $\varepsilon_{\text{ins}} = 0.002$, then it can be verified that $\frac{1}{4}$ W of steady heating would result in temperatures in the excess of 200°C.

Overheating is usually averted by using mechanical or solid-state thermostats (Chapter 6, Sec. VI) calibrated to turn the heater off and on at prescribed set temperatures. But the installation of mechanical thermostats or temperature sensors on a small-diameter tube is physically an awkward process that could result in lower effectiveness of insulation in the vicinity of mounting adaptors, wires, and other penetrations. Because heat conduction along the line is poor, a heater responding to a sensor environment characterized by an insulation emissivity equal, say, to 0.02, will overheat regions along the line where the effectiveness is much better. Using the preceding data, this could mean that some portions of the line would be at some 200°C while the temperature at the sensor's region is a mere 10°C.

Most of the difficulties disappear if the thermal conductivity of the line is improved. This has been done in practice by laying thin sheaths or wires made of high-conductivity material, such as copper, along the whole stretch of line. The pliability of thin, relatively soft copper makes it possible to form strands that remain continuous where they cover bends and other unwieldy regions.

If the farthest end of the line is to remain during shadow at 10°C solely by heat conduction from the body of the spacecraft, then the equation for the temperature difference along the fin with $U = 0$ gives

$$(293 - 283)/293 = (1 - 0.75\tau_m)(\cosh \gamma - 1)/\cosh \gamma$$

For best estimates, τ_m should be related to γ by

$$\tau_{mo} = 2\gamma_o \sinh \gamma_o/(\cosh^2 \gamma_o - \cosh \gamma_o + 1.5\gamma_o \sinh \gamma_o)$$

and hence, from the calculator,

$$\gamma_o = 0.5299, \qquad \tau_{mo} = 0.971$$

From the definitions,

$$\gamma_o^2 = 4\varepsilon A^r L\sigma T_0^3 \tau_{mo}^3/kA^k, \qquad A^r = \pi DL, \qquad A^k = (\pi/4)(D^2 - D_o^2)$$

where D and D_o are as indicated at left in Fig. 5.7b. Hence, with no solar heating, $\varepsilon = 0.02$, and ignoring heat conduction by the hydrazine and line material, a copper sheath of thermal conductivity $= 360$ W/m K (series 80,000) and diameter D yields

$$(0.5299)^2 = [4(0.02)\pi D(1)(5.67E-8)(20+273)^3(0.971)^3]/[360(\pi/4)$$
$$\times (D^2 - 0.01^2)]$$

$$D = 0.01228 \text{ m}$$

This translates into a sheath thickness t slightly more than 1 mm.

The cold case temperature specification is now satisfied. Moreover, a low-power heater–thermostat system can now be used for added margin. With increased line conduction, the risk of creating large temperature variations along the line is reduced significantly.

Thermal control in the hot case is accomplished by selecting the appropriate treatment on the outer surface of the insulation. If the outer surface is painted black (say, maximum $\alpha^s/\varepsilon = 1.14$), then steady-state temperature without heater power (open thermostat) is found from

$$\sigma T^4 = (\alpha^s/\varepsilon)(1/\pi)q^S$$

Using $q^S = 1400$ W/m^2,

$$T(\text{hot case}) = [(1.14/\pi)1400/5.67E-8]^{1/4} = 308 \text{ K} (35°C)$$

Thermal interaction with the propulsion module could affect this value, but the temperature should normally remain below the specified maximum 50°C in the module.

As a side benefit, the added bulkiness in hydrazine propellant lines when heater wires and conductive sheaths or strands are used has proven to aid immeasurably in the wrapping of a consistent and effective multilayered insulation. It also provides a more secure base for bonding thermostats and thermistors.

Example: Temperature Drop Across Honeycomb Panels–Comparison with Exact Solution

Honeycomb panels are widely used as spacecraft structure, including platforms for dissipating electronics and as substrates for solar cells.

Consider the honeycomb panel schematically shown in Fig. 5.8. As discussed in Chapter 2, Sec. III, a thermal representation may consist of two nodes located

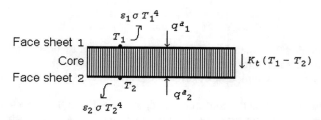

Fig. 5.8   Heat transfer in a honeycomb slab; a mean temperature is defined in the bulk heat balance $(\varepsilon_1 + \varepsilon_2)\sigma T_m^4 = q_1^a + q_2^a$.

opposite each other on the face sheets and coupled by a transverse conductance that can be estimated from

$$K_t = \frac{(\rho_c/\rho_m)k_m}{\delta}$$

where ρ is density, k conductivity, and the subscripts c and m refer, respectively, to the core and material from which the honeycomb is made. The thickness of the panel is denoted by δ.

Steady-state heat balances on nodes 1 and 2 give the two exact equations

$$q_1^a - \varepsilon_1\sigma T_1^4 - K_t(T_1 - T_2) = 0$$

$$q_2^a - \varepsilon_2\sigma T_2^4 + K_t(T_1 - T_2) = 0$$

The linearized representation is

$$(4\varepsilon_1\sigma T_m^3 + K_t)T_1 - K_t T_2 = q_1^a + 3\varepsilon_1\sigma T_m^4$$

$$K_t T_1 - (4\varepsilon_2\sigma T_m^3 + K_t)T_2 = -(q_2^a + 3\varepsilon_2\sigma T_m^4)$$

A natural definition for T_m in this case derives from an overall heat balance that equates the total energy absorbed to the heat radiated at the mean temperature; that is,

$$\sigma T_m^4 \equiv (q_1^a + q_2^a)/(\varepsilon_1 + \varepsilon_2)$$

Numerical data based on actual values from a tilted solar array in sunlight in LEO are as follows:

$$q_1^a = 1010.6 \text{ W/m}^2 \qquad q_2^a = 18.8 \text{ W/m}^2$$

$$\varepsilon_1 = 0.80 \qquad \varepsilon_2 = 0.83$$

$$K_t = 38.2 \text{ W/K per m}^2$$

$$\sigma T_m^4 \equiv (1010.6 + 18.8)/(0.80 + 0.83) = 631.5 \text{ W/m}^2 \ (T_m = 324.9 \text{ K})$$

Substituting into the nonlinear equation and solving numerically (iteration, hand calculator, MATHCAD solvers, etc.) give

$$T_1 = 57.7°C \qquad \text{and} \qquad T_2 = 45.5°C$$

The corresponding values from the simultaneous solution of the two linearized equations are

$$T_1 = 58.0°C \qquad \text{and} \qquad T_2 = 45.8°C$$

It is concluded that linearization with a mean temperature based on lump energy balance leads to satisfactory results. This is often the case when dealing with steady-state problems where temperature differences are not exessive ($\Delta T/T$ of order 0.1 or less).

It is useful to note that the configuration depicted in Fig. 5.8 applies to an exposed honeycomb panel (solar array or antenna dish), as well as a satellite equipment deck. In the latter case, with node 1 toward the satellite interior, $q_1^a - \varepsilon_1\sigma T_1^4$ is replaced by the net heat dissipation on the panel.

Example: Heat Pipe Spacing in Honeycomb Panels

Lateral heat distribution in honeycomb panels can be improved considerably by bonding heat pipes into the honeycomb (Chapter 6, Sec. IV is dedicated to heat pipes). The practice has now become standard for limiting thermally induced distortions in optical benches and enhancing the operation of battery groups or instruments. In most cases, the increase in weight is a modest penalty for overwhelming thermal benefits. (See Ref. 8 for a treatment on weight optimization.)

The configuration is represented in Fig. 5.9. The temperature at the panel interface in the direction of the pipes is nearly constant, but the fin effect (illustrated by the curve T vs x) between pairs of heat pipes can be large unless their spacing is kept moderately small.

With a sufficient number of heat pipes, a judicious placement of dissipating components will make all pipes on a platform run at about the same average temperature. The practice, however, is to minimize the number of pipes while remaining below a maximum allowable lateral temperature variation due to the fin effect. In finding the relation between temperature difference and heat pipe spacing, the analysis is usually performed assuming that there is no heat input along the fin other than that conducted from the heat pipes through the thickness of the honeycomb face sheets. Because only the temperature difference is being sought, heat exchange with the satellite is not included and the energy is assumed to radiate to deep space.

Consider the data given in Fig. 5.9 and an operation at 40°C (313 K). With noted symmetry and $U = 0$, the equation for γ^2 gives

$$\gamma = \left\{ \left[4(0.76)(1)(L)(L)(5.67E-8)(313)^3 \tau_m^3 \right] / \left[2(185)(1)(0.6\,E-3) \right] \right\}^{1/2}$$

$$= 4.880 \tau_m^{3/2} L$$

Using the optimization requirement

$$\tau_m = 2\gamma \sinh \gamma / (\cosh^2 \gamma - \cosh \gamma + 1.5\gamma \sinh \gamma)$$

iterations on L (that is, γ) and τ_m generate pairs of optimum values that are used to determine $(T_{hp} - T_L)$ as a function of $2L$ (the spacing) from the equation

$$(T_{hp} - T_L)/T_{hp} = (1 - 0.75\tau_{mo})(\cosh \gamma_o - 1)/\cosh \gamma_o$$

For example, selecting $\gamma = 0.5$, τ_m is calculated from the optimization equation as 0.9745. The equation for γ then gives $2L = 0.213$ m (21.3 cm) and the temperature difference is found as $(T_{hp} - T_L) = 9.5$ K. Continuing in this fashion, Fig. 5.9b is obtained. The reader may want to affirm that the number of heat pipes in a satellite radiator must be calculated based on the maximum operating temperature.

Example: Thermal Bending of Thin Tubular Elements

Tubular extendable elements (TEEs) carried on spacecraft as antennas, gravity gradient stabilizers, and instrument extension systems are typically made of thin, metallic ribbons (beryllium copper and stainless steel are common) that are rolled into longitudinal tubes and heat treated. They are then opened, flattened into ribbons again, and wound about their widths on a small-diameter spool that is driven by a

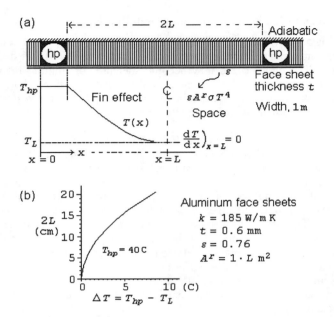

Fig. 5.9   Heat pipe spacing in honeycomb panels.

motor. When the element is extended, it takes the form of a tube with a near-circular cross section. Tabs and indents (called hingelocks or edgelocks) are cut along the ribbon edges to form on deployment a seam of interlocks that provide torsional rigidity. A three-dimensional sketch describing the shape during extension is shown in Fig. 5.10.

Element distortions induced by solar heating are of concern in many applications. The deflection happens because the longitudinal sections of the element expand and contract differently in response to the temperature variation around the perimeter when exposed to sunlight. The bent shape is the result of thermal moments that are calculated from the temperature distribution about the principal axes.[9] From circular symmetry, the net bending moment is about the Y axis and is given by

$$M_y = \int_a eETx \, \mathrm{d}a$$

where e is the coefficient of expansion of tube wall material and E its modulus of elasticity. Tip delection (in the minus-X direction) is

$$\delta = (M_y/2EI_y)L^2$$

where L is antenna length and I_y the moment of inertia that, for a thin cylindrical shell, is approximately $\pi R^3 t$. Substituting polar coordinates (Fig. 5.10),

$$\delta = \frac{1}{2} \frac{eL^2}{\pi R} \int_0^{2\pi} T(\phi) \cos \phi \, \mathrm{d}\phi$$

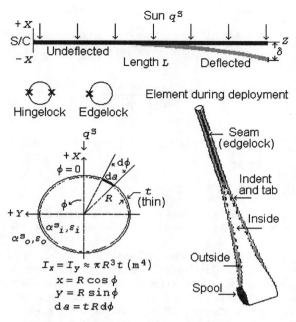

Fig. 5.10 Solar deflection of TEEs; bending moments are $M_y = \int_a eETx\,da$ and $M_x = \int_a eETy\,da$, which is zero due to symmetry; deflection $\delta_x = -(M_y/2EI_y)L^2$.

The thinness of a TEE and the high resistance to heat conduction at the interlocking seam make radiation the dominant mode of heat transfer. The temperature distribution for this case was obtained in Chapter 2, Sec. II as

$$\sigma T_\phi^4 = \frac{\alpha_o^s q^s}{\varepsilon_o \pi}\left[1 + \frac{\pi \cos\phi}{2[1 + (4/3)\rho]} - 2\sum_{n=1}^{\infty}\frac{(-1)^n \cos 2n\phi}{[1 + \rho + \rho/(16n^2 - 1)][4n^2 - 1]}\right]$$

where α_o^s is the solar absorptivity of the external surface and $\rho = \varepsilon_i/\varepsilon_o$ the ratio of inner to outer emissivities. The circumferential angle ϕ is designated 0 at maximum solar impingement.

The mean temperature associated with a cylinder in sunlight can be defined in the equation

$$\varepsilon_o \sigma T_m^4 = \frac{1}{2\pi}\int_{-\pi/2}^{\pi/2}\alpha_o^s q^s \cos\phi\,d\phi = \frac{\alpha_o^s q^s}{\pi}$$

and by rearranging the linearized expression $\sigma T^4 \sim 4\sigma T_m^3 T - 3\sigma T_m^4$ to read

$$T \sim (3/4)T_m + (1/4)\left(\sigma T^4/\sigma T_m^4\right)T_m$$

the temperature distribution can be written

$$T(\phi) = \frac{3}{4}T_m + \frac{1}{4}T_m\left[1 + \frac{\pi \cos\phi}{2[1 + (4/3)\rho]} - 2\sum_{n=1}^{\infty}\frac{(-1)^n \cos 2n\phi}{[1 + \rho + \rho/(16n^2 - 1)][4n^2 - 1]}\right]$$

Hence

$$\int_0^{2\pi} T(\phi) \cos \phi \, d\phi = \frac{\pi^2 T_m}{8[1 + (4/3)\rho]}$$

$$\delta = \frac{\pi T_m e L^2}{16R[1 + (4/3)\rho]}$$

Noting that

$$T(0) - T(\pi) \equiv \Delta T = \pi T_m/[4(1 + 4\rho/3)]$$

the expression can be written

$$\delta = \frac{e \Delta T}{4R} L^2$$

which is the usual linear form given for beam thermal deflection. On using this equation, the values of $T(0)$ and $T(\pi)$ may be calculated from the nonlinearized solution for T^4.

The analysis shows that deflections are lower for cooler elements (low T_m) with a high value of $\rho = \varepsilon_i/\varepsilon_o$. The element's mean temperature is proportional (fourth root) to α_o^s/ε_o and, therefore, a low solar absorptivity to high-emissivity treatment for the external surface would appear to help. However, a high external emissivity has an opposite and more dramatic effect on a desired high $\varepsilon_i/\varepsilon_o$. One satisfactory compromise has been to use a silver-plated finish (α_o^s/ε_o between 1 and 3) on the exterior surface and an anodize (ε_i about 0.7) or similar high-emissivity treatment on the inside.

For a numerical example, consider a cantilevered, 10-m-long, 2-cm-diam hinglocked (one seam) antenna made of 0.1-mm-thick beryllium copper ($e = 15E{-}6$ per K at $-50°C$ and $17E{-}6$ per K at $150°C$) with a silverized exterior (maximum α_o^s equals 0.12 and ε_o nominally equals 0.06). A chemical blackening treatment on the inside results in α_i^s/ε_i of 0.78/0.80. Using a value of 1400 W/m^2 for the solar constant, the maximum ($\phi = 0$) local temperature in the sun is

$$T(0) = \left\{ \{(0.12)(1400)/\pi(5.67E{-}8)(0.06)\}\{1 + (\pi/2)/[1 + (4/3) \right.$$

$$\times (0.78/0.06)] - 2 \sum_n (-1)^n/[1 + (0.78)(0.06)$$

$$\left. + (0.78)(0.06)/(16n^2 - 1)][4n^2 - 1]\} \right\}^{1/4} = 364.4 \text{ K}$$

Similarly, at the diametrically opposite point

$$T(\pi) = \left\{ \{(0.12)(1400)/\pi(5.67E{-}8)(0.06)\}\{1 - (\pi/2)/[1 + (4/3) \right.$$

$$\times (0.78/0.06)] - 2 \sum_n (-1)^n/[1 + (0.78)(0.06)$$

$$\left. + (0.78)(0.06)/(16n^2 - 1)][4n^2 - 1]\} \right\}^{1/4} = 349.6 \text{ K}$$

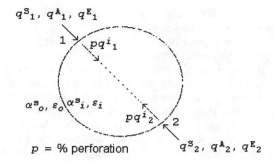

$$[\alpha^s_o(q^S_1 + q^A_1) + s_o q^E_1] + p[\alpha^s_i(q^S_2 + q^A_2) + s_i q^E_2]$$

$$= [\alpha^s_o(q^S_2 + q^A_2) + s_o q^E_2] + p[\alpha^s_i(q^S_1 + q^A_1) + s_i q^E_1]$$

$$p = \alpha^s_o/\alpha^s_i = s_o/s_i$$

Fig. 5.11 Perforation to eliminate thermal bending in TEEs.

The tip deflection is

$$\delta = (16\,E{-}8)(364.4 - 349.6)(10)^2/[4(1/100) = 0.592\,\mathrm{m}$$

which translates into an off-angle from the base of 3.4 deg.

The calculated distortion in this example is larger than the 1-deg pointing requirement usually imposed on extendable elements. Further improvement by lowering the exterior optical properties (especially solar absorptivity) or increasing the internal emissivity would only be marginal because the assumed values here are already near optima. If structural constraints permit, however, the deflection can be reduced or even eliminated by perforating the surface. The concept follows from the principle of balancing the thermal moment, as described in Fig. 5.11, by having diametrically opposite points of the element run at the same temperature.

In the numerical example, the deflection is theoretically eliminated if 15% ($p = \alpha^s_o/\alpha^s_i = 0.12/0.78$) of the surface is perforated. It must be noted, however, that uncertainties in the values of optical properties when they are low can have a marked effect on the result. For instance, it is possible to overestimate the value of solar absorptivity of the exterior surface and in actuality end up with a positive (toward the sun) deflection. It is left for an exercise to determine the sensitivity to properties and estimate the average temperature of a perforated element.

V. Transients

Much of the preceding relates to finding temperatures under steady-state conditions, defined either by orbit average values of dissipation and absorbed flux or as extended durations in a fixed orientation with respect to the heating source. It has been noted that averaging is often used in evaluating the thermal performance of platforms laden with massive electronics. The approach is suitable for predicting mission temperature limits and is very convenient in that solution routines do not involve stability or complicated convergence criteria.

Monitored thermistor data from orbiting satellites give credence to orbital averaging for component platforms where variations in the electronics dissipation is not too significant during the course of an orbit. Thermal designs of main canisters are generally directed toward reduced influences by environment fluxes, and in most cases of normal operation the mounting platform vascillates within ±2°C of the orbital average profile. These variations, and others that might occur momentarily during special events, can be predicted from greatly reduced models confined to the particular component and its immediate neighborhood, with the truncated surroundings usually replaced by sinks at constant orbital average temperature. This type of analysis gives compatible results with more extensive modeling when the temperature rate of change is not too radical and is relatively inexpensive and more efficient for evaluating instances of battery temperature change (Sec. V.B. of this chapter), electronics startups, failure modes, and the performance of instruments during periods of observation.

A similar approach is adopted in analyzing solitary electronics and mechanisms located outside the satellite's main body. These include antenna drive units, gimbals, remote sensors, and dampers. Thermal control of these items is achieved largely through the use of isolators and multilayered insulation, with only small areas exposed as radiators. As a result, variations in orbit temperature are generally not too steep and can be predicted from reduced models.

Critical transients occur in structures of large exposed area and low specific mass. Solar arrays and antenna dishes are specifically designed with these characteristics to obtain the most power or clearest signal with the least mass. Constructions include, among others, honeycomb and flexible roll-up substrates, composite materials, and mesh patterns. The dual requirement of large surface area and low mass limits the options in thermal control, and a potentially rapid temperature response to varying fluxes has to be contended with. The situation is most pronounced in cool-downs during occult and heat-ups as structures emerge from shadow into sunlight.

An illustration is given by the orbital temperature profile in Fig. 5.12 (see also Fig. 4.9). The curves are representative of the input submitted by thermal control for studying the thermal distortion of the TOPEX/Poseidon solar array and its effect on the satellite's pointing accuracy. Clearly, the temperature range and rates of change are quite significant here and an assessment would not have been complete without upgraded modeling and computational techniques. To this end, thermal modeling of the TOPEX array required a large number of coupled nodes representing different constructions in the panels. The profiles shown are only one set of some 100 obtained for 2.54-cm-square sections in the lighter region (2.8 kg/m^2) of the array located farthest from the satellite's main canister. As described in Fig. 5.12, each section consisted of three transverse nodes, with the two external nodes exchanging heat by radiation and conduction to the environment and other nodes in the overall network. In addition, to ensure consistency and convergence, comparisons were generally made using two computational routines from SINDA.

A. Analysis

As in steady state, simplified analytical methods are also used in transients to obtain general information and conduct parametric studies on the influence of the

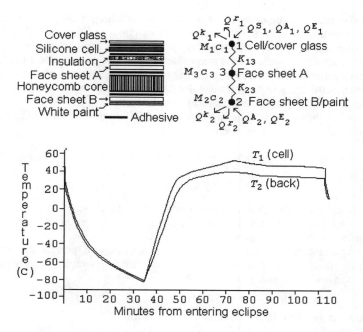

Fig. 5.12   Thermal model and one-orbit temperature profile of TOPEX/Poseidon solar array (cold case, beginning-of-life properties in a 1334-km, 63-deg inclination, $\beta = 0$ deg circular orbit; $S = 1287 \, \text{W/m}^2$, $f = 0.2$, $E = 189 \, \text{W/m}^2$).

thermal parameters. Typically, the approach considers a rather sizable region i as an isothermal entity whose interaction with regions j is given by the transient equations in Sec. V of this chapter, specifically,

$$(Mc)_i \frac{dT_i}{dt} = Q_i^d + (\alpha^s A^S q^S + \alpha^s A^A q^A + \varepsilon A^E q^E)_i$$

$$- \sum_j \Im_{ij} A_i^r (\sigma T_i^4 - \sigma T_{jr}^4) - \sum_j K_{ij}(T_i - T_{jk})$$

with initial condition

$$T_i(0^+) = T_0$$

It is important here to distinguish between the nearly repeating, periodic temperature profile of satellite structures in orbit and the transient behavior during a segment of the orbit when a particular event is taking place. The preceding initial condition can be made to define periodicity and give recurring results at each position in orbit, but the simpler and more interesting cases are those intervals where the temperature shows a marked response to sudden and significant changes in orbital heating.

In a first-order analysis, at least some of the conduction and radiation exchanges may be estimated, independently of a general solution of the preceding equations, by considering limiting or conservative values of extraneous heating. A fitting example on this is the interaction of a section of a flat solar array with adjacent connected sections. Because most array structures have negligible lateral conductance compared to transverse values (see example in Chapter 2, Sec. III), the isothermal section being analyzed may be assumed to have an adiabatic perimeter. Radiation exchange with the main satellite canister can also be simplified by considering the canister to be at a constant orbital average temperature, with an exchange factor equal to an average value calculated from known array orientation with respect to the sun at each position (time) in orbit.

Following these and similar simplifications, the single-node transient equation (with the subscript i now inferred) becomes

$$(Mc)\frac{dT}{dt} = Q - KT - \Im A'\sigma T^4$$

$$T(0^+) = T_0$$

where Q now includes absorbed environment heating, dissipation, and any extraneous input that is not a function of temperature. The factors K and $\Im$ account for the presence of coupled and surrounding systems.

An important special case that is frequently applied to large external structures is one in which the dominant mode of heat transfer is radiation. By ignoring conduction and introducing the parameters

$$\tau \equiv T/T_0, \qquad \theta \equiv \Im A'\sigma T_0^3/Mc, \qquad \eta^4 \equiv Q/\Im A'\sigma T_0^4$$

the transient problem takes the dimensionless form

$$\frac{d\tau}{d\theta} = \eta^4 - \tau^4$$

with

$$\tau(0^+) = 1.0$$

Generally, η is a function of time described by a specific value at each position in orbit. When it is constant, an exact closed-form solution (found by separation of variables) is given by

$$4\eta^3\theta = \ell n \left| \frac{(\eta+\tau)(\eta-1)}{(\eta-\tau)(\eta+1)} \right| + 2\left[\tan^{-1}\left(\frac{\tau}{\eta}\right) - \tan^{-1}\left(\frac{1}{\eta}\right) \right]$$

This equation can be shown to satisfy the final steady-state condition,

$$\tau_{ss} = \eta$$

or

$$T_{ss} = T_0\left[\frac{Q}{\Im A'\sigma T_0^4} \right]^{1/4} = \left[\frac{Q}{\Im A'\sigma} \right]^{1/4}$$

and give (by repeated application of L'Hospital's rule) in the limit as $\eta \to 0\,(Q = 0)$ the correct expression for perfect cool-down; that is,

$$\theta(\text{no heating input}) = (1/3)[(1/\tau^3) - 1]$$

or

$$T_{\text{no heating}} = T_0\left[1 + \frac{3\Im A'\sigma T_0^3 t}{Mc} \right]^{-1/3}$$

Because the differential equation and initial condition can be combined to give the initial rate of temperature change,

$$\left.\frac{d\tau}{d\theta}\right)_{\theta=0} = \eta^4(0^+) - 1$$

it follows that for constant heat input (which may be zero) there are available direct expressions for calculating the temperature at any time and the rate of change at the onset. As was noted earlier, this information is used in writing the specifications and requirements for qualifying solar arrays and antenna dishes in cryogenic thermal vacuum chambers.

Example: Solar Array Cool-Down in GEO

Because heating from planetary radiation is practically nonexistent at geosynchronous altitude, structures with a low ratio of mass to radiating area become very cold when shaded from the sun for extended periods. The predicted temperature at the end of a 72-min eclipse and the rate of change at the onset of cooling are often used as baseline for test verification of the structure's integrity.

Consider the following properties for an aluminum core honeycomb with graphite face sheets proposed as substrate for the Fairchild Leasecraft solar array: unit mass: 2.1 kg/m², across-section (transverse) conductance: 38.2 W/m²K, estimated bulk specific heat: 0.18 W-h/kg K, cell side (BOL): $\alpha^s_{cell}/\varepsilon_{cell} = 0.72/0.80$ (with cell efficiency taken into account), and back (BOL, white paint): $\alpha^s_{back}/\varepsilon_{back} = 0.26/0.83$.

In sun-pointing GEO, the average temperature just before going into occult is estimated from

$$\Im\sigma T^4 = (\varepsilon_{cell} + \varepsilon_{back})\sigma T^4 = \alpha^s_{cell}q^S$$

A cold case with solar constant equal to 1300 W/m² (contractor's requirement) gives

$$T = [(0.72)(1300)/(0.80 + 0.83)(5.67E{-}8)]^{1/4} = 317.2\,\text{K}(44.1°\text{C})$$

This value is used as the initial temperature in an eclipse cool-down analysis.

In the absence of all heating ($\eta = 0$) the initial (when $\tau = 1$) cool-down rate is

$$\left.\frac{d\tau}{d\theta}\right|_{\theta=0} = -1$$

or

$$\left.\frac{dT}{dt}\right|_{t=0} = -\Im A^r \sigma T_0^4 / Mc$$

Hence, based on a unit area, the steepest rate of temperature decrease is

$$\left.\frac{dT}{dt}\right|_{t=0} = -[(0.80 + 0.83)(1)(5.67E{-}8)(317.2)^4/(2.1)(0.18)](1/60)$$

$$= -41.3\,\text{K/min}$$

Also, with no heat input,

$$T = T_0\left[1 + 3\Im A^r \sigma T_0^3 t / Mc)\right]^{-\frac{1}{3}}$$

and the temperature at the end of 72 min is calculated as

$$T_{72\,min} = (317.2)[1 + 3(0.80 + 0.83)(1)(5.67E{-}8)(317.2)^3$$
$$\times (72/60)/(2.1)(0.18)]^{-\frac{1}{3}} = 103.1 \text{ K}$$

Example: Qualification of Solar Array Panels at Cold Temperatures

Whether an array can drop in temperature down to $103.1 \text{ K} (-170°C)$ when actually in orbit may be debatable. For, in the extremely cold environment of geosynchronous eclipse, even the slightest heating from a satellite's main body and other surrounding structure might be sufficient to restrain the rate of decrease and limit the temperature to a noticeably higher value.

In many satellites, the geometry that achieves simultaneously least impinging flux on the thermal radiators and maximum array exposure to sunlight is one in which electronics radiators have some view to the arrays. Consider as an example the symmetric 1000-W spacecraft shown in Fig. 5.13, where each set of main body radiators is found from a separate calculation to have a 10% (5% to the cell surface and 5% to the back) average view factor to a facing 5-m^2, Leasecraft type array. Because the fraction of IR heating intercepted by each wing is the view factor from

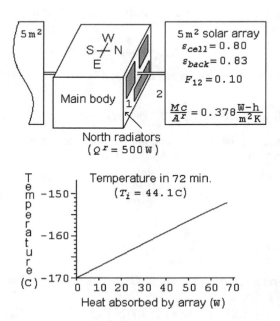

Fig. 5.13 Effect of satellite main body heating on a solar array temperature at the end of geosynchronous eclipse.

the radiator to the array, the energy incident on each array is approximately

$$Q^i(\text{IR}) = (0.10)(1000/2) = 50\,\text{W}$$

which, in an initial analysis, may be assumed equally and uniformly divided between the cell side and the back. The average IR energy absorbed by the array can then be calculated as

$$Q^a = \varepsilon_{\text{cell}}Q^i_{\text{cell}} + \varepsilon_{\text{back}}Q^i_{\text{back}} = (0.80 + 0.83)(50/2)) = 40.8\,\text{W}$$

Under normal sunlit conditions and a moderately large array size, this amount is a negligible perturbation in the overall energy balance. But in cool-downs it can translate into a significant temperature margin. Thus, with

$$\eta = \left[Q^a/\Im A^r \sigma T_0^4\right]^{\frac{1}{4}} = \{40.8/[(0.80 + 0.83)(5)(5.67\,E-8)(317.1)^4]\}^{\frac{1}{4}} = 0.3057$$

and, in 72 min,

$$\theta = \Im A^r \sigma T_0^3 t/Mc$$

$$= (0.80 + 0.83)(5)(5.67E-8)(317.1)^3(72/60)/(5)(2.1)(0.18) = 9.3640$$

the corresponding τ is found from the transient equation (with heating) to be 0.3592, giving a temperature of

$$T_{72\,\text{min}} = (0.3592)(317.1) = 113.9\,\text{K}(-159.1°\text{C})$$

vs $-170°\text{C}$ obtained when spacecraft heating is not considered. The difference ($11°\text{C}$) can lead to significant savings in test duration and cost.

A standard arrangement for a thermal qualification test is shown in Fig. 5.14a. The panels are suspended in a cryogenic vacuum chamber equipped with banks of IR heating lamps, whose size and spacing are made to produce the most uniformity in heating without causing excessive blockage to radiation exchange with the cryogenic walls. A thermal cycling test is now conducted by activating the lamps to heat up the panels to an uppermost temperature, then turning off the power to permit them to cool down to the required lowest temperature. The defining parameters of the test are the highest and lowest temperatures, dwell time at each temperature plateau, transition rate from one temperature level to the other, and the number of cycles.

Whereas rapid heating of the panels can be accomplished with relative ease by supplying large power from a ground power source, the walls of cryo chambers (shrouds with liquid or gaseous nitrogen), with $\varepsilon < 1$ and $T > 70$ K, are not a true simulation of space background and, consequently, cooling by radiation is slower than that which could take place in space. The situation is described in Fig. 5.14c. The curves are obtained from

$$Mc\frac{dT_p}{dt} = (\varepsilon_{\text{cell}} + \varepsilon_{\text{back}})_p \mathbf{F}\left(\sigma T_p^4 - \sigma T_w^4\right)$$

where p refers to panel and w to cryo wall. The canonical parameters are

$$\eta = \left[\sigma T_w^4/\sigma T_{p(0)}^4\right]^{1/4} = T_w/T_{p(0)}$$

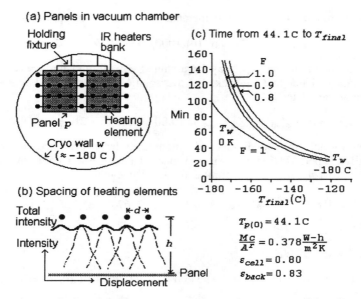

Fig. 5.14 Thermal qualification of solar array panels; heating elements are small to reduce blockage to cooling and h and d are adjusted for uniform intensity.

and

$$\theta = (\varepsilon_{\text{cell}} + \varepsilon_{\text{back}})\mathbf{F}\sigma T_{p(0)}^3 t / Mc$$

$T_{p(0)}$ is panel temperature at the start of cool-down and the factor $\mathbf{F}$ is included to account for inevitable blockage to radiation due to the presence of test appendages. Other corrections may be necessary to account for the relative size and emissivity of the shroud with respect to the specimen.

The cool-down time increases asymptotically as the target temperature approaches the limit set by the chamber wall. Despite a relatively low panel mass, it takes some 2 h to cool from 44.1 to $-170°C$ when $\mathbf{F} = 0.9$. For a target of $-159.1°C$ and $\mathbf{F} = 0.9$, the time is reduced to slightly over 70 min. The difference is invaluable when hundreds of cycles are to be exercised for qualification. (In its lifetime in LEO, an array can undergo some 30,000 high to low temperature cycles.)

B. Linearized Transient Solutions: Radiation Time Constant

When heating is not constant, the single-node problem may be considered in view of the linearization technique used for steady-state analysis. With η^4 a function of time and τ^4 replaced by $\tau_m^4 + 4\tau_m^3(\tau - \tau_m)$, the equations are written

$$\frac{d\tau}{d\theta} = \eta^4 - 4\tau_m^3\tau + 3\tau_m^4$$

$$\tau(0^+) = 1.0$$

and the solution (keeping τ_m constant) is

$$\tau(\theta) = 0.75\tau_m + (1 - 0.75\tau_m)e^{-4\tau_m^3\theta} + e^{-4\tau_m^3\theta} \int_0^\theta e^{4\tau_m^3\lambda}\eta^4(\lambda)\,d\lambda$$

Using the same arguments as before, best accuracy at a particular time θ_p is obtained from

$$\left.\frac{\partial \tau}{\partial \tau_m}\right)_{\theta_p,\eta} = 0$$

which yields

$$\theta_p \int_0^{\theta_p} e^{4\tau_m^3\lambda}\eta^4(\lambda)\,d\lambda - \int_0^{\theta_p} e^{4\tau_m^3\lambda}\lambda\eta^4(\lambda)\,d\lambda = \frac{1}{16\tau_m^3}\left[e^{4\tau_m^3\theta_p} - 1\right] - \theta_p(1 - 0.75\tau_m)$$

Pairs of τ_m and corresponding θ_p are found by numerical iteration. When η^4 equals a constant, the equations give

$$\tau(\theta) = 0.75\tau_m + (1 - 0.75\tau_m)e^{-4\tau_m^3\theta} + \frac{\eta^4}{4\tau_m^3}\left(1 - e^{-4\tau_m^3\theta}\right)$$

and

$$\frac{e^{4\tau_m^3\theta_p} - 1}{4\tau_m^3\theta_p} = \frac{4\tau_m^3 - 3\tau_m^4 - \eta^4}{\tau_m^4 - \eta^4}$$

Again, it must be pointed out that valid assessments are often possible by defining the linearization parameter τ_m based on physical considerations, rather than by taking partial derivatives that invariably entail complicated mathematical manipulations. A case in point is made by dealing with satellite component platforms. Because of generally good thermal coupling to massive components, the effective thermal mass of a platform is relatively large and the temperature in most cases remains within just a few degrees from an average orbital value, even as absorbed heating and components operation are varying. The result is that the ratio of the instantaneous temperature (Kelvin) to any other value during the orbit remains not far from unity, and selecting τ_m anywhere within the expected range should yield accuracies near the optimum value. In particular, for a suddenly applied constant heat input, by making τ_m coincide with the limiting steady-state value of τ, the accuracy improves at longer times while remaining restrained near the initial condition, which is always satisfied independently of τ_m.

Figure 5.15 includes some comparisons with the exact solution. Substituting $\tau_{ss}(=\eta)$ for τ_m, the results are seen to coincide near the initial and final conditions and are always well within acceptable accuracy when the heating input, as characterized by η, is not too great.

By substituting $\tau_{ss} = T_{ss}/T_0$ for both τ_m and η in the linear transient equation for τ (with radiation factor $\Im$) and using the definition of θ in terms of time t, one obtains

$$\frac{T - T_{ss}}{T_0 - T_{ss}} = \exp\left(-\frac{4\Im A'\sigma T_{ss}^3 t}{Mc}\right)$$

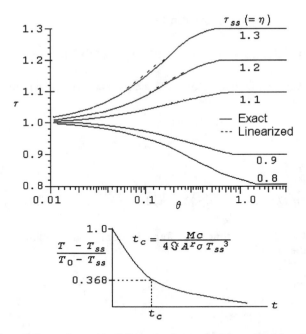

Fig. 5.15   Comparison of exact and linearized response to sudden heating; the time constant t_c is the time at which the temperature reaches 63.2% of the span between initial and steady-state values.

with T_{ss} found from

$$\sigma T_{ss}^4 = Q/\Im A^r$$

where Q is combined heating from dissipation, surrounding structure, and the environment. As in linear convection heat transfer,[4] the quantity $Mc/4\Im A^r \sigma T_{ss}^3$ may be viewed as a radiation time constant, signifying the sensitivity of temperature to new heating. At time

$$t_c = Mc/4\Im A^r \sigma T_{ss}^3$$

from the start of a new heat input, the temperature will have reached 63.2% $(1-1/e)$ of the difference between initial and final (mathematically unattainable) steady-state values.

Example: Thermal Control of Battery Radiators

Because of repeated charging and discharging, batteries have unusually, wide variations in orbital dissipation and show significant transients despite their relatively large mass.

The thermal conditions of satellite batteries are frequently moderated by automatic switch offs and switch ons and adjustments of recharge rates at preset temperatures. In addition, charge limits and voltage/temperature (V/T) charge control by ampere-hour integration may be regulated by onboard computers.[10] But management of the net heat output is normally accomplished by mounting to a thermally controlled platform.

Battery specifications require a cool mounting interface, but usually above 0°C. Various ranges have been specified for nickel–cadmium and nickel–hydrogen units, all with the intention of increasing life within reasonable engineering and cost. The range 0 to 20°C is considered standard, although −2 to 23°C is usually acceptable. Critical applications require 5 to 15°C and sometimes 5 to 10°C. The platform temperature difference is normally specified at ≤5°C.

High temperatures are avoided by sizing the radiator to accommodate hot case conditions while the battery is in extended discharge. Heaters, activated by thermostats, would then maintain the temperature above the lower limit during a no-charge or trickle-charge operation under cold case conditions. The radiator may even be oversized to lower the upper temperature and hasten cool-downs after discharge and possible overcharge, but for most cases of moderate to high power, oversized radiators face limitations both in available real estate and heater power.

A practical thermal design is usually based on transient considerations. However, despite seemingly adequate margins in analysis and design, orbiting battery platforms often deviate from predicted thermal performance, especially at the higher temperature range. This has been attributed to difficulties in characterizing actual heat dissipation as it varies with temperature, perturbations in consumed power, eclipse times, and charge efficiency, which often varies from one cell to another in the same batch. The controlled charge rate can also vary. In some NASA satellites, for example, higher charge rates are exercised during the short period immediately following eclipse when the solar array is cold and very efficient.[10] Making provisions for all possible influences within the given constraints on radiator size and heater power would require a nontraditional and certainly costly thermal design. Fortunately, commonly employed techniques have in most cases resulted in adequate performance for the intended duration even as temperatures occasionally exceed 20°C.

Predictability of the thermal behavior will improve by placing battery radiators where heating from the environment and surrounding structure is small, or at least known within small tolerances. An example of this is the relative location of the power system (which houses the batteries) described in Fig. 5.16. From the indicated geometry, in a low, Earth-pointing orbit with the solar vector in the roll–yaw plane, exposure of the batteries radiator to solar energy is generally momentary and not too great. For the TOPEX/Poseidon satellite, an additional improvement was to eliminate heat exchanges with the solar array by having a single wing placed on the opposite side of the batteries platform.

The power system depicted in Fig. 5.16 is patterned after the modular power subsystem (MPS) of NASA's MMS.[11] One feature of this design is that three batteries (NASA Standard 50 A-h Ni-Cd) are used to deliver the power normally available through only two. Consequently, the discharge rates per battery are lowered and life expectancy increased. Furthermore, heat pipes are embedded in the honeycomb mounting panel to limit spatial temperature variations to within 5°C (2.5°C lateral and about 3°C transverse are actually observed) and, hence, improve individual unit life by having nearly equal rates of charge and discharge among all three batteries. Figure 5.16 shows the relative location of the heat pipes and

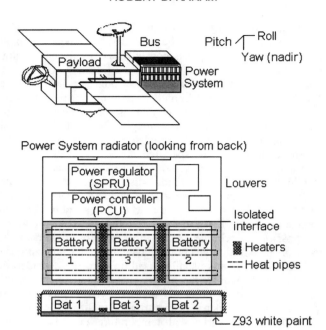

Fig. 5.16 Satellite power subsystem: three NASA 50 AH (22 cells) Ni Cd batteries, each 30 × 45 cm² base plate; the batteries section is isolated from the SPRU/PCU region.

heaters. From a thermal standpoint, temperature uniformity and the higher thermal constant of a basically two-battery system help to suppress variations in the orbital temperature.

Treated as a transient system, heating of a battery platform should be defined by the instantaneous incident flux and dissipation. Often, however, a periodic square wave (Fig. 5.17) representing constant heating during the sunlit and eclipse portions of the orbit is found to be a valid approximation. For a massive and properly oriented system, the deviations from a square function are mostly near when the spacecraft emerges from shadow into sunlight because, as was noted earlier, at this point the solar array is cold and the batteries are recharged at a relatively high rate to take advantage of the surge in available power. But this action lasts only as long as the array remains cool, and because of uncertainties in the transient variables during that period, thermal analysts retain the square wave assumption but use a worst-case constant dissipation while in discharge.

Hot case and cold case conditions are defined in terms of the orbit beta angle (Chapter 4, Sec. V.C). In the vicinity of Earth, a hot case occurs at 0-deg beta angle, when longest battery discharge occurs each time the satellite goes into maximum eclipse. The scenario is described in Fig. 5.17. The satellite enters eclipse (relative time a) with the radiator at temperature T_0, calculated based on maximum bias on heating during trickle charge in the sunlit portion of the orbit. As the batteries discharge during eclipse, the radiator temperature rises monotonically under maximum constant discharge dissipation and hot-biased Earth flux until the

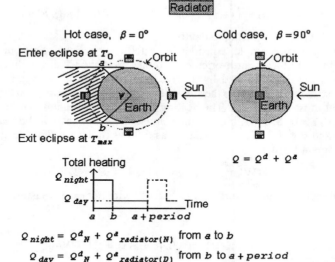

Fig. 5.17   Hot case and cold case senarios for batteries radiator.

maximum orbital value T_{max} is reached at end of eclipse (relative time b). At this point, the batteries are in trickle-charge mode again (maximum estimate in the hot case), and the temperature decreases until it reaches T_0 and the cycle starts again.

For cold case evaluation, it is noted that as beta angle passes the orbit plane beyond which the satellite remains fully sunlit, the batteries maintain a near full charge with low-average heat output in trickling. According to the nomenclature, the heating profile at beta near or equal to 90 deg is a degenerate case in which Q_{night} and Q_{day} coincide to give uniform heating throughout the orbit. The temperature for this scenario is calculated based on least values of trickle-charge dissipation and minimum incident fluxes.

The thermal problem is now defined as requiring that the radiator area be sufficiently large so that at the end of longest eclipse, following maximum heating due to discharge/overcharge in a hot case environment, the temperature will not exceed the specified upper limit. To complete the solution, the cold case must entertain sufficient heater power for the temperature to remain at or above the least allowable limit. In the following pages, parametric relations are obtained between specified acceptable temperature limits and the required radiating area and heater power.

The connection in the beta = 0 deg orbit between the temperature on entering eclipse T_0 and that on emerging into sunlight T_{max} after t_N hours of discharge is in the equation

$$T_{max} = T_{ss(N)} - \left(T_{ss(N)} - T_0\right) \exp\left(-4\varepsilon A\sigma T_{ss(N)}^3 t_N / Mc\right)$$

where the subscript N refers to nighttime or eclipse. In sunlight, in the same orbit, the roles of initial and final temperatures switch and the connection reads

$$T_0 = T_{ss(D)} - \left(T_{ss(D)} - T_{max}\right) \exp\left(-4\varepsilon A\sigma T_{ss(D)}^3 t_D/Mc\right)$$

with the subscript D referring to daytime.

For a numerical example, consider the satellite batteries radiator depicted in Fig. 5.16 in a 1112-km, Earth-oriented, circular orbit at 63.1-deg inclination. The solar arrays are being rotated by a drive, and a yawing maneuver is added to maintain normal orientation to the sun with negligible solar heating on the radiator throughout most of the orbit. The heating fluxes on the radiator can be obtained from TRASYS or SSPTA, but in this exercise the estimates will be made using Table 4.2 and Fig. 4.7. Hot case data are as follows (hot case, $\beta = 0$ deg).

Environment fluxes:

$$S = 1400 \, \text{W/m}^2, \qquad f = 0.46, \qquad E = 274 \, \text{W/m}^2$$

Normal incident fluxes (surface 1 in Fig. 4.7 with hot case adjustment):

$$q^S \approx 0$$

$$q^A = 22(1400/1350)(0.46/0.33) = 31.8 \, \text{W/m}^2$$

$$q^E = 42(274/236) = 48.8 \, \text{W/m}^2$$

Radiator coating properties (Z93 white paint):

$$\alpha^s/\varepsilon \, (\text{end of life}) = 0.28/0.84$$

Orbit period:

$$p = 2\pi[(R_E + a)^3/\mu]^{1/2} = 2\pi[(6340 + 1112)^3/1.437E9]^{1/2} = 106.6 \, \text{min}$$

Shadow angle (Fig. 4.9):

$$\psi = \pi - 2\cos^{-1}[R_E/(R_E+a)] = \pi - 2\cos^{-1}[6340/(6340+1112)] = 2.035 \, \text{rad}$$

with time in shadow

$$106.6(2.035/2\pi) = 34.5 \, \text{min} \, (0.58 \, \text{h})$$

and time in sunlight

$$106.6 - 34.5 = 72.1 \, \text{min} \, (1.20 \, \text{h})$$

Batteries dissipation (best estimate for the given mission load) in daylight:

$$Q_D^d = 15 \text{ to } 20 \, \text{W}$$

and in eclipse:

$$Q_N^d = 135 \text{ to } 140 \, \text{W}$$

Cold case ($\beta = 90$ deg) data are as follows.
Environment fluxes:

$$S = 1300 \, \text{W/m}^2, \qquad f = 0.2, \qquad E = 198 \, \text{W/m}^2$$

Normal incident flux (surface 1 in Fig. 4.7 with cold case adjustment):

$$q^S \approx 0$$

$$q^A = 22(1300/1350)(0.20/0.33) = 12.9 \text{ W/m}^2$$

$$q^E = 42(198/236) = 35.2 \text{ W/m}^2$$

Radiator properties:

$$\alpha^s/\varepsilon \text{ (beginning of life)} = 0.19/0.88$$

Orbit period:

106.6 min in sunlight

Batteries dissipation, trickle charge:

$$Q_D^d = 15 \text{ W}$$

The effective thermal mass (Mc) of the radiator, including batteries and mounting platform, has been estimated to be 29.7 W-h/K.

Using hot case data,

$$q_N^a = \varepsilon q^E = 0.84(48.8) = 41.0 \text{ W/m}^2$$

$$q_D^a = \alpha^s(q^S + q^A) + \varepsilon q^E = 0.28(0 + 31.8) + (0.84)(48.8)$$

$$= 49.9 \text{ W/m}^2$$

$$T_{ss(N)} = \left[(Q_N^d + q_N^a A)/\varepsilon\sigma A\right]^{1/4} = [(3 \times 140 + 41A)/0.84\sigma A]^{1/4}$$

$$= 171.29[(10.2439 + A)/A]^{1/4} \text{ K}$$

$$T_{ss(D)} = \left[(Q_D^d + q_D^a A)/\varepsilon\sigma A\right]^{1/4} = [(3 \times 20 + 49.9A)/0.84\sigma A]^{1/4}$$

$$= 179.91[(1.2024 + A)/A]^{1/4} \text{ K}$$

$$4\varepsilon A\sigma T_{ss(N)}^3 t_N/Mc = 4(0.84)A\sigma(171.29)^3[(10.2439 + A)/A]\}^{3/4}(0.58)/29.7$$

$$= 0.0187A(10.2439 + A)/A]^{\frac{3}{4}}$$

$$4\varepsilon A\sigma T_{ss(D)}^3 t_D/Mc = 4(0.84)A\sigma(179.91)^3[(1.2024 + A)/A]\}^{3/4}(1.20)/29.7$$

$$= 0.0341A(1.2024 + A)/A]^{\frac{3}{4}}$$

Defining

$$f_N(A) = [(10.2439 + A)/A]^{\frac{1}{4}}$$

$$f_D(A) = [(1.2024 + A)/A]^{\frac{1}{4}}$$

The transient equations become

$$T_{\max} = 171.29 f_N(A) - [171.29 f_N(A) - T_0] \exp\{-0.0187A[f_N(A)]^3\}$$

$$T_0 = 179.91 f_D(A) - [179.91 f_D(A) - T_{\max}] \exp\{-0.0341A[f_D(A)]^3\}$$

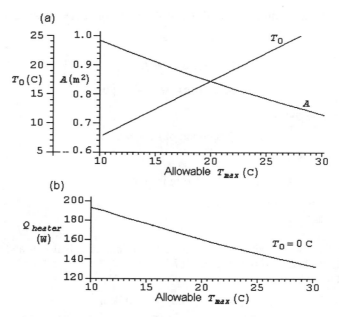

Fig. 5.18 Battery radiator size and heater power: a) radiator area as function of allowable maximum temperature in a hot case, $\beta = 0$ deg orbit and b) heater power required for $0°C$ operation in a cold case, $\beta = 90$ deg orbit.

For a specified maximum orbital temperature $T_{\max}$, the two equations can be solved numerically to determine the required radiator area A and the value of temperature T_0 just before entering eclipse. The results are shown graphically in Fig. 5.18a.

As expected, a larger area is needed to produce cooler temperatures. But the price for designing with large areas is an increase in heater power to sustain the lower temperature limit during cold case operation. From

$$Q_{\text{heater}} + Q_D^d = \left\{ \varepsilon_c \sigma T_{\min}^4 - \left[\alpha_c^s \left(q_c^S + q_c^A \right) + \varepsilon_c q_c^E \right] \right\} A$$

the heater power in a cold case beta = 90-deg orbit that will keep $T_{\min} \geq 0°C$ is at least

$$Q_{\text{htr}} = \{(0.88)(5.67E{-}8)(273)^4 - [(0.19)(12.9) + (0.88)(35.2)]\} A - (3)(15)$$

$$= 243.72A - 45$$

The results, as they depend on allowable maximum temperature, are plotted in Figure 5.18b. For example, if the platform specifications are 0 to 20°C, then, from Fig. 5.18a, the radiator must be made about 0.85 m² in area and, from Fig. 5.18b, 162.2 W must be available for heater power. A margin may have to be incorporated in the upper specification to compensate for spatial temperature variations.

The estimated heater power may be relatively high for a spacecraft with average power capability. But it must be noted that large consumptions in heater power take place only during fully sunlit orbits, or when the solar arrays are illuminated throughout the orbit and electrical power is abundant.

Still, if required, heater power can be eliminated or radically reduced by using thermal louvers (Chapter 6, Sec. V). But the added blockage to viewing space from louvers components not only slows down the cooling process following a discharge/overcharge mode, but a considerably larger area will be needed to retain the hot case temperature within the requirements. By way of comparison, the reader can perform similar calculations to those given here to determine the louvered area needed to remain below 20°C. It may be assumed that the louvers are calibrated (see Chapter 6, Sec. V.C) to be closed at 0°C (high radiation resistance with effective emissivity $\varepsilon_{\text{eff}} = 0.1$) and fully open at $\geq 20°C$ (low radiation resistance with $\varepsilon_{\text{eff}} = 0.7$). The heater power necessary to keep the temperature at least 0°C in the cold case is

$$Q_{\text{heater}} + Q_D^d = \left\{ \varepsilon_{\text{eff(closed)}} \sigma T_{0C}^4 - \left[\alpha_{\text{eff(closed)}}^s \left(q_c^S + q_c^A \right) + \varepsilon_{\text{eff(closed)}} q_c^E \right] \right\} A$$

with $\alpha_{\text{eff(closed)}}^s \approx 0$ and $\varepsilon_{\text{eff(closed)}} = 0.1$. Hence,

$$Q_{\text{heater}} = (0.1)(5.67E{-}8)(273)^4 A - (3)(15) = 31.5A - 45$$

This quantity is generally negative in this example, which implies that when louvers are used, the low battery power dissipation assumed during trickle charge is sufficient to maintain the temperature above the allowable minimum of 0°C.

Example: Heat-Up of Honeycomb Panels and Thermal Flutter

The transient temperature obtained from lump, single isothermal node analysis may be used as a time-vaying mean temperature in treating linearized multi-node transient problems. Application to the two-node honeycomb system reveals interesting behavior during initial heat-up.

The two-node configuration is shown in Fig. 5.19a. To simplify the analysis, the line separating the two nodes [center line (CL) in the sketch] is situated such that it divides the panel into two sections of equal thermal masses. Also, following the linearization procedure, T^4 is replaced by the first two terms of a Taylor expansion about a parameter T_m that is considered constant in the mathematical manipulations.

Writing the nonsteady energy balance for each section,

$$(Mc/A)_1 \frac{dT_1}{dt} = q_1^a - \varepsilon_1 \sigma T_1^4 - K_t(T_1 - T_2)$$

$$(Mc/A)_2 \frac{dT_2}{dt} = q_2^a - \varepsilon_2 \sigma T_2^4 + K_t(T_1 - T_2)$$

$$T_1(0) = T_{01}, \qquad T_2(0) = T_{02}$$

On linearization,

$$(Mc/A)_1 \frac{dT_1}{dt} = q_1^a + 3\varepsilon_1 \sigma T_m^4 - \left(4\varepsilon_1 \sigma T_m^3 + K_t \right) T_1 + K_t T_2$$

$$(Mc/A)_2 \frac{dT_2}{dt} = q_2^a + 3\varepsilon_2 \sigma T_m^4 - \left(4\varepsilon_2 \sigma T_m^3 + K_t \right) T_2 + K_t T_1$$

$$T_1(0) = T_{01}, \qquad T_2(0) = T_{02}$$

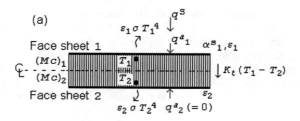

(a)

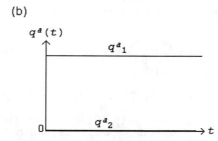

(b)

Fig. 5.19   Two-node honeycomb panel under sudden heating; center line is selected such that $(Mc)_1 = (Mc)_2$; for maximum temperature difference, $q_2^a = 0$.

Locating CL such that $(Mc/A)_1 = (Mc/A)_2 = (Mc/A)$, the time derivative of either linear equation results in

$$\frac{d^2 T_{1,2}}{dt^2} + P\frac{dT_{1,2}}{dt} + Q T_{1,2} = R_{1,2}$$

with

$$T_1(0) = T_{01} \quad \text{and} \quad T_2(0) = T_{02}$$

The subscript 1, 2 stands for node 1 or 2, and

$$P = [A/Mc]\big[2K_t + 4(\varepsilon_1 + \varepsilon_2)\sigma T_m^3\big]$$

$$Q = [A/Mc]^2\big[4(\varepsilon_1 + \varepsilon_2)K_t\sigma T_m^3 + 16\varepsilon_1\varepsilon_2\big(\sigma T_m^3\big)^2\big]$$

$$R_{1,2} = \{A/Mc\}^2\big\{ K_t\big[(q_1^a + q_2^a) + 3(\varepsilon_1 + \varepsilon_2)\sigma T_m^4\big]$$
$$+ 4\big(q_{1,2}^a + 3\varepsilon_{1,2}\sigma T_m^4\big)\varepsilon_{1,2}\sigma T_m^3\big\}$$

On subtracting, the equation for the temperature difference $\Delta = T_1 - T_2$ is obtained as

$$\frac{d^2\Delta}{dt^2} + P\frac{d\Delta}{dt} + Q\Delta = \frac{4\big(q_1^a\varepsilon_2 - q_2^a\varepsilon_1\big)\sigma T_m^3}{(Mc/A)^2}$$

with

$$\Delta(0) \equiv \Delta_0$$

The general solution for the case when the coefficients are constant is

$$\Delta = C_1 e^{mt} + C_2 e^{nt} + D$$

with

$$D = \frac{q_1^a \varepsilon_2 - q_2^a \varepsilon_1}{(\varepsilon_1 + \varepsilon_2) K_t + 4 \varepsilon_1 \varepsilon_2 \sigma T_m^3}$$

and m, n the roots of

$$p^2 + Pp + Q = 0$$

That is,

$$m, n = 0.5[-P \pm (P^2 - 4Q)^{0.5}]$$

C_1 and C_2 are determined by applying the initial conditions and are found as

$$C_1 = (na - b)/(n - m)$$
$$C_2 = (b - ma)/(n - m)$$

where

$$a = \Delta_0 - D$$

$$b = \frac{(q_1^a - q_2^a) + 3\sigma T_m^4(\varepsilon_1 - \varepsilon_2) - 4\sigma T_m^3(\varepsilon_1 T_{01} - \varepsilon_2 T_{02}) - 2K_t \Delta_0}{Mc/A}$$

The results may be stored into a QBasic or Fortran program or an advanced hand calculator for rapid determination of temperature difference as function of time. The values of T_m used in the equations are from the heat-up analysis of a single, combined isothermal system.

For a numerical application, assume the slab in Fig. 5.19 in nearly uniform temperature ($\Delta_0 = 0$) at $-170°C$ when emerging from an extended geosynchronous eclipse. An incident solar flux $q^S = 1400$ W/m^2 suddenly impinges and remains normal to the cell side while the opposite side is receiving no heating. This constitutes a worst-case temperature difference profile. Now consider the following data:

$$(Mc/A)_1 = (Mc/A)_2 = (Mc/A) = (2.1)(0.18)/2 = 0.189 \text{ W-h/m}^2 \text{ K}$$

$$K_t = 38.2 \text{ W/m}^2 \text{ K}$$

$$\alpha_1^s = 0.72 \text{(solar cells)}$$

$$\varepsilon_1 = 0.80 \text{(solar cells)}$$

$$q_1^a = (0.72)(1400) = 1008 \text{ W/m}^2$$

$$q_2^a = 0$$

$$\varepsilon_2 = 0.83 \text{ (white paint)}$$

A time-varying T_m is obtained from pairs of τ and θ in the exact transient equation for the combined sections; thus,

$$\theta = (\varepsilon_1 + \varepsilon_2)\sigma T_0^3 t / (2Mc/A)$$

$$= (0.80 + 0.83)(5.67E-8)(-170 + 273)^3 / [(2.1)(0.18)(t/3600)$$

$$= (7.4214E-5)t$$

where t is in seconds, and

$$\eta = \left[q_1^a / (\varepsilon_1 + \varepsilon_2)\sigma T_0^4 \right]^{0.25}$$

$$= [(0.72)(1400)/(0.80 + 0.83)(5.67E-8)(-170 + 273)^4]^{0.25} = 3.1375$$

For τ equals, say, 1.5,

$$T_m = 1.5(-170 + 273) = 154.5\,\mathrm{K}(-118.5°\mathrm{C})$$

From the transient temperature equation

$$4\eta^3\theta = \ln|[(\eta + \tau)(\eta - 1)]/[(\eta - \tau)(\eta + 1)]| + 2[\tan^{-1}(\tau/\eta) - \tan^{-1}(1/\eta)]$$

dimensionless time is found as

$$\theta = [1/[4(3.1375)^3]]\{\ln|[(3.1375 + 1.5)(3.1375 - 1)]/(3.1375 - 1.5)$$

$$\times (3.1375 + 1)| + 2[\tan^{-1}(1.5/3.1375) - \tan^{-1}(1/3.1375)]\} = 5.305E-3$$

and, hence,

$$t = 5.305E-3/7.4214E-5 = 71.48\,\mathrm{s}$$

The parameters for the linear two-node system when $T_m = 154.5\,\mathrm{K}(\sigma T_m^3 = 0.2091)$ are

$$P = \{[2(38.2) + 4(0.80 + 0.83)(0.2091)]/[(1.05)(0.18)]\}/3600 = 0.1143/\mathrm{s}$$

$$Q = \{[4(0.8 + 0.83)(38.2)(0.2091)$$

$$+ 16(0.8)(0.83)(0.2091)^2]/[(1.05)(0.18)]^2\}/(3600)^2 = 1.1350E-4/\mathrm{s}^2$$

$$m = \{-0.1143 + [(0.1143)^2 - (4)(1.135E-4)]^{1/2}\}/2 = -1.0018E-3/\mathrm{s}$$

$$n = \{-0.1143 - [(0.1143)^2 - (4)(1.135E-4)]^{1/2}\}/2 = -0.1133/\mathrm{s}$$

$$D = [(1008)(0.83) - (0)(0.80)]/[(0.8 + 0.83)(38.2) + 4(0.8)(0.83)(0.2091)]$$

$$= 13.3178\,\mathrm{K}$$

$$a = 0 - 13.3178$$

$$b = \{[1008 + 3(0.2091)(154.5)(0.8 - 0.83)$$

$$- 4(0.2091)(103)(0.8 - 0.83)]/[(1.05)(0.18)]\}/3600 = 1.481\,\mathrm{K/s}$$

Hence,

$$C_1 = [(-0.1133)(-13.3178) - (1.481)]/[(-0.1133) - (-1.0018E{-}3)]$$
$$= -0.2485 \text{ K}$$

$$C_2 = [(1.481) - (-1.0018E{-}3)(-13.3178)]/[(-0.1133) - (-1.0018E{-}3)]$$
$$= -13.0693 \text{ K}$$

and

$$\Delta_{71.48\,\text{s}} = -0.2485e^{(-1.0018E{-}3)(71.48)} - 13.0693e^{(-0.1133)(71.48)} + 13.3178$$
$$= 13.0825 \text{ K}$$

Table 5.2 shows the results of similar calculations at other times using a QBasic program.

Evidently, the usual charateristics of a monotonic decay as implied by the negative exponential in Δ does not materialize. When T_m is continually adjusted to reflect the instantaneous average temperature, the nonlinear nature of the problem is recaptured as can be seen from the small, but definite, overshoot in the period within 57.0 and 71.5 s. The response is clarified by the enlargement in Fig. 5.20.

The shudder known as thermal flutter that is observed in orbiting satellites as large solar panels, antennas, or long booms emerge into sunlight may well be related to the overshoot in the temperature difference across the structure. A periodic spike in the thermal moment will appear as illustrated in Fig. 5.20. Thus, if the dampening coefficients do not completely dissipate the motion in the course of an orbital period, the repeated jolts could cause a buildup in the disturbance. It is suspected that this is more likely to show itself in LEOs where the spikes repeat in intervals of less than 2 h. (For another view on this problem, see Ref. 12. For induced disturbances in the Hubble Space Telescope, see Ref. 13).

Table 5.2   Results from QBasic program

$\tau(= T_m/T)$	T_m, K	t, s	Δ, K
1.00	103.00	0	0
1.05	108.15	7.0	7.2055
1.10	113.30	14.1	10.4673
1.20	123.60	28.3	12.5975
1.40	144.20	57.0	13.0855
1.45	149.35	64.3	13.0880
1.50	154.50	71.5	13.0825
1.55	159.65	79.0	13.0753
1.80	185.40	116.9	13.0167
2.50	257.50	257.9	12.7803
3.13	322.39	323.1	12.4230

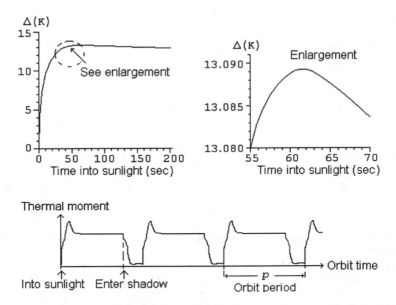

Fig. 5.20 Temperature differential and induced thermal moment in a solar array following repeated sudden heating.

References

[1]Morton, K. W., and Mayers, D. F., *Numerical Solution of Partial Differential Equations*, Cambridge Univ. Press, Cambridge, England, UK, 1994, pp. 6–59.

[2]Press, W. H., Teukolsky, S. A., Vetterling, W. T., and Flannery, B. P., *Numerical Recipes in Fortran*, 2nd ed., Cambridge Univ. Press, Cambridge, England, UK, 1992, pp. 818–880.

[3]Zwillinger, D., *Handbook of Differential Equations*, Academic, San Diego, CA, 1989, pp. 630–635.

[4]Incropera, F. P., and DeWitt, D. P., *Fundamentals of Heat Transfer*, Wiley, New York, 1981, pp. 146–165, 214–229.

[5]Gilmore, D. G., and Collins, R. L., "Thermal Design Analysis," *Satellite Thermal Control Handbook*, edited by D. G. Gilmore, The Aerospace Corp. Press, El Segundo, CA, 1994, Chap. V, pp. 5-21–5-61.

[6]Gaski, J. D., *System Improved Numerical Differencing Analyzer, SINDA 1987/ANSI*, Network Analysis Associates, Fountain Valley, CA.

[7]Karam, R. D., and Eby, R. J., "Linearized Solution of Conducting-Radiating Fins," *AIAA Journal*, Vol. 16, No. 5, 1978, pp. 536–538.

[8]Curran, D. G. T., and Lam, T. T., "Weight Optimization for Honeycomb Radiators with Embedded Heat Pipes," *Journal of Spacecraft and Rockets*, Vol. 33, No. 6, 1996, pp. 822–828.

[9]Gatewood, B. E., *Thermal Stresses*, McGraw–Hill, New York, 1957, pp. 9–11.

[10]Chetty, P. R. K., *Satellite Technology and Its Applications*, 2nd ed., TAB Professional and Reference Books, Blue Ridge Summit, PA,1991, pp. 97–171.

[11] "Multimission Modular Spacecraft (MMS) Power Subsystem," NASA S-700-16, 1981.

[12] Thornton, E. A., and Kim, Y. A., "Thermally Induced Bending Vibrations of a Flexible Rolled-Up Solar Array," *Journal of Spacecraft and Rockets*, Vol. 30, No. 4, 1993, pp. 438–448.

[13] Foster, C. L., Tinker, M. L., Nurre, G. S., and Till, W. A., "Solar-Array-Induced Disturbances of Hubble Space Telescope Pointing System," *Journal of Spacecraft and Rockets*, Vol. 32, No. 4, 1995, pp. 634–644.

Chapter 6

Thermal Control Hardware

I. Introduction

A S must have been noticed by now, satellites must employ certain thermal control hardware to produce the required temperatures and provide margins to compensate for uncertainties and tolerances.

Flight-proven thermal hardware, generally considered standard equipment, consists of multilayered insulation (MLI); coatings; heat pipes and related systems; louvers; and heaters with associated thermostats or solid-state controllers. Also included are incidentals such as isolators, conductive straps, and interface fillers. There have been few uses of thermoelectric and stored-coolant cryogenic systems (Chapter 4, Sec. VI) and, for sometime now, there has been excitement over the capillary pumped loop (CPL), a two-phase device that has been proposed for use on the space station.

II. Multilayered Insulation

Insulation is used to minimize the thermal interaction of a system with its surroundings and, hence, reduce the expenditure in the energy needed to maintain it at a selected temperature. The thermal design of most satellite canisters is based on the principle of maximizing coverage by insulation while directing waste heat to thermally controlled radiators for disposal into space.

Insulating from surrounding space is generally accomplished with fluffy stacks of very thin (from ≈ 7.5 μm) sheets of Kapton or Mylar® that are made reflective by vapor-deposited aluminum (gold has also been used) on one or both surfaces. Fluffiness is seen to improve when using crinkled or embossed sheets separated by nonmetallic material such as a Dacron mesh. A somewhat thicker (≈ 25 μm) end sheet and still heavier (≈ 50 μm to ≈ 125 μm) aluminized Kapton or silvered Teflon outer cover enclose the stack to form what is known as a thermal blanket (or MLI blanket or insulation blanket), which is held together by stitching with non-metallic thread or intermittent taping along the edges (every 10–15 cm for large blankets) and sometimes by using nonmetallic snap buttons (say, one for every 25×25 cm^2 of surface area.) Outer covers also serve as a substrate for preferred surface coatings that may be electrically conductive and resistant to the effects of atomic oxygen (Sec. II.E of this chapter).

A schematic of a typical lay-up is shown in Fig. 6.1a. The assembly is secured to the satellite structure by bonding or using Velcro® strips, and the grounding straps shown in Fig. 6.1b are often added to reduce accumulation of electrostatic charge on the surface during orbit (Chapter 4, Sec. VI).

(a)

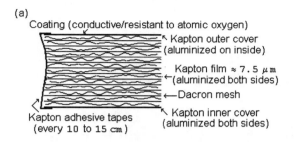

(b)

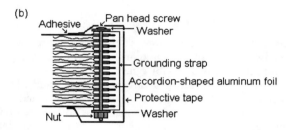

Fig. 6.1   Typical construction of MLI: a) typical lay-up and b) electrical grounding.

Construction and installation of thermal blankets have been described in numerous industry reports and engineering drawings. There is also much information in the publications of the AIAA series of Thermophysics Conferences, notably between 1970 and 1980, as well as in many satellite and heat transfer journals. NASA 905-G-407 (Ref. 1) is a documented procedure that continues to be consulted, and NASA CR-72605 (Ref. 2) contains many of the details still in practice. But, unquestionably, the truest appreciation of MLI can only be gotten by working with the actual hardware.

A.   Effective Emissivity and Effective Absorptivity of MLI

Because aluminum (or gold) has low emissivity ($\varepsilon < 0.03$), the stacked layers of a thermal blanket form a large resistance to radiation heat transfer. According to the equations for script F (Fig. 3.3 and Table 3.1), a shield of 15 or so separated sheets, aluminized on both sides, diminishes the ability to radiate by at least an order of magnitude. Resistance to heat conduction across the stack is also high because it occurs only at intermittent, small contact areas between sheets and becomes negligible when interspacers are used and the thermal blanket is made to remain fluffy. Therefore, in theory, MLI should make a formidable insulator.

In practice, however, the benefits suffer from edge and fin effects, cutouts, penetrations, and the unavoidable abuses from bending and molding to fit corners and irregularly shaped surfaces. A measure of this demotion is given by the effective emissivity, ε_{ins}, as it appears in the radiation heat transfer equation

$$Q_{ins} = \varepsilon_{ins} A\left(\sigma T^4 - \sigma T_\infty^4\right)$$

where Q_{ins} is heat rate to a sink at temperature T_∞ through the layers of a thermal blanket covering a radiator of area A and at temperature T. Following standard terminology, the lower the effective emissivity, the more effective is the MLI.

The sink temperature in orbit, with impinging solar, albedo, and Earth flux is given by

$$\sigma T_\infty = (\alpha^s_{oc} / \varepsilon^s_{oc})(q^S + q^A) + q^E$$

where oc refers to outer cover. Inclusion of T_∞ in the definition is necessary by virtue of the limiting case that requires an insulated nondissipator in thermal equilibrium to assume the temperature of the outer cover. It is shown in Sec. II.B of this chapter that the formalism is consistent with the notion that an insulated radiator with dissipation q^d may be treated as a standard radiator of emissivity ε_{ins} and solar absorptivity (called effective solar absorptivity) α^s_{ins} that satisfy the equation

$$q^d + \alpha^s_{ins} (q^S + q^A) + \varepsilon_{ins} q^E = \varepsilon_{ins} \sigma T^4$$

Also, α^s_{ins} is related to the other, measurable properties by

$$\alpha^s_{ins} / \varepsilon_{ins} = \alpha^s_{oc} / \varepsilon_{oc}$$

Although an actual thermal blanket deviates from a purely radiative system, the concept of effective emissivity and solar absorptivity simplifies the computations by combining all modes of heat transfer across the MLI into an equivalent radiation process with pseudo-optical properties. In practice the approach works best when the values of ε_{ins} (and related α^s_{ins}) are obtained from tests.

Some thermal engineers prefer to include the outer cover (but not the innermost sheet) in analyzing insulated radiators. In this case, an effective emissivity ε^* is used in the context of the equation

$$q^d = Q^d / A = \varepsilon^* (\sigma T^4 - \sigma T^4_{oc})$$

The values of ε^* and ε_{ins} are often used interchangeably. But if the MLI is to be viewed as a purely radiating system, then the correct connection between the two definitions must be obtained from a heat balance on the outer cover. Thus, when the interaction is with the space environment,

$$\sigma T^4_{oc} = q^d / \varepsilon_{oc} + (\alpha^s_{oc} / \varepsilon_{oc})(q^S + q^A) + q^E$$

and, on substituting into the ε^* equation,

$$q^d = \frac{\varepsilon^*}{1 + \varepsilon^* / \varepsilon_{oc}} \left\{ \sigma T^4 - \left[\left(\frac{\alpha^s_{oc}}{\varepsilon_{oc}} \right)(q^S + q^A) + q^E \right] \right\}$$

Noting that $[(\alpha^s_{oc} / \varepsilon_{oc})(q^S + q^A) + q^E] \equiv \sigma T^4_\infty$ and comparing with the definition of ε_{ins},

$$\varepsilon_{ins} = \varepsilon^* / [1 + (\varepsilon^* / \varepsilon_{oc})]$$

or

$$\varepsilon^* = \varepsilon_{ins} / [1 - (\varepsilon_{ins} / \varepsilon_{oc})]$$

Generally, the emissivity of the outer cover is very large compared to the insulation effective emissivity, which makes ε_{ins} and ε^* practically equal.

Another measure of MLI effectiveness is in the definition of effective conductance κ_{stack}, which satisfies the conduction equation

$$Q_{\text{ins}}/A = \kappa_{\text{stack}}(T_{\text{ic}} - T_{\text{oc}})$$

where ic refers to inner cover. The factor κ_{stack} is sometimes equated to an insulation conductivity divided by the thickness of the stack. Obviously, this definition is complicated because of the need to know the temperatures of the innermost sheet and outer cover (monitored during test) and because large MLI blankets often have fairly uniform effectiveness despite variations in their thickness.

B. Thermal Modeling of Insulated Radiators

The definitions may be extended to include lateral conduction along the outer cover and perhaps other MLI sheets and accessories. But the mathematics would undoubtedly become very complicated as it would have to represent a large collection of interacting, two-dimensional conducting–radiating fins. Some insight into lateral influences may be gained from Ref. 3, which features results from tests designed to evaluate the distribution of heat loss due to joints and penetrations and the effects of patching, overlapping, and stitching. A more recent evaluation is found in Ref. 4.

But most analyses are conducted based on one-dimensional radiation through the stack of layers, with losses by other modes and along different heat paths considered inherent in the test value of the effective emissivity (hence, the word "effective"). The validity of this approach has been indirectly verified by results from many satellite thermal vacuum tests.

Based on this and the properties of materials in standard insulation blankets, thermal modeling of radiators covered by MLI usually adheres to the following assumptions:

1) Each insulated radiator node i exchanges heat by radiation across the insulation either with corresponding sinks or a single corresponding outer cover node $oc(i)$ located on the external surface.

2) Heat transfer among outer cover nodes in the same plane is negligible.

3) The thermal mass of an outer cover node is so negligibly small (arithmetic node) that its temperature responds instantly to heating without change in stored energy.

Hence, a heat balance on an insulated radiator node gives

$$(Mc)_i \frac{\mathrm{d}T_i}{\mathrm{d}t} = Q_i^d - \varepsilon^* A_i \left(\sigma T_i^4 - \sigma T_{\text{oc}(i)}^4\right) - \sum_j \mathfrak{I}_{ij} A_i \left(\sigma T_i^4 - \sigma T_j^4\right)$$

$$- \sum_j K_{ij}(T_i - T_j)$$

and for the corresponding outer cover node

$$\alpha_{\text{oc}(i)}^s \left(A^S q^S + A^A q^A\right) + \varepsilon_{\text{oc}(i)}(A^E q^E) + \varepsilon^* A_i \left(\sigma T_i^4 - \sigma T_{\text{oc}(i)}^4\right) = \varepsilon_{\text{oc}(i)} A_{\text{oc}(i)} \sigma T_{\text{oc}(i)}^4$$

Here $A_{\text{oc}(i)}$ is the radiating area of the outer cover node (corresponding to internal node i) and j are regular (internal) interacting surfaces. The two equations are part of a network that enables the determination of T_i and $T_{\text{oc}(i)}$ at all locations.

From the second of the last two equations,

$$\sigma T_{oc(i)}^4 = \left[\alpha_{oc(i)}^s(A^S q^S + A^A q^A) + \varepsilon_{oc(i)}(A^E q^E) + \varepsilon^* A_i \sigma T_i^4\right] / \left[\varepsilon_{oc(i)} A_{oc(i)} + \varepsilon^* A_i\right]$$

Substitution into the first gives, after some manipulation,

$$(Mc)_i \frac{dT_i}{dt} = Q_i^d + f\alpha_{oc(i)}^s(A^S q^S + A^A q^A) + f\varepsilon_{oc(i)}(A^E q^E) - f\varepsilon_{oc(i)} A_{oc(i)} \sigma T_i^4$$

$$- \sum_j \Im_{ij} A_i (\sigma T_i^4 - \sigma T_j^4) - \sum_j K_{ij}(T_i - T_j)$$

with f given by

$$f = \varepsilon^* / [\varepsilon_{oc(i)}(A_{oc(i)}/A_i) + \varepsilon^*]$$

A heat balance on the radiator i that incorporates the implication that it is insulated can now be written

$$(Mc)_i \frac{dT_i}{dt} = Q_i^d + \alpha_i^s(A^S q^S + A^A q^A) + \varepsilon_i(A^E q^E) - \varepsilon_i A_{oc(i)} \sigma T_i^4$$

$$- \sum_j \Im_{ij} A_i (\sigma T_i^4 - \sigma T_j^4) - \sum_j K_{ij}(T_i - T_j)$$

where

$$\alpha_i^s \equiv f\alpha_{oc(i)}^s \qquad \text{and} \qquad \varepsilon_i \equiv f\varepsilon_{oc(i)}$$

When $A_{oc(i)} = A_i$, the values reduce to α_{ins}^s and ε_{ins} as defined earlier, and the formulation need not contain the outer cover in the nodal description. That is, the radiator's pseudo-optical properties, α_i^s and ε_i, now serve as the modulating factors for absorbing and emitting energy. The results do not yield the temperature of the satellite blanket surface, but that is generally evaluated in a separate study that includes tests designed to confirm the integrity of insulation materials under extreme heating and cooling.

C. Determining the Effective Emissivity by Test

MLI effective emissivity can be determined by tests in vacuum chambers evacuated to below 10^{-6} torr (mm of Hg), with sufficient durations to eliminate convection by gaseous residuals. However, testing of individual MLI blankets is not generally a requirement, and verification of the overall effectiveness is usually deferred until the satellite thermal balance test (Chapter 7, Sec. II) when the blankets are in their final installed configuration, with penetrations, cutouts, and attachments in place.

An arrangement is shown in Fig. 6.2. Heater power is applied to the insulated radiator while the chamber wall is kept at cryogenic temperature T_w to minimize extraneous heating. Ideally, the radiator, the MLI, and the distribution of heat would closely match actual satellite conditions.

In the case shown, the net heat going through the insulation can be determined by subtracting from the measured electric power (EI) the energy radiating from the back surface whose emissivity ε_b is known from an independent measurement. The insulation effective emissivity is then obtained as an average value from a series

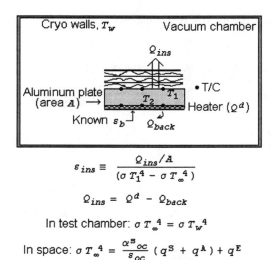

$$\varepsilon_{ins} \equiv \frac{Q_{ins}/A}{(\sigma T_1^4 - \sigma T_\infty^4)}$$

$$Q_{ins} = Q^d - Q_{back}$$

In test chamber: $\sigma T_\infty^4 = \sigma T_w^4$

In space: $\sigma T_\infty^4 = \frac{\alpha^s{}_{oc}}{\varepsilon_{oc}}(q^S + q^A) + q^E$

Fig. 6.2   Determining the effective emissivity of MLI by test in cryo vacuum.

of calculations at various power levels and corresponding temperatures using the equation

$$\varepsilon_{ins} = (Q_{ins}/A)/(\sigma T_1^4 - \sigma T_w^4)$$

with

$$Q_{ins} = Q_{heater} - \varepsilon_b A(\sigma T_2^4 - \sigma T_w^4)$$

Other arrangements designed to reduce errors in determining Q_{ins} include using guard heaters, testing with the same type of insulation on both sides of the radiator, and performing simultaneous tests on two identical sets suspended back to back.

D.   Systems Level Evaluation of Thermal Blankets

Because the effectiveness of multilayered insulation cannot be detected by visual inspection (some of the best-looking thermal blankets have the worst performance) it is not possible to determine, outside of an actual performance test, how effective an assembled insulation will be. Pointers for reducing potential problems include the following:

1) Experience is the only known efficacious avenue for making reliable thermal blankets. It must extend the entire range of fabrication, handling, and installation. Depending on size and shape, MLI with this background should yield an effective emissivity (test value) lower than 0.02, and thermal engineers can justifiably specify 0.02 as an upper limit for performance, with no exceptions. The value $\varepsilon_{ins} = 0.02$ may then be assumed in cold case analyses whereas 0.001 or less may be assumed in the hot case. Hence, the specifications for MLI effective emissivity are, for manufacturing: $\varepsilon_{ins} \leq 0.02$ and for analysis: $0.0 \leq \varepsilon_{ins} \leq 0.02$.

2) Edges and seams are a drawback to good MLI performance. Improvements can be made by overlapping (by at least 10 cm in large blankets) individual sections, and by free wrapping (one sheet of core material at a time) surfaces that are small or have awkward geometries.

3) Multiplicity of pins, tabs, and buttons must be discouraged. These are usually used to keep a blanket in place and make it look nice and fluffy. But beyond a nominal number, these items can introduce a relatively significant heat loss. A rule of thumb is that their spacing be such that, in final installation, the thickness of a 15-layer thermal blanket with covers and interspacers will not be less than 1.5 cm (2.0 cm is nominal for large blankets).

4) The number of cutouts and penetrations should also be minimized. A serious effort should be made to collect into as few bundles as possible the wires and cables that must penetrate the insulation. Special insulation boots can then be constructed to cover those areas.

5) On installation, the innermost layer and the stiff outer cover must not be allowed to squeeze or crush the body of the blanket. Parts of these sheets can be cut and removed, or tailored to fit the desired contours without seriously affecting overall performance.

6) Venting is aided when perforated material is used. Venting is important for expediting the disposal of residual air and other gases when in test chambers or space vacuum.

E.   Effect of Atomic Oxygen on MLI

Space weathering is presenting difficult challenges in selecting and developing thermal control materials that can endure long-term exposures. Attack by atomic oxygen in LEO on polymers used in making thermal blankets is of particular concern.[5] Much remains to be understood and done in this area, and so far no definite solutions have been proposed.

Interaction with atomic oxygen is described as an oxidation process with sputtering.[6] It is significant in the presence of solar radiation on surfaces leading in the direction of satellite motion, particularly at altitudes from 200 to 500 km. The origin of the atoms is mostly dissociation by the sun's UV radiation of molecular oxygen and water vapor at the edge of the atmosphere. The density is placed at the order of 10^8 atoms/cm^3, and fluences as high as 10^{22} atoms/cm^2 along the leading edge have been reported at the lower elevations with a translational energy per atom estimated about 5 eV on striking a surface at 8 km/s relative velocity.

Most of the quantitative conclusions on long-term effects of atomic oxygen derive from studies of material samples on the Long Duration Exposure Facility (LDEF),[7-9] which remained in LEO for 69 months before being retrieved. Other data have been provided by returned components on the Solar Max Mission[10] and materials that had undergone short but repeated exposures on the STS space shuttle.

Kapton samples from LDEF reveal severe damage, with surface recessions exceeding 250 μm in 69 months of exposure.[5,11] Silvered Teflon (silverized fluoroethylyne propylene Teflon) with the Teflon facing out also shows physical alteration,[12] but its erosion (2.5 μm from 125 μm) does not appear as rapid as Kapton, and its optical properties seem only slightly affected despite some deformation and a considerable change in color. Comparison with erosions from accumulated STS flights confirm the extreme susceptibility of Kapton and Mylar and a slower, but definite effect on silvered Teflon.

The material and thickness of MLI outer covers must, therefore, be evaluated in view of mission duration and orientation in low orbit and plans for possible orbital replacements and repairs. Protection of nearby sensitive instruments and mechanisms from escaping debris is also of concern. In this regard, material studies should not exclude durable surface treatments that might have adverse optical properties but could be made to apply by a suitable thermal design.

Finally, the aggravation may be lessened by reducing, or perhaps eliminating, surface cracks, openings, and sharp edges. Studies indicate undercutting and possible hastening of erosion due to surface defects.[11] Modifications on MLI blanket constructions should consider unexposed, tucked-in borders and venting provisions that do not include exposed surface openings. This may be done by perforating the interior stack to allow gas flows either radially along edges that are covered (but not sealed) with flaps or through access holes in the satellite structure.

III. Surface Coatings and Finishes

Surfaces can either naturally or after treatment be made to absorb and emit radiant energy at specific rates corresponding to the spectrum of radiation. Some of these properties and their directional and spectral behavior are discussed in Chapter 3.

Thermal control coatings (TCCs) have been classified as follows:

1) Solar reflectors, which have a low solar absorptivity but high emissivity, are useful in a solar or albedo environment as they reflect much of the impinging energy while retaining the high IR emissivity needed for efficient rejection of spacecraft waste heat. They include white paints, with as-applied ratio of α^s/ε as low as 0.2, and optical solar reflectors (OSRs) that work on the principle of the second-surface mirror, where solar radiation penetrates a thin transparent high-emissivity material (Teflon, glass, etc.) then reflects off a metallic substrate, effectively producing a low-α^s, high-ε surface. A schematic is given in Fig. 6.3a. A popular OSR is the flexible silvered Teflon ($\approx$127 μm thick with the Teflon facing outward toward the environment), which can be made with an adhesive backing for direct application on a surface. Some measurements on the 127-μm silvered Teflon show as-applied

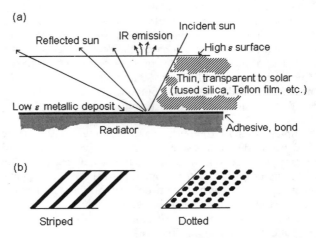

Fig. 6.3   Composite optical systems: a) second surface mirrors and b) patterned surfaces; on the average, for a pattern of 1 and 2 of areas A_1 and A_2, $\alpha^s_{av} = (\alpha^s_1 A_1 + \alpha^s_2 A_2)/(A_1 + A_2)$ and $\varepsilon_{av} = (\varepsilon_1 A_1 + \varepsilon_2 A_2)/(A_1 + A_2)$.

values of α^s/ε as low as 0.07/0.79. Quartz mirrors (with average measured as-applied $\alpha^s/\varepsilon = 0.08/0.78$) have also seen extensive use. They come in the shape of wafers (2–4 cm squares, $\approx$150–200 μm thick) that are bonded to surfaces in the same manner as solar cells. The singular advantage of quartz mirrors is their resistance to a chemical reaction and the long-term stability of their properties in a solar environment.

2) Flat coatings, which reflect and absorb nearly equally in the solar and IR spectra, include black paints, with high values (>0.8) of α^s and ε, and metallic paints, with α^s between 0.25 and 0.50 and approximately the same for ε. Black paints are widely used inside satellite canisters, including the exteriors of the electronics covers, to enhance heat sharing by radiation.

3) Solar absorbers, which have high α^s/ε, are seldom used on satellite surfaces exposed to the sun because of the high and unpredictable temperatures associated with them. Among the better-known ones are silverized finishes and vapor-deposited aluminum (VDA) applied on berrylium copper or stainless steel extendable elements. Measured data of VDA have not been consistent, and reported values of α^s/ε have ranged from 0.08/0.03 to 0.17/0.035. Solar absorbers are sometimes also used in combination patterns with other coatings to obtain tailored properties. Examples are shown in Fig. 6.3b.

A. Coatings Degradation: Systems Level Evaluation

Unlike the fabrication of MLI, the manufacturing and application of surface coatings have become largely automated and less prone to human imperfections. In addition, cleaning, handling, protection, and storage practices have improved to a level where the possibility of physical damage or contamination on the ground has been greatly reduced. Ground requirements for each particular coating are usually contained in the coating's own process control and application specifications.[13]

Despite careful and successful management on the ground, problems have persisted in space with no apparent full resolutions. Among the outstanding ones are degradation of properties, erosion from atomic oxygen, and delamination of bonded material.

Two causes are often cited for degradation of properties, particularly the solar absorptivity. One is contamination from spacecraft debris and volatiles, and the other is corruption from exposure to ultraviolet radiation and charged elementary particles. Contamination affects almost any surface finish, but it varies with temperature and proximity and shape factor to the contaminating source. Degradation of absorptivity from exposure to ultraviolet radiation and bombardment by charged particles seems especially pronounced in Kapton (polyimide film) and white paints with organic binders.[14,15]

As noted in Sec. II.E of this chapter, material erosion due to reaction with atomic oxygen is most prevalent in a low Earth orbit between 200 and 500 km and is particularly significant on sunlit surfaces leading along the satellite's velocity vector.

Delamination and adhesive deterioration of coatings that have bonded layers or are applied with adhesive films appear to be caused mainly by adhesive migration from cracked metallic backing and residual gas (air) pockets formed during application.[12,16]

Although no complete answers have been found, the effects can be reduced or partially circumvented by observing the following guidelines:

1) A vigorous effort should be made to place radiators with critical coatings in regions of the satellite where uv radiation and contamination fluxes are not excessive. Sometimes this may require awkward radiator orientation or specialized satellite maneuvers, but the benefits can be significant when translated into narrower temperature margins for thermal qualification and reduced heater power.

2) Application of adhesive coatings, particularly silvered Teflon tape, on radiators that will undergo severe temperature vascillations in orbit should be carried out in vacuum (tools already exist) to avoid trapping air. Air pockets in adhesive tapes are known to be a nucleus for potential cracking and delamination. The risk is further reduced by perforation (typically, about 1-mm-diam holes spaced some 2.5 cm apart), and visible pockets and wrinkles should be worked out by puncturing and applying pressure with a roller.

3) As discussed in Sec. II.E of this chapter, exposed Kapton in LEO should be avoided as this material is seriously damaged from interaction with atomic oxygen. However, the erosion may be acceptable in locations that are off the leading direction and away from solar impingement. In any event, the heavier, thicker film should be preferred.

4) Discoloration and some erosion also appear in silvered Teflon, especially when used as floating sheets or bonded to flexible surfaces. The optical properties of the 127-μm silvered Teflon, however, do not seem radically affected, and the material is considered an acceptable coating on hard surface radiators in LEO.

5) Quartz mirrors and white paints Z93 (zinc oxide in potassium silicate binder) and YB71 (zinc orthotitanate in potassium silicate binder) appear stable in a solar environment and resistant to reaction with atomic oxygen. Black paint Z306 (polyurethane binder) is highly susceptible to erosion from atomic oxygen and should be used in that environment only on interior surfaces and external radiators along the trailing edge or hidden from solar radiation. White paints S13G-LO (zinc oxide in silicone binder) and A276 (titanium dioxide in a polyurethane binder) also show considerable effects from solar radiation and atomic oxygen, and their use should be limited to small areas or regions with minor direct solar impingement (back of sun-facing solar array, for instance). In general, potassium silicate binders are stable, whereas organic binders are not. But nonorganic binders are more difficult to apply and repair and can crack and chip off a flexible substrate. They are not recommended as surface treatment on polymer MLI outer covers.

6) Procedures for handling and protecting thermal control coatings on the ground must be strictly enforced. Protective covers for white paints and quartz mirrors and removable benign latex film sprays for silvered Teflon are effective and will not generally interfere with activities around the satellite during ground work and operations. Flight radiators exposed in vacuum testing must be cleaned, and their properties confirmed before being covered again.

7) The program should require that the solar absorptivity and emissivity of primary radiators be monitored following prolonged periods of ground assembly and functional testing. Portable, noncumbersome instruments such as Gier Dunkle MS-251 and DB-100 reflectometers (Chapter 3, Sec. IV.C) have been used to make periodic checks to verify consistency.

8) The use of new, exotic coatings as prime thermal control systems should be discouraged when their history is based mainly on laboratory data. There is no clear evidence that conditions of simulated degradation and accelerated testing are accurate representations of what actually occurs in space. It is safest to rely on

coatings with flight background, even when the thermal design must accommodate the worst of their observed degradation.

B.   Common Satellite Coatings and Their Properties

Some of the coatings that have been used with various degrees of satisfaction in missions lasting at least 5 years are listed in Table 6.1 with their optical properties. Background sources include, among others, Ref. 17, spacecraft charging at high altitudes (SCATHA) experiments,[14,15,18] flight thermistor data from ATS-6 solar arrays,[19] and reports from the LDEF Post-Retrieval Symposiums.[7-9] It must be stated, however, that there can be no claims of a general consensus on these or any other reported values as tolerances and degradation ultimately depend on the parameters of each particular application (orbit altitude, surface orientation, intensity of solar radiation, accumulated hours in sunlight, inherent and surrounding material outgassing, etc.) Nonetheless, the ranges given here are sufficiently wide to encompass most published and privately conveyed information.

It will be noted that no attempt is made to apportion the degradation of solar absorptivity on individual causes. However, one would expect that the effects of contamination show themselves early in the mission, when the rates of escaping satellite volatiles are highest, and degradation from UV radiation increases with time in sunlight (solar and albedo). The effect of atomic oxygen are most prominent in LEO in the presence of UV radiation on surfaces whose normals are in the direction of the velocity vector.

No reliably stable flat coatings with medium properties have been advanced, although there have been occasional uses of aluminum paints (with as applied $\alpha^s/\varepsilon \approx 0.3/0.3$) when the application is not too critical. Also, combinations in narrow stripes or dots of black paint on aluminum, as shown in Fig. 6.3b, have been applied to obtain in-between properties. Similarly, the striping or dotting of polished aluminum with white paint tends to reduce the temperature while in the sun because average emissivity is increased without a radical change in solar absorptivity. The pattern can be made either by masking off portions of the surface while the rest is being treated, or by locally laying an adhesive tape treated with the desired supplemental coating. In doing this, however, one must maintain as much uniformity per unit area as is practical to avoid large variations in temperature because of localized differences in surface properties.

Certain ranges of properties are also obtained either by varying the thickness of the surface finish or by time-controlled chemical treatment. Figure 6.4a summarizes measured data on the effect of Teflon film thickness on average normal solar absorptivity and emissivity of silvered Teflon.[17,20] Figure 6.4b is typical of the effect of anodize time on the properties of 2024-T3 aluminum alloy.

Figure 6.5 shows the change in solar absorptivity of fused silica in GEO. Contamination is evidently the major contributor to the degradation of quartz mirrors, as evidenced by the sharp increase in α^s when a contaminant-prone conductive coating is used. It is noted that this occurs in the first days of a mission, when the outgassing rate of spacecraft volatiles is relatively high.

Dependence of surface properties on the in-plane angle φ has not been a major factor in thermal analysis or design as it is found to be small for most thermal

Table 6.1   Thermal control coatings in general use

Z93 white paint: as applied[a] $\alpha^s = 0.17 \pm 0.01$, $\varepsilon = 0.86 \pm 0.02$
   Near-Earth[b]
      Heavy exposure to uv: first yr, $\alpha^s = 0.19$; 5 or more yr, $\alpha^s = 0.25$
      Light exposure to uv[c]: first yr, $\alpha^s = 0.19$; 5 or more yr, $\alpha^s = 0.22$
      Resistant to atomic oxygen
   Geosynchronous
      First yr, $\alpha^s = 0.23$; 2 yr, $\alpha^s \approx 0.30$ (ATS-6 solar array flight temperature data)
YB71 white paint: as applied $\alpha^s = 0.15 \pm 0.01$, $\varepsilon = 0.86 \pm 0.02$
   Near-Earth
      Heavy exposure to uv: first yr, $\alpha^s = 0.18$; 5 or more yr, $\alpha^s = 0.25$
      Light exposure to uv: first yr, $\alpha^s = 0.17$; 5 or more yr, $\alpha^s = 0.22$
      Resistant to atomic oxygen
   Geosynchronous
      First yr, $\alpha^s = 0.25$; 2 yr, $\alpha^s = 0.35$
S-13G-LO (RTV 602 silicone binder) white paint: as applied $\alpha^s = 0.20 \pm 0.02$, $\varepsilon = 0.84 \pm 0.02$
   Near-Earth
      Heavy exposure to uv: first yr, $\alpha^s = 0.30$; 5 or more yr, $\alpha^s = 0.40$
      Light exposure to uv: first yr, $\alpha^s = 0.30$; 5 or more yr, $\alpha^s = 0.37$
      Severe erosion from atomic oxygen between 200 and 500 km
   Geosynchronous
      First yr, $\alpha^s = 0.33$; 5 or more yr, $\alpha^s = 0.68$
127-μm silvered Teflon: as applied $\alpha^s = 0.10 \pm 0.02$, $\varepsilon = 0.77 \pm 0.02$
   Near-Earth
      First yr, $\alpha^s = 0.17$; 5 or more yr, $\alpha^s = 0.28$
      Light, acceptable erosion from atomic oxygen when used on hard surfaces
   Geosynchronous
      First yr, $\alpha^s = 0.17$; 3 yr, $\alpha^s = 0.23$; after 3 yr, $\Delta\alpha^s = +0.03$/yr
ITO-coated 127-μm silvered Teflon: as applied $\alpha^s = 0.14 \pm 0.02$, $\varepsilon = 0.82 \pm 0.02$
   Near-Earth (1334 km)
      First yr, $\alpha^s = 0.20$; 3 yr, $\alpha^s = 0.32$
Uncoated quartz mirrors (152 μm): as applied $\alpha^s = 0.08 \pm 0.005$, $\varepsilon = 0.78 \pm 0.02$
   Near-Earth
      First yr, $\alpha^s = 0.09$; after 1 yr, $\Delta\alpha^s = +0.004$/yr[d]
   Geosynchronous
      First yr, $\alpha^s = 0.09$; after 1 yr, $\Delta\alpha^s = +0.004$/yr
Conductively coated quartz mirrors (127 μm): as applied $\alpha^s = 0.09 \pm 0.01$, $\varepsilon = 0.78 \pm 0.02$
   Near-Earth
      First yr, $\alpha^s = 0.11$; after 1 yr, $\Delta\alpha^s = +0.005$/yr
   Geosynchronous
      First yr, $\alpha^s = 0.11$; after 1 yr, $\Delta\alpha^s = +0.005$/yr
Chemglaze Z306 black paint: as applied $\alpha^s = 0.95$, $\varepsilon = 0.85$
      $\alpha^s/\varepsilon = 1.12 \pm 0.02$, no significant degradation
      Severe erosion from atomic oxygen between 200 and 500 km
D111 (carbonaceous pigment in glass binder) black paint: as applied $\alpha^s = 0.98$, $\varepsilon = 0.93$
      Stable; resistant to atomic oxygen; recommended for heating flux calibration coupons
Uncoated 5-mil (127-μm) aluminized Kapton (Kapton side): as applied $\alpha^s = 0.48 \pm 0.02$, $\varepsilon = 0.81 \pm 0.01$
   Near-Earth
      Severe erosion by atomic oxygen between 200 and 500 km
   Geosynchronous
      After 3 yr, $\alpha^s = 0.60$; maximum $\alpha^s = 0.67$

(cont.)

Table 6.1 (Continued)

ITO-coated 50-μm Al Kapton (Kapton side): as applied $\alpha^s = 0.40 \pm 0.02$ $\varepsilon = 0.71 \pm 0.01$
Near-Earth
    Severe erosion by atomic oxygen between 200 and 500 km
    Geosynchronous
       $\Delta\alpha^s = +0.05/\text{yr}$; maximum $\alpha^s = 0.60$
Polished aluminum (properties depend on finish): $\alpha^s = 0.08$ to 0.15, $\varepsilon = 0.03$ to 0.04
    $\Delta\alpha^s = +0.001/\text{yr}$
Solar cells (properties depend on type, cell efficiency not included): $\alpha^s = 0.70$ to 0.91, $\varepsilon = 0.76$ to 0.81
    $\Delta\alpha^s = +0.005/\text{yr}^{\text{d}}$

[a]Emissivities are hemispherical.
[b]Below 1500-km altitude.
[c]Surface mostly hidden from direct sun but with some exposure to albedo.
[d]Degradation due mainly to contamination.

treatments. One striking departure was the behavior noted in the so-called black Kapton used on the TOPEX/Poseidon instrument module as a conductive outer cover for the satellite's insulation blankets. The material has significant specularity[21] and the mesh used for reinforcement leaves the surface with a ridgelike roughness that apparently affects solar reflection in the direction of φ. But because of its low reflectivity, black Kapton did not cause a dramatic change in the TOPEX heating flux profile; however, it was important to place it only along orientations that did not result in reflections toward sensors and other instruments.

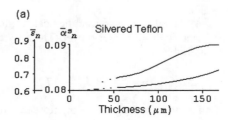

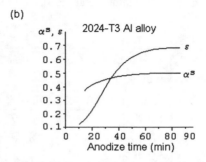

Fig. 6.4 Surface properties control: a) normal properties of silvered Teflon as function of film thickness[17] and b) total average properties of anodized 2024-T3 aluminum alloy as function of anodize time (after FSC IOC TH-RK-024, 1973).

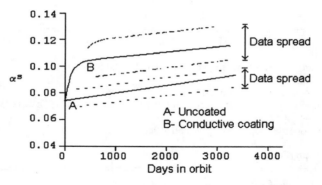

Fig. 6.5   Degradation of fused silica mirrors in GEO.[18]

C.   Calorimetric and On-Orbit Measurements

Calorimetry refers to determining properties from measurements of temperature and heat (heater power). The method is used in ground tests to obtain total or average hemispherical emissivity and can be applied to finding optical properties in orbit.

By knowing the radiating area and monitoring steady-state temperatures and electric heater power Q^d, the surface emissivity of an isothermal sample in a relatively large vacuum chamber is calculated from

$$\varepsilon = (Q^d/A^r)/(\sigma T^4 - \sigma T_w^4)$$

where T is sample temperature and T_w the temperature (usually a cryogenic average) of the chamber walls.

If the thermal mass Mc is also known, another calculation can be made by monitoring the cooling rate with no heater power. Thus, by equating the rate of loss in internal energy with the rate of radiation,

$$Mc\frac{dT}{dt} = -\varepsilon A^r(\sigma T^4 - \sigma T_w^4)$$

and the emissivity is found by estimating slopes or, more accurately, by integrating from time zero (when the sample's temperature is T_0) to time t (temperature T_t). Noting that

$$\frac{1}{T^4 - T_w^4} = \frac{1}{2T_w^2}\left[\frac{1}{T^2 - T_w^2} - \frac{1}{T^2 + T_w^2}\right]$$

integration by separation of variables gives

$$\varepsilon = \frac{1}{t}\left\{\frac{Mc}{4A^r\sigma T_w^3}\left[\ln\left|\frac{(T_t + T_w)(T_0 - T_w)}{(T_t - T_w)(T_0 + T_w)}\right|\right.\right.$$
$$\left.\left. + 2\tan^{-1}\left(\frac{T_t}{T_w}\right) - 2\tan^{-1}\left(\frac{T_0}{T_w}\right)\right]\right\}$$

The experiment is also useful in finding the thermal mass when the emissivity is known.

Thermal engineers sometimes exercise these procedures in piggyback experiments during cryogenic vacuum testing of a satellite or a subsystem. Small samples of coated substrates are instrumented with heaters and thermocouples then suspended in reserved, out-of-the-way locations inside the vacuum chamber for independent testing. Tests are also conducted in high-vacuum chambers designed specifically for calorimetric measurements. One such facility is available at NASA Goddard Space Flight Center (GSFC). The cool-down test in the NASA chamber is initiated after heating the sample by an X 25 solar simulator through a quartz port.

Contrary to their apparent simplicity, calorimetric tests are difficult to process because of uncertainties in measurements and effects from a nonuniform temperature distribution around the chamber wall. In addition, an accurate determination requires that the samples be very nearly isothermal and free of edge and instrumentation heat losses that are difficult to determine, especially when the total amount of heat involved is very small.

There are two, seemingly contradictory methods for approaching isothermal conditions. One is to apply a coating on a somewhat thick (0.50-cm), highly conductive substrate, such as aluminum or copper, of relatively large surface area (5.0-cm-diam disk). The gradients caused by transverse and edge effects are, thus, overwhelmed by radiation from the exposed surface and may be further reduced by isolating the rest of the sample from the surroundings with low-emissivity treatment (or insulation) and low-conduction suspensions or mounts. The other method is to strive for temperature uniformity in a localized region, away from edges and their effect, by using a substrate of very low conductance. Either approach is adequate when properly calibrated.

Coupons of the first variety[22] have been installed on spacecraft for assessing surface properties in orbit. The interest has been mostly in monitoring the degradation of solar absorptivity. An important contribution to this activity has been the SCATHA program on the P-78 spacecraft, including a modified experiment (ML12) that essentially avoids contaminants and yields data specifically on the influence of solar radiation (see Ref. 15 for summary and updates). The P-78 operates at or near geosynchronous altitudes and data collection has continued since 1979.

A coupon with low lateral conductance, instrumented by a heater and a temperature sensor, is shown in Fig. 6.6. Distinguishing features of this miniature high-tech unit include high accuracy and negligible heat loss along instrumentation leads.[23] Also, the dual insulation stand (skirt) makes radiation exchange exclusively between the surface and the environment, with the only relatively significant heat loss occurring at low conductivity standoffs. (Figure 6.6 depicts two thin Fiberglas® G-10 stubs with mounting tabs at the farthest edge from where data are monitored.) When installed on a spacecraft, the skirt may be surrounded by MLI with selected outer treatment to avoid excessive temperatures during exposure to sunlight.

The results of a numerical (SINDA) model confirm the insensitivity of the region farthest from the mounts to all influences other than those of the heater and impinging energy. Calculations were repeated for spacecraft interface temperature at the standoffs of −50, 0, and 60°C with only minor (second decimal place) effect noted on the temperature at the location of the sensor. A typical map is given in Fig. 6.6.

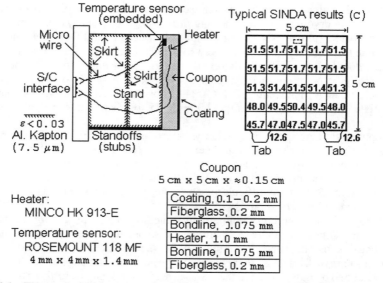

Coupon
5 cm x 5 cm x ≈0.15 cm

Heater:
  MINCO HK 913-E

Temperature sensor:
  ROSEMOUNT 118 MF
  4 mm x 4 mm x 1.4 mm

| Coating, 0.1 – 0.2 mm |
| Fiberglass, 0.2 mm |
| Bondline, 0.075 mm |
| Heater, 1.0 mm |
| Bondline, 0.075 mm |
| Fiberglass, 0.2 mm |

Fig. 6.6   Thermal coating coupon for calorimetric measurements of surface proper-ties in orbit; SINDA results are for 1400 W/m² solar input on a silvered Teflon coating ($\alpha^s = 0.2$, $\varepsilon = 0.76$) with 0.5-W heater power and 0°C spacecraft temperature; the ideal equilibrium temperature without losses is 51.7°C.

Data reduction is simpler for orbits where Earth radiation is negligible. In this case, the energy balance on a one-sided isothermal surface of area A in a fixed, well-defined orientation with respect to the sun is

$$(Mc/A)\frac{dT}{dt} = \alpha^s q^S - \varepsilon\sigma T^4$$

where the value of normal incident solar flux q^S is usually available either from a separate source [Earth Radiation Budget Satellite (ERBS), NOAA, etc.] or from accompanying calibration coupons of known properties (D111 black coating has known and stable properties). Because the motion of a satellite at very high altitudes (particularly GEO) is slow with respect to solar direction, there occur sufficiently long intervals during the orbit when the temperature and solar heating may be con-sidered constant. Hence, the ratio of solar absorptivity to emissivity is immediately obtained from

$$\alpha^s/\varepsilon = \sigma T^4/q^S$$

If, as observed in many coatings, the emissivity is not measurably affected by space environment, then α^s can be calculated using the as-applied value of emissivity, and it would not be necessary to employ an active heater to determine both properties. Furthermore, a confirmation of the value of ε can be obtained by monitoring the temperature during eclipses and applying the transient cool-down equation.

Temperature profiles of the sensor region for three coupons with different coat-ings in various orientations are shown in Fig. 6.7. The heating fluxes are for a $\beta = 0$ circular orbit at 1334-km elevation. Sun-oriented coupon S is a solar cell sample

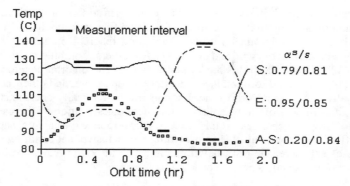

Fig. 6.7 Coupon temperature in 1334-km, $\beta = 0$ circular orbit; S is sun-oriented solar cell with 0.3-W heater activated during eclipse; E is Earth-oriented stable black with constant 0.3-W heater power; and A-S is antisun white coating with constant 0.3-W heater power.

with its stand mounted to the cell side of a rotating solar panel that remains normal to the sun during daytime. A 0.3-W heater is available and may be activated during the eclipse portion of the orbit to reduce the amplitude of the temperature cycle and help maintain the integrity of coupon materials, especially the adhesives. The curve designated by E refers to an Earth-oriented sample of D111 black coating used as calibration monitor for Earth IR radiation and albedo. Antisun coupon A-S is S-13G-LO white paint. It is mounted to the anticell side of a solar panel as representative of the coating on the back of the array.

The time-varying energy balance on an isothermal location reads

$$\frac{Mc}{A}\frac{dT}{dt} = \alpha^s(q^S + q^A) + \varepsilon q^E + q^d - \varepsilon\sigma T^4$$

where q^d is heater power per unit area in the region of analysis. But, as seen from the curves, there are intervals during the orbit (pointed out by dark short lines) when the time rate of change of temperature is very small and may be neglected. Because, generally, Mc/A is also small, the heat balance during these periods is reasonably accurately given by

$$\alpha^s(q^S + q^A) + \varepsilon q^E + q^d - \varepsilon\sigma T^4 = 0$$

which does not contain the usually elusive uncertainties in rates of change and thermal mass. Thus, if the heater power can be turned off and on by command, then for two separate zero slope orbit positions, 1 and 2,

$$(q^S + q^A)_1\alpha^s + (q^E - \sigma T^4)_1\varepsilon = q^d$$
$$(q^S + q^A)_2\alpha^s + (q^E - \sigma T^4)_2\varepsilon = 0$$

If incident heating fluxes at the two positions are known from onboard radiometers or calibration coupons of stable properties, then measurements of temperature and heater power will give the coefficients of the two equations, which can then be solved simultaneously for α^s and ε.

Actually, environment sources that cause changes in surface properties consist of more than impinging flux and include, in addition to contamination, effects from charged particles, micrometeorite bombardment, inherent outgassing products, reaction with atomic oxygen, and probably a few others. Combined with this is the fact that albedo and Earth flux are variables that depend on orbit position, time of day and of year, cloud formation, air moisture, temperature, and other factors that continually change with orbit time. Lacking an impractically enormous amount of data, it is impossible to establish a consistent classification of the causes and effects of property changes. In this respect, recoverable experiments, such as the LDEF (Ref. 24), are invaluable in providing a conclusive, net result after certain periods of exposure. On-orbit determinations, on the other hand, are important when continual changes need to be known.

IV. Heat Pipes

Heat pipes greatly improve heat sharing in a satellite at a relatively small increase in weight. Their operation involves change of phase, with large heat rates transferred at an almost constant temperature. Hence, they dampen temperature differences in mounting and radiating panels and moderate the temperature in regions of concentrated heat sources or sinks.

The basic heat pipe is schematically shown in Fig. 6.8. It is a tube in which an interior capillary wick (grooves, mesh, sintered porous material, etc.) is soaked with liquid and the container sealed at saturation conditions. As heat is applied to one end (the evaporator), a differential pressure is created and the vapor is driven to the cooler end (the condenser), where it condenses back into the wick. But because of liquid loss at the evaporator, the meniscus there depresses, resulting in a capillary head that drives the condensate back, which forces the process to repeat.

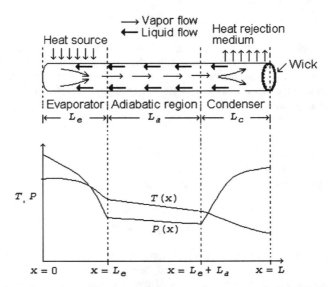

Fig. 6.8 Basic heat pipe: the temperature is nearly constant in the adiabatic region and may be calculated from $\frac{1}{L_e+L_a}[\int_0^{L_e} T(x)\,dx + \int_{L_e+L_a}^{L} T(x)\,dx]$.

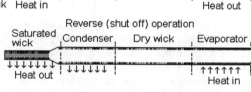

Fig. 6.9 Basic variable conductance heat pipe and the liquid trap diode.

As the driving potential is capillary action, a heat pipe is a natural for microgravity applications without assistance from mechanical pumps or other moving devices.

Variations on the concept include the variable conductance heat pipe [(VCHP) (Sec. IV.H of this chapter)] and the heat pipe diode. As shown in Fig. 6.9, the condenser end of the VCHP entails a reservoir of noncondensable gas whose front moves into and out of the active section following pressure changes with heat input. When heating is high, the increase in pressure forces the front toward the reservoir, leaving more length for vapor condensation. As heating decreases, the gas compensates by occupying more of the condenser. In this way, large variations in heat input result in small changes in temperature. In the liquid trap diode, one end has a disconnected wick that leads to trapping or blocking the working fluid (here in the liquid phase) and thus shut off the operation if condenser and evaporator reverse roles by changes in heat distribution.

Engineering books on heat pipes and related devices include *Heat Pipe Theory and Practice,*[25] *Heat Pipes,*[26] *Heat Pipe Science and Technology,*[27] and *The Physical Principles of Heat Pipes.*[28] A list combining all the reported references in these works would probably cover most of the truly incredible number of papers written on the subject through 1994. Many of these can be found in the proceedings of the International Heat Pipe Conference (since 1973) and the journals and conference publications of AIAA, the Society of Automotive Engineers (SAE), the American Institute of Chemical Engineers (AIChE), and the American Society of Chemical Engineers (ASME). The earlier works (1965–1975) seem the more interesting, especially to thermal engineers who lived through the difficulties of transforming the industrial heat pipe into the presently reliable, long life satellite component.

A. Grooved Heat Pipe with Ammonia as Working Fluid

Most heat pipes in U.S. satellites have the axial groove wick typified by the ATS-6 configuration shown in Fig. 6.10. Although its operation has generally been

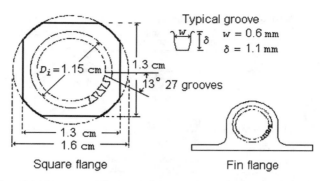

Fig. 6.10   Grooved wick extrusions with square (ATS-6) or fin (MMS) flange.

confined to moderate temperatures (-40 to $80°C$) with relatively low-heat loads (less than 80 W), and despite its sensitivity to gravity effects in ground testing, the grooved heat pipe remains more qualified than any other for space operation and is closest to being an off-the-shelf satellite item. It is relatively easy to fabricate and clean, and the fin shape of the grooves conforms to the desired low resistance to heat flow across the wick, thus reducing the susceptibility to vapor formation in the liquid passages that can seriously hamper the operation (Sec. IV.D of this chapter).

Familiar units are extruded square or finned aluminum tubes ranging from about 0.5 to 2.0 cm in diameter, with lengths usually less than 2.0 m. Rectangular or trapezoidal grooves are uniformly distributed around the inner wall (Fig. 6.10), and a flat or flanged exterior is often incorporated to improve heat transfer into and out of the pipe by increasing the contact area with the external medium. The pipes may be bent (4-cm bend radius in the 1.15-cm square or flanged extrusions) into various shapes to cover large areas with few individual units.

Summary reviews of grooved heat pipes, including test performance data using different groove geometries and various working fluids, are given in Refs. 29 and 30. Reference 31 details the cleaning procedure that subsequently became the basis for NASA's required practice.

Ammonia is almost always the working fluid in satellite grooved heat pipes. Its heat transport capability is excellent in the temperature range usually encountered, and the techniques for purifying it are now at a stage that makes it completely compatible with its aluminum container. This is absolutely necessary to avoid even the remotest interactions that might generate noncondensable gases that can cut back on the working length and instigate bubble formation. Of course, there is always concern about the effects of an ammonia leak, but today's heat pipe is virtually secured as a leak-proof pressure vessel with very advanced welds around the end caps and fill valves.

Ammonia's ability to transport heat can be judged by its liquid transport factor (sometimes called figure of merit M with units of W/m^2). In zero gravity and laminar flow, the factor is represented by (Sec. IV.C, this chapter)

$$M = \rho_l \sigma \lambda / \mu_l$$

where ρ_l, σ, and μ_l are, respectively, liquid density (kg/m^3), surface tension (N/m) and viscosity (N s/m^2), and λ the latent heat of vaporization (J/kg). On reviewing the tables of properties, it can be confirmed that M for ammonia at 20°C is at least an order magnitude larger than that of water or methanol, two other fluids considered compatible at the temperature range of satellite operations.

Pertinent properties of ammonia are given in Table 6.2. Vapor specific heat at constant volume is based on the polyatomic nature of ammonia gas that suggests $\gamma = c_p/c_v = 1.33$. The gas constant R_v is assumed 489 J/kg K in accordance with the universal gas constant relation

$$R_v = 8314/\text{molecular weight}$$

B. Specifications and Test Verification

Table 6.3 is a typical listing of performance specifications for grooved heat pipes. The data apply to various flanged shapes (straight, C, L, U, Z) used in some of the command and data handling modules (C&DH) manufactured by Orbital Sciences Corporation [(OSC) (previously Fairchild Space Company)]. The cross section is identical to that of ATS-6.

Usually the stated heat loads in heat pipe specifications exceed the levels expected during satellite operation (2.5 times is common) and, depending on the program, verification by a ground test may or may not be required on every flight and spare unit.

Testing is usually done in air, with the evaporator raised against gravity at a specified tilt for added margin. Gravitational resistance to capillary flow may also be viewed as a contingency against unplanned advantageous deviations from horizontal leveling in the thermal balance test (see last item in Chapter 7, Sec. II.E). A schematic of the basic apparatus is given in Fig. 6.11. Also shown is an arrangement with heaters placed on separate sections along the pipe to emulate an actual heating profile in the satellite. This constitutes several evaporation and condensing regions, as is usually the case in most operations.

Results from tilt tests on an ATS heat pipe are given in Fig. 6.12. The test value of heat transport capability is seen to be larger at high-tilt angles than what is predicted by theory. This is consistent with other test data and has been attributed to drainage effects.[30] It could also be due to the assumed character of flow with liquid overfills and an overestimate of the effects of liquid-vapor shear interaction.

Less emphasis is being placed on burn-out and restart tests nowadays because of established confidence in heritage heat pipes. A similar attitude seems to prevail on the requirements for life tests, where representative units undergo extended ground operation (sometimes accelerated by higher heat loads) with periodic checks on the amount of generated noncondensable gases.

C. Mathematical Theory of Heat Pipe Operation: Capillary Limit

Although the principle of operation of a heat pipe is not difficult to perceive, a mathematical description of the physical process taking place inside is rather elusive. Not only are there very many parameters involved, but the analysis ultimately depends on operating regimes that are unknown but whose existence must be assumed to determine the conditions that will produce them.

But even with defined operations and assumptions of one-dimensional flow, consistent geometry, exact fill, and ideal wicking (capillary) and fluid characteristics, the fluid dynamics of heat pipes is complex and useful engineering results have been obtained only by assuming average behavior at each cross section with

Table 6.2   Properties of saturated ammonia (NH_3, molecular weight, 17; boiling point = 240 K)

Temperature, K	230	250	270	290	310	330	350	370
Pressure, N/m^2	6.06E4	1.65E5	3.82E5	7.75E5	1.425E6	2.422E6	3.87E6	5.88E6
Latent heat, J/kg	1.40E6	1.34E6	1.27E6	1.20E6	1.12E6	1.02E6	0.90E6	0.74E6
Surface tension, N/m	0.0352	0.0306	0.0264	0.0202	0.0176	0.0132	0.009	0.0047
Liquid density, kg/m^3	694	669	642	615	584	550	512	465
Vapor density, kg/m^3	0.55	1.40	3.10	6.00	11.00	18.90	31.00	52.00
Liquid viscosity, N s/m^2	2.2E−4	2.0E−4	1.7E−4	1.5E−4	1.2E−4	1.0E−4	8.2E−5	6.4E−5
Vapor viscosity, N s/m^2	0.7E−5	0.8E−5	0.9E−5	1.0E−5	1.1E−5	1.2E−5	1.3E−5	1.4E−5
Liquid conductivity, W/m K	0.62	0.58	0.54	0.49	0.45	0.40	0.36	0.32
Vapor conductivity, W/m K	0.017	0.020	0.022	0.025	0.030	0.037	0.047	0.061
Liquid c_p, J/kg K	4.43E3	4.51E3	4.59E3	4.65E3	4.86E3	5.07E3	5.40E3	5.86E3
Vapor c_p, J/kg K	1.92E3	2.32E3	2.69E3	3.04E3	3.44E3	3.90E3	4.62E3	6.21E3
Vapor c_v, J/kg K	1.44E3	1.74E3	2.02E3	2.29E3	2.59E3	2.93E3	3.47E3	4.67E3

Table 6.3   ATS-6 and OSC C&DH heat pipe specifications

Straight and U heat pipes
    Heat Load: 60 W
    Input/output: 30.5-cm evaporator, remainder of heat pipe as condenser
    Transport capability (maximum): 38 W-m
C, L, and Z heat pipes
    Heat load: 20 W
    Input/output: 15-cm evaporator and 15-cm condenser
    Transport capability (maximum): 31.75 W-m
All heat pipes
    ΔT (maximum evaporator outer wall to minimum condenser
        outer wall): $\leq 5.5°C$
    ΔT (within evaporator): $\leq 1.7°C$
    ΔT (within condenser): $\leq 1.7°C$
    Tilt: evaporator elevated 2.54 mm
    Operating temperature: 0 to 40°C
    Qualification temperature: -20 to 65°C
    Ammonia purity: 99.998%
    Proof pressure: no detectable deformation at 100 atm (10^7 N/m^2) and 120°C
    Burst pressure: no structure failure at 137 atm and 120°C
    Straightness: 0.0833 cm/m
    Design life: 7–10 yr
    Noncondensable gas generation: $\leq 1.5E-4$ g-mole/m/7 yr
    Leak rate: $\leq 10E-8$ scc/s helium at 15.3 atm (1.55E6 N/m^2)
    Mass: depends on exterior design, typically, 0.25–0.35 kg/m

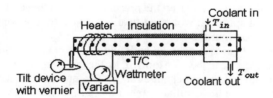

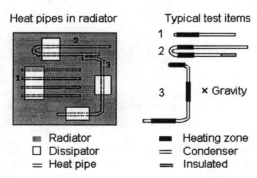

Fig. 6.11   Verification tests of satellite heat pipes: in radiator pipe test, the evaporators in 1 and 2 are tilted against gravity (out of paper plane) and item 3 is tested horizontally.

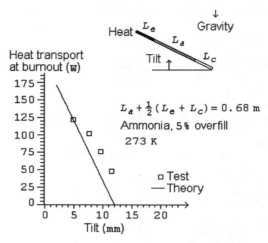

Heat transport at burnout (w)

$L_a + \frac{1}{2}(L_e + L_c) = 0.68$ m

Ammonia, 5% overfill

273 K

□ Test
— Theory

Tilt (mm)

Fig. 6.12   Tilt test on ATS grooved heat pipe.[30]

mathematically well-defined transitions along the length. This seems to work well in satellites where heat rates are made relatively low by distributing the load among a number of heat pipes. The resulting flows then tend to be predictable by standard viscous flow theories and, because of usually massive satellite systems, an analysis in steady state is often justified.

The seminal work in heat pipe theory is Cotter's "Theory of Heat Pipes."[32] Subsequent analytical efforts, including the present treatment, have mostly elaborated on Cotter's original approach with modifications added to fit a nonideal condition or a specific wick configuration.

Basically, the analysis is concerned with finding the maximum heat rate applied in the evaporator that can be transported to the condenser with the heat pipe functioning in a normal, phase-change mode. The operation is said to be within the capillary limit when the capillary pressure is sufficiently high to replenish the liquid in the evaporator as rapidly as it is being lost by evaporation. For an exact fill, this condition of connected (vs dried out) flow is guaranteed if the pressure remains continuous in an arbitrary section (x_1–x_2 in Fig. 6.13) according to the identity

$$[P_v(x_1) - P_v(x_2)] + [P_v(x_2) - P_l(x_2)] + [P_l(x_2) - P_l(x_1)] + [P_l(x_1) - P_v(x_1)] = 0$$

where P_v and P_l are vapor and liquid pressures at their common interface. In a sense, this equation provides an equivalent one-dimensional (along x) fluid flow description to a situation in which there is mass exchange in the radial direction between liquid and vapor. This is a considerable simplification of an otherwise extremely complex problem.

Recalling the meaning of capillary pressure $P_c(x)$ as the difference $[P_v(x) - P_l(x)]$ that causes the curvature of the meniscus at the liquid–vapor interface, pressure continuity is rewritten

$$P_c(x_2) - P_c(x_1) = [P_v(x_2) - P_v(x_1)] - [P_l(x_2) - P_l(x_1)]$$

In terms of integrated variations,

$$\int_{x_1}^{x_2} \frac{\mathrm{d}P_c}{\mathrm{d}x}\,\mathrm{d}x = \int_{x_1}^{x_2} \frac{\mathrm{d}P_v}{\mathrm{d}x}\,\mathrm{d}x - \int_{x_1}^{x_2} \frac{\mathrm{d}P_l}{\mathrm{d}x}\,\mathrm{d}x$$

Now the task is to find expressions for the pressures that, when substituted into the preceding equation, will yield a relation between the heating input and parameters of geometry and flow.

In anticipation of finding the maximum capability for transporting heat, it is assumed that the surface of the liquid in the wick at initial evaporator position ($x = 0$) is so depressed as to yield maximum capillary action, whereas it is so flat at the condenser end ($x = L$) that capillary force there is negligible. Hence, based on the coordinate direction in Fig. 6.13,

$$\int_0^L \frac{\mathrm{d}P_c}{\mathrm{d}x}\,\mathrm{d}x = -P_c(0)$$

with the negative sign indicating that capillary pressure is driving the liquid in the $-x$ direction.

Gravity ($g = 9.81$ m/s^2 on Earth and approximately 0 m/s^2 in orbit) can have a positive or negative effect depending on the direction of the normal to the meniscus in each capillary cavity. This may be included by referring to a net capillary pressure

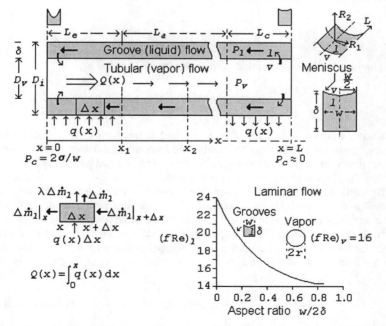

Fig. 6.13 Heat and mass transfer in grooved heat pipes: maximum capillary action is for meniscus angle at wall $= \pi/2$; the principal radii of curvature are $R_1 = w/2$ and R_2 (parallel to groove wall) $= \infty$; Laplace–Young equation gives $P_c = 2\sigma/w$.

given by

$$P_{c\,net} = P_c(0) \pm G\rho_l g \cos\theta$$

where G is a factor defining the extent of gravitational communication among the pores and θ the horizontal inclination. By proper construction, G can be made negligibly small (it is considered zero for grooved heat pipes) and the Laplace–Young equation for a general-shaped cavity gives

$$P_{c\,net} = P_c(0) \quad (=P_{c,\max}) = 2\sigma/r_c$$

where r_c is the effective capillary radius. It is indicated in Fig. 6.13 that one principal radius of curvature of the meniscus in a rectangular channel is half the width w, whereas the other (along the length) is infinite. Thus,

$$(1/r_c)_{\text{rectangular channel}} = (1/R_1 + 1/R_2) = 2/w$$

and

$$P_{c\,net\,per\,groove,0-g} = P_c(0) = 2\sigma/w$$

In this work, the desired expressions for liquid and vapor pressure are found by utilizing the concept of friction factor (Chapter 2, Sec. IV) that puts the equation of motion for viscous flow in terms of Reynolds number and conduit shape. Thus, for liquid flow with gravity as possibly the only body force

$$\frac{dP_l}{dx} = 2f_l\rho_l\frac{u_l^2}{D_l} \pm \rho_l g \sin\theta$$

Introducing the liquid Reynolds number $Re_l \equiv \rho_l u_l(x)D_l/\mu_l$,

$$\frac{dP_l}{dx} = 2(f_l Re_l)u_l(x)\frac{\mu_l}{D_l^2} \pm \rho_l g \sin\theta$$

Subscript l refers to liquid and f, ρ, u, μ, and D are, respectively, friction factor, density, speed (positive, m/s), viscosity, and hydraulic diameter, which is defined as four times the cross-sectional area divided by the wetted perimeter. The second term on the right is gravitational contribution.

The relation between flow rate and heat transport is contained in the conservation of mass and energy during a phase change. In keeping with the quest to find the maximum heat transport capability, it must be assumed that local external heating (or cooling), $q(x)$ (W/m), be completely expended into a change of phase in the corresponding location of working fluid. A mass balance on a liquid element Δx (see the elemental volume in Fig. 6.13 and recall that u_1 is considered positive) gives

$$q(x)\Delta x = \lambda[(\dot{m}_l)_{x+\Delta x} - (\dot{m}_l)_x] = \lambda[\rho_l A_{l\,\text{flow}} u_l|_{x+\Delta x} - \rho_l A_{l\,\text{flow}} u_l|_x]$$

$$= \lambda \rho_l A_{l\,\text{flow}}\left(\frac{du_l}{dx}\right)\Delta x$$

where $\dot{m}_l$ is rate of mass flow, $A_{l\,flow}$ the cross section of liquid flow, and λ the latent heat of evaporation. Hence,

$$\frac{du_l}{dx} = \frac{q(x)}{\lambda \rho_l A_{l\,flow}}$$

Assuming constant properties and negligible liquid motion at the pipe's end in the evaporator $[u_l(0) = 0]$,

$$u_l(x) = \frac{1}{\lambda \rho_l A_{l\,flow}} \int_0^x q(x)\,dx$$

It is understood that internal local axial heat flow, $Q(x)$ (W), is accumulated along positive x according to

$$Q(x) = \int_0^x q(x)\,dx$$

Hence,

$$u_l(x) = Q(x)/\lambda \rho_l A_{l\,flow}$$

and the equation for liquid pressure gradient becomes

$$\frac{dP_l}{dx} = F_l Q(x) \pm \rho_l g \sin\theta$$

where F_l is the liquid friction coefficient given by

$$F_l = 2(f_l Re_l)\mu_l / D_l^2 \lambda \rho_l A_{l\,flow}$$

Analogous relations for vapor pressure are obtained by utilizing the conservation of mass

$$\dot{m}_l = -\dot{m}_v$$

Thus, in the vapor region (subscript v)

$$u_v(x) = \frac{1}{\lambda \rho_v A_{v\,flow}} \int_0^x q(x)\,dx = \frac{Q(x)}{\lambda \rho_v A_{v\,flow}}$$

and

$$\frac{-dP_v}{dx} = F_v Q(x) \pm \rho_v g \sin\theta + \Gamma$$

with the negative sign indicating a pressure drop along increasing x and Γ a factor accounting for the effect on vapor motion due to radial influx. The vapor friction coefficient F_v is

$$F_v = 2(f_v Re_v)\mu_v / D_v^2 \lambda \rho_v A_{v\,flow}$$

where

$$Re_v \equiv \rho_v u_v(x) D_v / \mu_v$$

and Γ (Ref. 25, pp. 43–47) is given by

$$\Gamma(x) = \left(1.33/A_{v\,\text{flow}}^2 \rho_v \lambda^2\right) \frac{\mathrm{d}Q^2}{\mathrm{d}x}$$

Incompressibility is examined by noting that the speed of sound at T_v is

$$u_s = (\gamma_v R_v T_v)^{1/2}$$

giving the local Mach number as

$$M_v(x) = u_v/u_s = Q(x)/\left[\lambda \rho_v A_{v\,\text{flow}} (\gamma_v R_v T_v)^{1/2}\right]$$

where $\gamma_v \, (= c_p/c_v)$ is the ratio of vapor specific heats and R_v (J/kg K) the gas constant.

Substituting the new expressions for individual pressure gradients in the original pressure continuity equation

$$P_c(0) = \frac{2\sigma}{r_c} = \int_0^L (F_l + F_v)Q(x)\,\mathrm{d}x + \int_0^L [\pm(\rho_l + \rho_v)g\sin\theta + \Gamma]Q(x)\,\mathrm{d}x$$

By using the preceding equation for Γ while contending that $Q(x)$ is zero at $x = 0$ and $x = L$,

$$\int_0^L \Gamma Q(x)\,\mathrm{d}x = 0$$

and in zero gravity

$$P_c(0) = \frac{2\sigma}{r_c} = \int_0^L (F_l + F_v)Q(x)\,\mathrm{d}x$$

Because in most satellite applications of interest

$$\left|\frac{\mathrm{d}P_l}{\mathrm{d}x}\right| \gg \left|\frac{\mathrm{d}P_v}{\mathrm{d}x}\right|$$

there is little error in writing

$$\frac{2\sigma}{r_c} = \int_0^L F_l Q(x)\,\mathrm{d}x$$

The heat transport factor HTF (W–m), is defined by

$$HTF \equiv \int_0^L Q(x)\,\mathrm{d}x$$

Therefore, when the geometry is consistent and fluid properties constant,

$$HTF = (2\sigma/r_c)/(F_l + F_v) \approx 2\sigma/r_c F_l$$

which is functionally independent of length. It is also seen from the definition of F_l that for uniform geometry and constant properties the heat transport factor is proportional to $\sigma \lambda \rho_l/\mu_l$. As noted earlier, this combination of liquid properties is a measure of the effectiveness of the working fluid to transport heat. It is relatively high for ammonia.

The integral form used throughout is essential to establish a basis for treating situations in which heat absorption and rejection are not uniform and may occur at intermittent locations along the pipe. It is important to note, however, that for a standard heat pipe consisting of one evaporator, perhaps followed by an inactive (adiabatic) section, and one condenser, *HTF* can be calculated without reference to length. Hence, once obtained for a given cross section and working fluid, it will serve to judge the performance of all heat pipes of the same cross section and working fluid in terms of whether or not they can transport a particular heat load along a given length. A numerical example that includes the calculation of *HTF* is given later. The result can be used as a baseline for comparing the performance of heat pipes of different lengths.

Maximum capillary pressure was inserted in the analysis to find the maximum heat transport Q_{max} possible within capillary limitations. In the one-evaporator–one-condenser configuration, this is the integrated axial heat flux in the evaporator; that is,

$$Q_{max} = \int_0^{L_e} q(x) \, dx$$

where L_e is evaporator length. For uniform heating [$q(x)$ a constant], axial heat flow can be put in terms of maximum heat transport as follows:

$$Q(x) = (Q_{max}/L_e)x, \qquad \text{for } 0 \le x \le L_e$$

$$Q(x) = Q_{max}, \qquad \text{for } L_e \le x \le (L_e + L_a)$$

$$Q(x) = (Q_{max}/L_c)(L - x), \qquad \text{for } (L_e + L_a) \le x \le L$$

where L_c and L_a are the lengths of the condenser and adiabatic sections. Hence, maximum heat transport factor HTF_{max}, is

$$HTF_{max} = \int_0^L Q(x) \, dx = \int_0^{L_e} \frac{Q_{max}}{L_e} x \, dx$$

$$+ \int_{L_e}^{L_e+L_a} Q_{max} \, dx + \int_{L_e+L_a}^L \frac{Q_{max}}{L_c}(L - x) \, dx$$

$$= (0.5L_e + L_a + 0.5L_c)Q_{max}$$

This equation contains a constraint, dictated by HTF_{max}, on the manner of distribution of maximum heat load along a heat pipe of given length. Requirements on both Q_{max} and HTF_{max} are usually given in heat pipe specifications. Note that when a heat pipe consists of just an evoparator and a condenser, without an adiabatic section,

$$HTF_{max} = Q_{max} L/2$$

Applying the results to a grooved heat pipe with N rectangular channels (see Fig. 6.13)

$$D_l = 4w\delta/(w + 2\delta)$$

$$A_{l \, \text{flow}} = Nw\delta$$

Table 6.4   Friction factor and Reynolds number for vapor flow in axially grooved heat pipes

$Re_v < \approx 2300$, $M_v < \approx 0.3$
$f_v Re_v = 16$ (Moody diagram)

$Re_v < \approx 2300$, $M_v > \approx 0.3$
$f_v Re_v = 16\left[1 + (\gamma_v - 1)M_v^2/2\right]^{-\frac{1}{2}}$

$Re_v > \approx 2300$, $M_v < \approx 0.3$
$f_v Re_v^{\frac{1}{4}} = 0.079$ (Moody diagram)

$Re_v > \approx 2300$, $M_v > \approx 0.3$
$f_v Re_v^{\frac{1}{4}} = 0.079\left[1 + (\gamma_v - 1)M_v^2/2\right]^{-\frac{3}{4}}$

$$D_v = D_i - 2\delta$$

$$A_{v\,\text{flow}} = \pi D_v^2/4$$

$$F_{l,0-g} = (f_l Re_l \mu_l/8\lambda\rho_l)[(w + 2\delta)^2/N(w\delta)^3]$$

$$F_{v,0-g} = (8 f_v Re_v \mu_v/\pi\lambda\rho_v)/(D_i - 2\delta)^4$$

Also, in most satellite applications of interest the liquid in the grooves is in laminar flow with $f_l Re_l$ obtained from the curve in Fig. 6.13. The character of vapor flow is in accordance with Table 6.4. It should be noted that normal operation may still be possible with turbulent or compressible vapor flow.

Finally, in zero gravity and dominating liquid pressure gradient the heat transport factor for a grooved heat pipe is

$$\int_0^L Q(x)\,dx\bigg|_{\text{grooved}/0-g} = \frac{16\sigma\lambda\rho_l N w^2\delta^3}{(fRe)_l\mu_l(w + 2\delta)^2}$$

which suggests the existence of functional relations among properties and groove dimensions that would maximize the performance.[33]

Example: Numerical Calculations for Grooved Heat Pipes

Consider the following data for a 1-m aluminum (6061-T6, $k = 168$ W/m K) heat pipe of circular cross section and rectangular axial grooves with ammonia as working fluid: tube inside diameter (includes groove depth), $D_i = 1.15$ cm; number of grooves, $N = 27$; groove depth, $\delta = 1.1$ mm; groove width, $w = 0.6$ mm; vapor core diameter, $D_v = D_i - 2\delta = 0.93$ cm; aspect ratio, $w/2\delta = 0.273$; evaporator length, $L_e = 15$ cm; and condenser length, $L_c = 15$ cm. It is required to find the maximum uniform (evaporator and condenser) heat load that can be transported at 17°C in zero gravity.

Ammonia properties at 17°C (290 K) are as follows: latent heat of vaporization, $\lambda = 1.2E6$ J/kg; surface tension, $\sigma = 0.0202$ N/m; liquid density, $\rho_l = 615$ kg/m³; vapor density, $\rho_v = 6.0$ (kg/m³); liquid viscosity, $\mu_l = 1.5E{-}4$ Ns/m²;

vapor viscosity, $\mu_v = 1.0E-5$ Ns/m^2; ratio of specific heats, $\gamma_v \approx 1.33$ (poly-atomic); and vapor (gas) constant, $R_v \approx$ universal gas constant/molecular weight $= 8314/17 = 489$ J/kg K.

The calculations are performed in accordance with uniform heating and assuming laminar, incompressible flow for both liquid and vapor. The final results are checked to confirm these assumptions.

Maximum capillary pressure:

$$P_{c(\max)} = P_c(0) = 2\sigma/w = (2)(0.0202)/6.0E-4 = 67.3 \text{ N/m}^2$$

Liquid friction coefficient $(fRe)_l = 18$ (from the curve in Fig. 6.13 with $w/2\delta = 0.273$):

$$F_l = (fRe)_l \mu_l (w + 2\delta)^2/8\lambda\rho_l N(w\delta)^3$$

$$= (18)(1.5E-4)[0.6E-3 + 2(1.1E-3)]^2/8(1.2E6)$$

$$\times (615)(27)(0.6E-3)^3(1.1E-3)^3$$

$$= 0.462 \text{ N/m}^2 \text{ per W-m]}$$

Vapor friction coefficient $(fRe)_l = 16$ (laminar flow in circular cross section):

$$F_v = 8(fRe)_v \mu_v/\pi\lambda\rho_v(D_i - 2\delta)^4$$

$$= (8)(16)(1.0E-5)/(\pi)(1.2E6)(6.0)(1.15E-2 - 2.2E-3)^4$$

$$= 0.0076 \text{ N/m}^2 \text{ per W-m}$$

(Note that $F_v \ll F_l$.)

Maximum heat transport factor:

$$HTF_{\max} = \int_0^L Q_{\max}(x)\,dx$$

$$= P_c(0)/(F_l + F_v) = 67.3/(0.462 + 0.0076) = 143.3 \text{ W-m}$$

Maximum heat transport:

$$Q_{\max} = HTF_{\max}/(0.5L_e + L_a + 0.5L_c)$$

$$= 143.3/[(0.5)(0.15) + 0.70 + (0.5)(0.15)]$$

$$= 168.6 \text{ W}$$

Laminar flow and incompressibility are confirmed as follows:

$$Re_v \equiv \rho_v u_v D_v/\mu_v$$

Maximum u_v occurs at entrance of adiabatic section where $Q(x) = Q_{\max}$

$$Re_{v(\max)} = D_v Q_{\max}/A_{v\,\text{flow}}\lambda\mu_v$$

$$= (1.15E-2 - 2.2E-3)(168.6)(4)/[(\pi(1.15E-2 - 2.2E-3)^2$$

$$\times (1.2E6)(1.0E-5)]$$

$$= 1924 < 2300$$

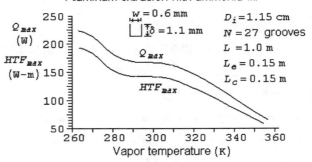

$$Q(x) = \int_0^x q(x)\,dx, \quad Q_{max} = \int_0^{L_e} q(x)\,dx, \quad HTF = \int_0^L Q(x)\,dx$$

Aluminum extrusion with ammonia fill

Fig. 6.14   Maximum heat transport capability (Q_{max}) and maximum heat transport factor (HTF_{max}) for grooved heat pipe with uniform heating.

Hence, the flow is laminar ($u_v = 0.345$ m/s) and

$$M_{v(max)} = Q_{max} / \left\{ A_{v\,flow} \lambda \rho_v (\gamma_v R_v T_v)^{\frac{1}{2}} \right\}$$

$$= (168.6)(4) / \left\{ \pi(1.15E{-}2 - 2.2E{-}3)^2 (1.2E6) \right.$$

$$\left. \times (6)[(1.33)(489)(290)]^{\frac{1}{2}} \right\}$$

$$= 8.0E{-}4 < 0.3$$

Hence, the flow is incompressible ($u_s = 434.3$ m/s).

Therefore, the calculated data are valid estimates of the heat pipe's heat transport capability as it performs at its capillary limit. Similar calculations can be made at other temperatures and the results plotted as shown in Fig. 6.14.

D.   Sonic, Entrainment, and Boiling Limits

It is natural to inquire about processes that might hamper a heat pipe from performing at its capillary limit. One such hindrance occurs when vapor speed (maximum at evaporator exit) exceeds the local speed of sound at heat loads lower than what is permitted by the capillary limit. The heat input at which this happens is known as the sonic limit, and it results in a large temperature gradient along the pipe. The situation is usually compared to flow in a coverging–diverging nozzle, with the process of a heat pipe's variable mass passing through a constant cross-sectional area made equivalent to a nozzle's constant mass flow through a changing area (see pp. 79–84 in Ref. 25.)

Although the phenomenon associated with the sonic limit is of some significance in industrial applications involving large-heat loads and high temperatures, it is not known to happen in normal satellite operation except, possibly, during a heat pipe startup at cold temperatures. The numerical results in the preceding example show that with ammonia the heat load at normal temperatures will have to be on the order of kilowatts for the Mach number to reach or exceed 1. On the other hand, from the Mach number equation

$$M_v = Q/A_{v,\text{flow}}\lambda\rho_v(\gamma_v R_v T_v)^{\frac{1}{2}}$$

it is seen that a sudden application of heat at the evaporator when the whole system is at a cold temperature might create a large pressure difference between evaporator and condenser that could result in sonic conditions. Although this is rare in common satellite operations, a contingency plan might be considered to uniformly warm up a cold heat pipe before applying an unbalanced heat load.

Another conceivable limitation, referred to as entrainment limit, is due to speeding vapor tearing droplets off the liquid surface in the wick, thus causing a reduced supply of liquid to the evaporator. This seems to be of particular concern with the axially grooved design because of the relatively large unpartitioned and exposed liquid surface in the axial groove.

A measure of entrainment is given by the Weber number We, which relates the shearing force of the vapor to the capillary force that holds the liquid in the wick. A simple defintion is

$$We \equiv \kappa\left(\rho_v u_v^2 A_p/\sigma C_p\right)$$

where κ is a proportionality factor and A_p and C_p are surface area and wetted perimeter of the wick pore, respectively. In terms of heat flux,

$$We = \kappa(\rho_v A_p/\sigma C_p)[Q(x)/\lambda\rho_v A_{v\,\text{flow}}]^2$$

Entrainment is considered significant when $We \geq 1$.

For a grooved heat pipe of length L,

$$A_p = wL \qquad \text{and} \qquad C_p = 2L + 2w \approx 2L$$

Hence A_p/C_p is about half the groove width (capillary radius), and if κ is on the order of one[34] then, using the numerical example as a typical case, a heat load on the order of 100 W at a nominal temperature (say, 17°C) gives

$$We = \kappa(\rho_v w/2\sigma)[Q(x)/A_{v\,\text{flow}}\lambda\rho_v]^2$$

$$= (1)[(6)(6.0E{-}4)/2(0.0202)][100(4)/\pi$$

$$\times (1.15E{-}2 - 2.2E{-}3)^2(1.2E6)(6)]^2$$

$$= 3.7E{-}3$$

This indicates that entrainment is not likely to occur in normal satellite operations. According to this calculation, the problem would not arise unless heat transport is in the range of 10^4–10^5 W.

Perhaps least predictable is a heat pipe boiling limit, which is defined as the heat input at which nucleate boiling ensues at the evaporator's wall/wick interface. Among the influences that initiate and direct the process are minute crevices and indentations in the surface, cleanliness (wettability), and the deposition and release of noncondensable foreign gases. Irreproducibility of these and other factors in experiments has made it difficult to establish consistent parametric classifications,[35,36] and a tractable treatment necessarily ignores all effects other than uniform heat transfer across the evaporator wall.

As external heat is conducted radially across the wall to the evaporator's liquid-soaked wick, the temperature of the liquid at the wall/wick is higher than at the liquid–vapor interface. And because liquid pressure is lower, by an amount that equals surface tension, than the saturation pressure at the liquid–vapor interface, the saturation pressure at the wall/wick is higher than the liquid pressure, which leads to vapor bubble formation. Unless the bubbles collapse or detach and open at the liquid surface, they will be a resistance to heat transfer across the pipe wall causing hot spots and eventual dry-out with a sharp increase in evaporator temperature.

In a gravitational field, buoyancy and stirring forces might force the bubbles to detach and move to the liquid surface unless they are trapped by wick construction (by its inherently high thermal conductivity and openness, a groove wick will fascilitate the motion). In zero gravity, however, once bubbles form they will remain at the wall/wick unless they collapse.

To find the conditions for bubble collapse, a spherical embryonic bubble of radius r_b is considered in equilibrium at the evaporator's wall/wick, as indicated in Fig. 6.15a. Noting the nomenclature, with saturation pressure denoted by P_w^s, the net normal outward force due to the pressure difference between inside and

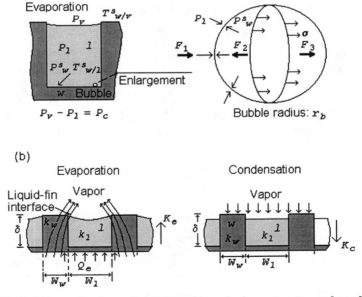

Fig. 6.15 Bubble mechanics and radial heat transfer in evaporator and condenser: a) bubble equilibrium (existence) requires $F_2 - F_1 = F_3$ and b) conductance per unit heat pipe length.

outside the bubble is

$$F_2 - F_1 = \pi r_b^2 (P_w^s - P_l)$$

The resisting force of surface tension acting around the circumference (imagine a bubble made of two hemispheres) is

$$F_3 = 2\pi r_b \sigma$$

For a stable, intact bubble, $F_2 - F_1 = F_3$, or

$$\pi r_b^2 (P_w^s - P_l) = 2\pi r_b \sigma$$

As noted earlier, liquid pressure is pressure at the liquid–vapor interface less surface tension. Hence,

$$\pi r_b^2 (P_w^s - P_v + P_c) = 2\pi r_b \sigma$$

or

$$P_w^s - P_v = 2\sigma/r_b - P_c$$

The wick structure is bounded by the pipe wall and the vapor core. Denoting saturation temperature at the wall–wick interface with the liquid by $T_{w/l}^s$, and by $T_{w/v}^s$ at the wick interface with the vapor, the Clausius–Clapeyron equation (along the saturation temperature–pressure line) gives approximately

$$\frac{\mathrm{d}P}{\mathrm{d}T} = \lambda \rho_v / T_v \approx (P_w^s - P_v) / (T_{w/l}^s - T_{w/v}^s)$$

On substituting the earlier result,

$$T_{w/l}^s - T_{w/v}^s = (T_v/\lambda \rho_v)[(2\sigma/r_b) - P_c]$$

The path of heat transfer due to a radial temperature difference is shown in Fig. 6.15b to be made up of a liquid section in parallel with solid wick wall. For uniform heating, the total input Q_e (W) in the evaporator is

$$Q_e = K_{\mathrm{eff}}(T_{w/l}^s - T_{w/v}^s)$$

where K_{eff} (W/K) is the effective radial conductance of the wick along the entire evaporator. For cylindrical geometry, this may be put in terms of a radial evaporator conductivity k_e by the relation (Chapter 2, Sec. III)

$$K_{\mathrm{eff}} = k_e (2\pi L_e)/\ell n(D_i/D_v)$$

with D_i the inner diameter and D_v the vapor core diameter, as shown in Fig. 6.13. Hence,

$$Q_e = [2\pi L_e k_e T_v/\lambda \rho_v \ell n(D_i/D_v)][(2\sigma/r_b) - P_c]$$

is the amount of heat into the evaporator that will create the thermodynamic conditions of a stable bubble (of radius r_b). If the applied heat is

$$Q_e^- < [2\pi L_e k_e T_v/\lambda \rho_v \ell n(D_i/D_v)][(2\sigma/r_b) - P_c]$$

then the bubble will collapse. On the other hand, if

$$Q_e^+ > [2\pi L_e k_e T_v / \lambda \rho_v \ell n (D_i/D_v)][(2\sigma/r_b) - P_c]$$

then the bubble will grow to a correspondingly larger equilibrium radius.

There are various methods for estimating an evaporator radial conductivity that accounts for both liquid and solid heat paths in the wick. Chi[25] treats k_e as a material property based on the relative volume content of solid and liquid and a correlated film coefficient. Thus (see Fig. 6.15b), the conductance per unit length at the evaporator, K_e is the sum of the conductances through the liquid (K_l) and through the wall, with heat transfer to the vapor at the solid-vapor interface (K_{w-v}). That is,

$$K_e = K_l + K_{w-v}$$

$$1/K_{w-v} = 1/K_w + 1/C_{w-v}$$

where K_l (W/m K) $= k_l W_l/\delta$, K_w (W/m K) $= k_w W_w/\delta$, and C_{w-v} the conductance at the liquid–fin interface. In terms of unit evaporator length, Chi's correlation is written

$$C_{w-v} = [k_l/0.185 W_w] W_w$$

Defining evaporator conductivity k_e by the equation

$$K_e \equiv k_e(W_w + W_w)/\delta$$

substitutions result in

$$k_e = [k_l k_w W_w \delta + k_l W_l (k_l \delta + 0.185 k_w W_w)]/[(W_l + W_w)(k_l \delta + 0.185 k_w W_w)]$$

A similar treatment for the condenser gives condenser conductivity k_c as

$$k_c = (k_l W_l + k_w W_w)/(W_l + W_w)$$

For the grooved heat pipe in the numerical example, W_w is calculated as

$$W_w = (\pi D_i - Nw)/N = [11.5\pi - (27)(0.6)]/27 = 0.74 \, \text{mm}$$

and, hence,

$$
\begin{aligned}
k_e = & \{(0.74E{-}3)(0.49)(168)(1.1E{-}3) \\
& + [(0.6E{-}3)(0.49)][(0.185(0.74E{-}3)(168) \\
& + 0.49(1.1E{-}3)]\}/\{[(0.6E{-}3) \\
& + (0.74E{-}3)][(0.185(0.74E{-}3)(168) + (0.49)(1.1E{-}3)]\} \\
= & \ 2.34 \, \text{W/m K}
\end{aligned}
$$

Also, in a grooved heat pipe, if $P_c = 2\sigma/w$ is conceded to be much smaller than $2\sigma/r_b$, as macroscopic w is compared to microscopic r_b, then

$$
\begin{aligned}
Q_e(W) \approx & \{[2\pi(0.15)(2.34)(290)]/[(1.2E6)(6) \, \ell n(1.15/0.93)]\}\{2(0.0202)/r_b\} \\
= & \ 1.690E{-}5/r_b
\end{aligned}
$$

Now the question is, what value should be assigned to the embryonic radius when it could vary by orders of magnitude depending on any one of various influencing factors? Reported or recommended figures[25,27,37,38] have r_b ranging from 10^{-8} to 10^{-5} m. As is readily seen, when r_b is on the order of 10^{-5} m or greater, the heat input must be limited to less than a couple of watts so as not to exceed the boiling limit. On the other hand, values on the order of 10^{-7} m or smaller would place the restriction strictly in the capillary limit domain calculated earlier as 168.6 W. It is also clear that, even for the same order of magnitude, tolerances can have a marked effect on the predicted value of heating limit. Using the given numerical data, whereas $r_b = 10^{-7}$ m would make the heat pipe borderline capillary limited, the case of $r_b = 0.5 \times 10^{-7}$ m would make the operation definitely capillary limited.

To avoid pointless fencing, many thermal engineers view heating limits in the context of test data that reflect the overall operation of a heat pipe as it is intended to be used in a satellite. Faced with weight and cost limitations to deliver at nominal temperatures large quantities of external heat into a single heat pipe, the radial heat input to grooved heat pipes in NASA satellites has generally been limited to less than 1 W/cm^2 of evaporator area. Deviations from capillary limited operations at this level of heating seem to have been rare and may be traced to a cleaning (wettability) or manufacturing deficiency. The results of a survey (Table 3.3, p. 146 in Ref. 28) testify that adequately cleaned and filled heat pipes of various wicks, using ammonia as working fluid, can sustain heat fluxes in the excess of 5 W/cm^2 of evaporator area while operating in normal heat pipe mode at nominal temperatures.

E.   Thermal Modeling of Satellite Panels with Heat Pipes

Generally, the interior workings of a heat pipe, including startups and other transients, are a separate study and do not enter into thermal modeling. Nodal networks for satellite panels with heat pipes often represent the vapor in each pipe as an isothermal arithmetic node or a set of such nodes connected by very high conductance. The network conductors then trace the heat paths along the electronics mounting platform, across heat pipe bond lines or clamped surfaces; along flanges, tube wall, and grooves; across the wetted surface; into the vapor (evaporation); and out (condensation) to the cooling medium. The analysis assumes that the heat pipe is capable of transporting the heat loads involved and that it is functioning as intended. If required, failures and reduced performance can be simulated by removing the vapor as an active node or by modeling only part of the length as a working heat pipe. (Incidently, some NASA specifications require that the number of heat pipes in a satellite be determined based on permitting a failure of every other heat pipe, or some similar criterion, without a detrimental reduction in the thermal performance.)

Because of uncertainties in the uniformity of bond lines and groove structure and in the value of the convection coefficient of a two-phase fluid flow, tests are often performed on the actual hardware to obtain conductance data to use in the analysis. Usually, the test value is an average of several measurements made at different heat loads and temperatures.

Table 6.5 lists conductances from unpublished results obtained in controlled tests (mostly between 0 and 40°C) on the configurations shown in Fig. 6.16. Most of the data relates to ATS-6 heat pipes. Later tests and monitored temperatures from

Table 6.5   Conductance values from tests on ATS- 6 heat pipes bonded in honeycomb panels,[a] Fig. 6.16

Two heat pipes bonded with 0.075-mm epoxy interface (free or embeded in honeycomb)
  Conductance from vapor node 1 to vapor node 2:

$$K_{v1-v2} = 21.2 \text{ W/K per heat pipe meter}$$

Heat pipe bonded (slightly off center) in 2.0-cm honeycomb panel with 0.63-mm aluminum face sheets; epoxy bond thickness approximately 0.075 mm
  Conductance from face sheet node to diametrically opposite vapor node:

$$K_{fs-v} = 20.4 \text{ W/K per heat pipe meter}$$

Heat pipe partially saddled (centered using 0.075-mm epoxy bond) by $15 \times 3.8 \times 1.3 \text{ cm}^3$ aluminum block
  Conductance from bond line to vapor:

$K_{bl-v} = 9.0$ W/K per $3.8 \times 15 \text{ cm}^2$ of contact area ($= 1579$ W/K per m^2 of contact area)

Block/pipe bonded with 0.075-mm epoxy (slightly off center) into 2.0-cm, 0.50-mm honeycomb panel (aluminum face sheets)
  Conductance from face sheet node to diametrically opposite vapor node:

$K_{fs-v} = 8.4$ W/K per $3.8 \times 15 \text{ cm}^2$ of contact area ($= 1474$ W/K per m^2 of contact area)

Block/pipe saddle bonded to face sheet (thermal grease interface, approximately 0.15 mm thick. Some designs also use bolts) diametrically opposite saddled heat pipe bonded (0.075-mm epoxy bond) into honeycomb
  Conductance from vapor node 1 to vapor node 2 across flanges/blocks/face sheet/bond lines:

$$K_{v1-v2} = 3.9 \text{ W/K over } 15 \times 3.8 \text{ cm}^2 \text{ contact area}$$

Effective conductivity along adiabatic length (vapor node):

$$\mathcal{O}(10^5) \text{ W/m of pipe length per degree Kelvin}$$

Film coefficients:
  At the evaporator: $5500 < h_e < 7000 \text{ W/m}^2 \text{ K of evaporator internal (wet) area}$
  At the condenser: $8000 < h_c < 10000 \text{ W/m}^2 \text{ K of condenser internal (wet) area}$

---

[a]Extruded grooved aluminum heat pipe with square flange; flange: $1.30 \times 1.30 \text{ cm}^2$; tube: 1.15 cm diameter; and 27 grooves with average width $= 0.6$ mm and depth $= 1.1$ mm.

satellites using the same design (NASA's MMS, TOPEX/Poseidon, the command and data handling modules manufactured by OSC, etc.) show general consistency. There is also agreeable correlation with values determined by heat transfer analyses, although analytical data inevitably depend on assumptions relating to properties and the thickness and uniformity of bond lines.

Application: Thermal Interaction Between a Heat Pipe Radiator and a Coolant Fluid Header

Future satellites, with electrical power in the kilowatts range, are envisioned carrying large, deployable radiators with intricate fluid networks for distributing and rejecting waste heat. A number of ideas in promoting such systems have been advanced,[39-42] all proposing heat pipes and other devices that profit from phase

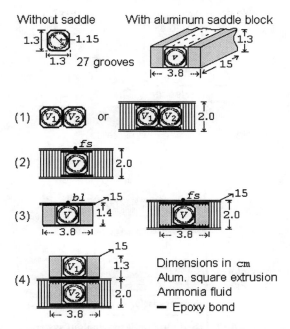

Fig. 6.16   ATS-6 heat pipes: saddled configurations and bonding in honeycomb panels (Table 6.5).

change heat transfer and capillary pumping. One of the greater challenges in this effort has been the mechanical design of stowable systems with flexible joints and interfaces that sustain continuity of heat and fluid flow after deployment.

Some of the principles and limitations associated with these concepts may be revealed by considering the heat pipe radiator–coolant system schematically shown in Fig. 6.17 (also Fig. 6.18). The sketches show a coolant fluid being continuously pumped through a header as it collects heat from a satellite and transfers it to the radiator heat pipes. The discussion is a follow-up on previous work,[43] and the object is to obtain relationships among flow and geometric parameters as a basis for sizing the radiator to satisfy a temperature specification at a given heat load and allowable pressure.

Grooved, ammonia-filled ATS-6 heat pipes are centrally bonded into the honeycomb radiator as shown in Fig. 6.17, with their bent protruding ends saddled and mated to a header adaptor. Spacing is as close as the saddles will permit so that the interface is nearly a continuous line of saddle blocks.

The header design and the method of interfacing with the satellite and radiator are a separate study, which takes into account such factors as available space, compatibility of coolant with wall material, pumping requirements, and the system's pressure and temperature. Also, in addition to having to be consistent with the available pressure head and rate of heat removal, the sizing of the conduit must result in an adequate arithmetic product of the convection heat transfer coefficient and the associated area that will yield the desired temperatures. It should be noted that the finned (4 fins) circular cross section selected in this example may not be the optimum design, and it is understood that the conduit diameter cannot be made

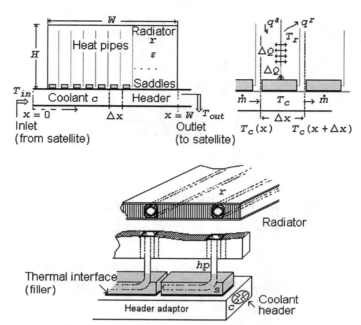

Fig. 6.17   Heat pipe radiator–coolant system.

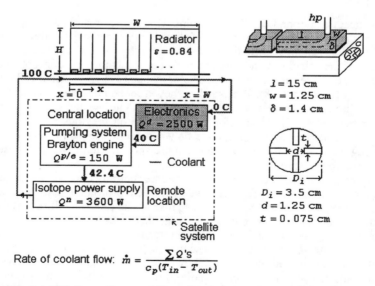

Fig. 6.18   Satellite thermal control by a circulating coolant with heat rejection to a heat pipe radiator; data for numerical example.

too large because of weight and space restrictions and because heat transfer suffers as circumferential temperature gradients increase with diameter (a discussion on optimizing the diameter in terms of entropy generation is found in Ref. 44).

In performing the analysis, steady-state conditions and normal heat pipe operation are assumed, with the saddle region acting as evaporator and the remainder of the pipe as condenser. Both surfaces of the radiator are considered to have the same temperature profile, which is actually the case when the radiator has uniform optical properties and identical heating on both sides, and is nearly always true for thin panels. Also, as is often the approach in this type of analysis, the region defined by Δx in Fig. 6.17 is chosen sufficiently large to encompass at least one whole saddle, but small enough so that a formulation in differential calculus is possible. This technique is exact when saddles are negligibly small, and inaccuracies increase progressively as saddle lengths grow. Comparison with finite element solutions[43] suggests the validity of the procedure when the saddle length is about 1/50 or less of total header length.

Tracing the path of energy packet ΔQ from the coolant c to the radiator r across the interface enclosed by Δx,

$$\Delta Q = -(\dot{m}c_p)\Delta T_c = K(T_c - T_r)\,\Delta x$$

where $\dot{m}$ is coolant mass rate of flow, c_p coolant specific heat at constant pressure, and K (W/K per meter of header length) total conductance from the coolant to the radiator. In differential form

$$\frac{dT_c}{dx} = -[K/(\dot{m}c_p)](T_c - T_r)$$

This is the first of two equations needed to calculate coolant and radiator temperatures. The boundary condition is contained in the heat balance, which equates the total heat to be rejected Q to that released by the fluid as its temperature drops from T_{in} to T_{out} along the header length; that is,

$$Q = \int_0^W dQ = \dot{m}c_p(T_{in} - T_{out})$$

Now, each heat pipe is nearly isothermal throughout its length and, on neglecting edge effects, the isotherms in the radiator become parallel to the heat pipes. Hence, for uniform radiator cross section and no transverse gradients, the lateral temperature distribution is one dimensional and is described by (Chapter 5, Sec. IV.B.)

$$\frac{d^2T_r}{dx^2} + \frac{A^a q^a(x)}{k A^k W} - \frac{\varepsilon A^r \sigma T_r^4}{k A^k W} = 0$$

where W is radiator length along the temperature gradient and the superscripts a, k, and r refer, respectively, to absorbed, conducted, and radiated energy.

A solution to the preceding equation following the linearization procedure discussed in Chapter 5, Sec. IV.B will prove cumbersome due to the unconventional boundary conditions defined by the set of intermittent temperature lines (or heat sources) representing the location of the heat pipes. To reduce the difficulty, an assumption is made to the effect that lateral conduction in the radiator is weak and, therefore, the heat released by each heat pipe is radiated mostly in its own immediate vicinity. This is not a radical departure from actual conditions in honeycomb panels that are lightweight and constructed in separate sections hinged together along lines of large thermal resistance. Accordingly, the differential equation is replaced by

$$K(T_c - T_r) + A^a q^a(x)/W = \varepsilon A^r \sigma T_r^4/W$$

This is the second equation that will allow solving for T_c and T_r in terms of the heat to be rejected and the system's flow and geometric parameters.

Solutions in simple, compact form are found by substituting

$$T_r^4 \approx T_m^4 + 4T_m^3(T_r - T_m)$$

with T_m conveniently defined by

$$\sigma T_m^4 \equiv \frac{1}{\varepsilon A^r}\left[Q + \int_0^W q^a(x)\, dA^a\right]$$

The results for a rectangular radiator of length W (along the header) and height H (approximately the length of heat pipe; $dA^a = H\, dx$) with uniformly absorbed environment flux U are

$$T_c = C_1 + (T_{in} - C_1)\exp(-C_2 x)$$

and

$$T_r = C_3 + C_4 T_c$$

where

$$C_1 = \left(U + 3\varepsilon\sigma T_m^4\right)/4\varepsilon\sigma T_m^3$$

$$C_2 = [K/(\dot{m}c_p)]\left[4\varepsilon\sigma T_m^3/\left(4\varepsilon\sigma T_m^3 + K/H\right)\right]$$

$$C_3 = \left(U + 3\varepsilon\sigma T_m^4\right)/\left(4\varepsilon\sigma T_m^3 + K/H\right)$$

$$C_4 = (K/H)/\left(4\varepsilon\sigma T_m^3 + K/H\right)$$

and

$$\varepsilon\sigma T_m^4 \equiv [Q + UWH]/WH$$

Emissivity ε and environment heating U must reflect the radiator's shape factors to space. Thus, in calculating the mean temperature, U may refer to heating on one side of the radiator whereas ε is the sum of the emissivities if both sides are viewing space.

Total conductance K from coolant fluid to radiator is calculated according to the combination rules described in Chapter 2, Sec. V; that is,

$$K = 1/[(K_{c-s})^{-1} + (K_{s-hp})^{-1} + (K_{hp-r})^{-1}]$$

where c, s, hp, and r are coolant, thermal interface between header and saddle, heat pipe vapor, and the radiator surface. As required by the definition of conductance, the separate quantities must have consistent definitions in terms of geometry, which means that when K is in W/K per meter of header (or radiator) length then, with no peripheral temperature variation in the header adapter,

$$K_{c-s} = hA^c$$

where h is internal convection coefficient and A^c the wetted perimeter over 1 m of header length. Similarly, K_{s-hp} and K_{hp-r} must have values for the particular saddled heat pipe configuration but must be modified to indicate the interface geometry along header length. For the configuration and data in Table 6.5, an isothermal saddle of length 15 cm along the header yields

$$K_{s-hp} \approx K_{bl-v}(100/15) \text{ W/K per meter of header length}$$

where, according to the case of a heat pipe partially saddled in Table 6.5, $K_{bl-v} = 9.0$ W/K over a length of 15 cm and width equals 3.8 cm. Also,

$$K_{hp-r} \approx 2K_{fs-v}(100/15)H \text{ W/K per meter of header length}$$

where H is heat pipe length (heat pipes are assumed to be 15 cm apart) and, in accordance with the case of a heat pipe bonded slightly off center in Table 6.5, $K_{fs-v} = 20.4$ W/K per meter of heat pipe length. The factor 2 on the right-hand side comes from the assumption that the heat pipe is coupled equally to both face sheets of the honeycomb radiator. Therefore,

$$K = 1/[1/hA^c + 1/60 + 1/272H]$$

Figure 6.18 shows the application to the cooling of a nuclear-powered communications satellite, in which the circulating working fluid (helium/xenon mixture) of the radioisotope thermodynamic cycle is used for the thermal control of a shelf on which standard electronics are mounted. The satellite is in GEO, and precision pointing maintains the radiator nearly free of solar impingement. The coolant is known to leave the nuclear source at 100°C, and it is required that the electronics mounting platform be maintained between 0 and 40°C. The operating pressure is $2.62E5$ N/m^2 and pressure loss external to the satellite system must not exceed 25% of that value, or $6.55E4$ N/m^2. The coolant properties as function of temperature are given in Table 6.6.[45]

The flow loop in Fig. 6.18 satisfies the temperature requirements for the noted power allocation. One would expect that electronics of higher dissipation would be placed on a shelf nearer to incoming flow, whereas low and nondissipators would be placed farther downstream. In a more general situation, the isotope power supply and the electronics may have conflicting temperature requirements, which could possibly be resolved by incorporating bypass conduits in which the coolant is made to flow at different rates in each channel. But in this example there is one continuous mass flow rate that must satisfy the energy conservation

$$\dot{m} = \sum Q_i/c_p(T_{in} - T_{out}) = (2500 + 150 + 3600)/[(247.86)(100)]$$

$$= 0.252 \text{ kg/s}$$

Table 6.6   Properties of He/Xe mixture (mol wt 83.8)

T, °C	c_p, J/kg K	ρ, kg/m^3	k_f, W/m K	μ, Ns/m^2	Pr, $\mu c_p/k_f$
−17.8	247.86	6.08	0.0216	$2.08E{-}5$	0.239
40.0	247.86	6.08	0.0260	$2.57E{-}5$	0.245
96.1	247.86	6.08	0.0299	$2.98E{-}5$	0.247
200	244.22	5.92	0.0353	$3.67E{-}5$	0.254

For the finned circular header in Fig. 6.18, the following quantities are calculated.

Flow area:

$$A_{\text{flow}} = (\pi/4)(3.5/100)^2 - 2(3.5 - 1.25)0.075/(100)^2 = 9.28E{-}4\,\text{m}^2$$

Wetted perimeter:

$$\pi(3.5/100) + 4(3.5 - 1.25)/100 = 0.20\,\text{m}$$

Convection surface area:

$$A^c = 0.20\,\text{m}^2 \text{ per meter of header length}$$

Hydraulic diameter:

$$D_h = 4(9.28E{-}4)/0.20 = 0.0186\,\text{m}$$

Coolant average speed:

$$u = \dot{m}/\rho A_{\text{flow}}$$
$$= 0.252/\{6.08[(\pi/4)(3.5/100)^2 - 2(3.5 - 1.25)0.075/(100)^2]\}$$
$$= 44.7\,\text{m/s}$$

Reynolds number at average temperature equal to 50°C:

$$Re = \rho u D_h/\mu = (6.08)(44.7)(0.0186)/2.64E{-}5$$
$$= 191479 > 100{,}000\,(\text{turbulent flow})$$

Speed of sound at the given system's pressure and average temperature:

$$u_s = (P/\rho)^{1/2} = (2.62E5/6.08)^{1/2} = 207.6\,\text{m/s}$$

Mach number:

$$M = u/u_s = 44.7/207.6 = 0.22 < 0.3\,(\text{incompressible flow})$$

Friction factor (from Table 2.4):

$$f = 0.046Re^{-1/5} = 0.046(191479)^{-0.2} = 0.004$$

Pressure drop:

$$\Delta P/W = 2f\rho u^2/D_h = 2(0.004)(6.08)(44.7)^2/0.0186 = 5225.1\,\text{N/m}^2$$

Allowable header length:

$$W_{\text{max}} = \Delta P_{\text{max}}/5225.1 = 6.55E4/5225.1 = 12.5\,\text{m}$$

Nusselt number for the given order of Prandtl number (see Table 2.4):

$$Nu = 5 + 0.016\,Re_{D_h}^c\,Pr^d$$
$$c = 0.88 - 0.24/(4 + Pr)$$
$$d = 0.33 + 0.5e^{-0.6Pr}$$

At 50°C,

$$c = 0.88 - 0.24/(4 + 0.246) = 0.823$$

$$d = 0.33 + 0.5e^{-(0.6)(0.246)} = 0.761$$

$$Nu = 5 + 0.016(191479)^{0.823}(0.246)^{0.761} = 127.4$$

The Nusselt number remains approximately constant in the temperature range of 0–100°C.

Average convection coefficient:

$$h = Nu\,k/D_h = (127.4)(0.0267)/(0.0186) = 182.9 \text{ W/m}^2 \text{ K}$$

Conductance from the coolant to saddle:

$$K_{c-s} = hA^c = 182.9(0.2) = 36.6 \text{ W/K per meter of header length}$$

Conductance from the coolant fluid to the radiator:

$$K = 1/[1/36.6 + 1/60 + 1/272H]$$

$$= 272/(11.97 + 1/H) \text{ W/K per meter of header length}$$

With no environment heating ($U = 0$) and full exposure to space at uniform emissivity $\varepsilon = 0.84$ (Z93 white paint):

$$\sigma T_m^4 \equiv [Q + UWH]/2\varepsilon WH$$

$$= (2500 + 150 + 3600)/(2)(0.84)LH$$

$$= 3720.24/WH \text{ (W/m}^2)$$

$$C_1 = \left(U + 3\varepsilon\sigma T_m^4\right)\big/4\varepsilon\sigma T_m^3 = (3/4)T_m$$

and

$$C_2 = [K/\dot{m}c_p]\left[4\varepsilon\sigma T_m^3\big/\left(4\varepsilon\sigma T_m^3 + K/H\right)\right]$$

$$= \{272/[(11.97 + 1/H)(0.252)(247.86)]\}\{4(0.84)\sigma T_m^3\big/[4(0.84)\sigma T_m^3$$

$$+ 272/(11.97 + 1/H)H]\}$$

or, by rearranging,

$$C_2 = 4.35H\sigma T_m^3\big/[(1 + 11.97H)\sigma T_m^3 + 80.95] \text{ per meter of header length}$$

Hence, from

$$T_c = C_1 + (T_{in} + C_1)\exp(-C_2 x)$$

the temperature (in Kelvin) at the header outlet is

$$273 = (3/4)T_m + [373 - (3/4)T_m]\exp(-C_2 L)$$

This equation is the constraint connecting W (header or radiator length) to H (heat pipe length beyond 15-cm evaporator). When one is defined, the other is uniquely determined to satisfy the temperature requirements. The restriction remains that W be less than 12.5 m in accordance with the specifications for pressure loss.

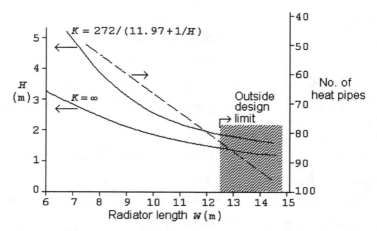

Fig. 6.19 Heat pipe length (radiator height) vs header length (radiator length), which ensures 0°C coolant into satellite electronics compartment; pressure constraints limit the header length to 12.5 m; and the number of heat pipes is based on 15-cm intervals.

The results are summarized in Fig. 6.19. The number of heat pipes at 15-cm intervals is indicated on the right-side abscissa. The limitations of transporting heat at the interfaces are pointed out by comparing with the case where the conductance from the coolant to the radiator is infinitely large, or

$$C_{2,K\to\infty} = 4\varepsilon H\sigma T_m^3 / \dot{m}c_p = 4(0.84)H\sigma T_m^3 / [0.252(247.86)]$$

$$= 0.0538 H\sigma T_m^3 \text{ per meter of header length}$$

It is concluded that a conventional construction with nominal conductances and using heat pipes of moderate lengths (1–2 m) will require that the header length be almost equal to the permitted upper limit. This leaves little or no margin for accommodating a higher satellite dissipation or perturbations from environment heating. Longer heat pipes may be considered, but then one must contemplate the possibility that they may not be capable of transporting the required amount of heat within the capillary limit.

For example, suppose mechanical and packaging constraints limit the header length to a maximum of 8 m. Then, from the curves, some 53 heat pipes, each having about 3.16-m condenser length would be needed. In this case,

$$\sigma T_m^4 = Q/2\varepsilon W H = \{6250/[(2)(0.84)(8)(3.16)]\} = 147.16 \text{ W/m}^2$$

$$T_m = (Q/2\varepsilon\sigma W H)^{1/4} = \{6250/[(2)(0.84)(5.67E{-}8)(8)(3.16)]\}^{1/4} = 225.7 \text{ K}$$

$$\sigma T_m^3 = 0.652 \text{ W/m}^2 \text{ K}$$

$$C_1 = (3/4)T_m = (3/4)(225.7) = 169.3 \text{ K}$$

$$C_2 = 4.35 H\sigma T_m^3 / \big[(1 + 11.97H)\sigma T_m^3 + 80.95\big]$$

$$= 4.35(3.16)(0.652)/\{[1 + (11.95)(3.16)](0.652) + 80.95\}$$

$$= 0.0844 \text{ per meter of header length}$$

$$K = [272/(11.97 + 1/H)] = 272/(11.97 + 1/3.16)$$

$$= 22.14 \text{ W/K per header meter}$$

$$K/H = 22.14/3.16 = 7.0 \text{ W/K m per header meter}$$

$$C_3 = 3\varepsilon\sigma T_m^4 / (4\varepsilon\sigma T_m^3 + K/H)$$

$$= 3(0.84)(147.16)/[(4)(0.84)(0.652) + 7.0] = 40.35 \text{ K}$$

$$C_4 = (K/H) / (4\varepsilon\sigma T_m^3 + K/H) = 7.0/[(4)(0.84)(0.652) + 7.0]$$

$$= 0.7616$$

and the temperature distributions are

$$T_c = 169.3 + (373 - 169.3)\exp(-0.0844x)$$

$$T_r = 40.35 + 0.7616T_c$$

Results are shown in Fig. 6.20a. Heat pipe vapor temperature is calculated from

$$K(T_c - T_r) = K_{\text{hp}-r}(T_{\text{hp}} - T_r)$$

with $K = 22.14$ W/K per header meter and

$$K_{\text{hp}-r} = 272H = (272)(3.16) = 859.52 \text{ W/K per header meter}$$

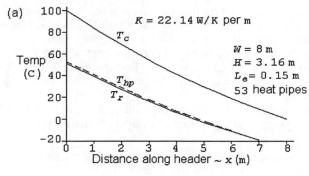

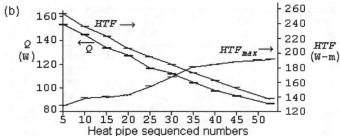

Fig. 6.20   Thermal performance of heat pipe radiator–coolant system: a) temperature distribution based on coolant inlet and outlet at 100 and 0°C, respectively and b) heat transport and heat transport factor of individual heat pipes.

The vapor is within 1.3°C (52.7°C vs 51.4°C at $x = 0$) to 0.6°C (−24.1°C vs −24.7°C at $x = 8$ m) of local radiator temperature.

Heat transport in each heat pipe is obtained by defining the header section for heat transfer as the pipe's saddle length with average temperatures at that location. That is, from

$$\Delta Q = K(T_c - T_r)\Delta x$$

the input to each heat pipe with saddle between $x = a$ and $x = a + 0.15$ m may be calculated from

$$\Delta Q|_{x=a \text{ to } x=a+0.15} = 22.14(T_c - T_r)_{x=a+0.075}(0.15)$$

If the pipes are sequenced from 1 to 53 from inlet to outlet, then heat transport to heat pipe n is

$$\Delta Q|_{\text{hp number } n} = 22.14(T_c - T_r)_{x=0.075+0.15(n-1)}(0.15)$$

The results are shown in Fig. 6.20b, where the curves are drawn in a fashion to indicate that only integer values apply along the ordinate. The heat transport factor is obtained from

$$HTF|_{\text{hp number } n} = 0.5(L_e + L_c)\Delta Q|_{\text{hp number } n}$$

$$= 0.5(0.15 + 3.16)\Delta Q|_{\text{hp number } n} = 1.655\Delta Q|_{\text{hp number } n}$$

Comparison is made with the results of the numerical example, which provides data on the expected maximum heat transport capability of an ammonia heat pipe having the same cross section as that described here. It is discovered that the calculated heat transport factor for the present length (3.31 m total, of which 15 cm is evaporator) exceeds the limiting values in all pipes sequenced 1–33. The remaining units (34–53) operate below the potential limit. Table 6.7 is a partial list of the calculations. The limiting values of the heat transport factor (HTF_{max}) for each heat pipe are found from applying the procedure in the numerical example for each known T_{hp}.

An interesting upgrade would be to shape the radiator to accommodate heat pipes of increasing lengths along header flow direction. But nonuniformity entails

Table 6.7   Calculations of limiting values of heat transport factors

Heat pipe no.	Location x, m	T_c, K	T_r, K	T_{hp}, K	HTF, W-m	HTF_{max}, W-m
1	0.075	371.7	323.4	324.6	265.5	116.1
10	1.425	349.9	306.8	307.9	236.8	139.4
20	2.925	328.4	290.5	291.5	208.4	143.0
30	4.425	309.5	276.1	277.0	183.5	167.7
33	4.875	304.3	272.1	272.9	177.0	175.4
34	5.025	302.6	270.8	271.6	174.8	177.8
40	5.925	292.8	263.4	264.2	161.5	184.1
50	7.425	278.2	252.2	252.9	142.8	190.6
53	7.875	274.1	249.1	249.7	137.4	192.0

complexities in analysis and test and additional expense in tooling and manufacturing. Moreover, other limits will inevitably emerge with increased power demands. The more direct approach is to develop heat pipes or other devices with larger heat transport capabilities.

F. Systems Level Evaluation of Heat Pipes

The following are among the factors that influence the decisions on selecting and integrating satellite heat pipes:

1) Adequate facilities and personnel experience must rank high in the process of selecting a supplier. Only experience offers full awareness of the intricacies and subtleties that go into the making of heat pipes, especially the crucial tasks of cleaning and filling.

2) Similar considerations apply when heat pipes are integrated into satellite panels. The critical factors here are bond thickness and continuity and the straightness and flatness of the heat pipe layout. Flatness is a prelude for reducing gravity effects during ground testing.

3) Embedding heat pipes into radiator panels removes most of the difficulties associated with the application of thermal control coatings and the mounting of dissipators on flat surfaces.

4) That a heat pipe (or any other device that operates on phase-change principles) can internally pump enormous amounts of heat may not be as significant as efficient transport of much lesser amounts from and into the pipe. Unlike a test configuration where heat is supplied by wires or heater sleeves wrapped around the evaporator and removed by direct coupling to cooling fluids, electronics heating and radiation in satellites are usually spread over a large area, and exchanging heat with a heat pipe can present serious design and weight problems. It may prove simpler and more effective to distribute component heating among more than one heat pipe, each carrying a nominal heat load, and with a spacing consistent with the required temperature distribution.

5) A frequent NASA requirement is that a heat pipe layout in a satellite, especially when the grooved wick is being used, must conform to testing in Earth's gravity. It is conceivable that certain options for a good thermal design would have to be rejected to meet this requirement. It is recommended in such situations to determine whether noncritical test objectives should be compromised and permit testing even though some heat pipes would be inactive or operating at a reduced level. If the effect on the overall operation is not too radical, then intended flight performance in zero gravity may be extrapolated by analysis.

G. The Capillary Pumped Loop (CPL)

A solution to the problem of using capillary action to transport large amounts of phase change heat over long distances may be found in the capillary pumped loop (CPL), shown in Fig. 6.21. The operation has been confirmed in ground tests and, to some degree, in the STS space shuttle; but despite its great promise and the many improvements made since the early 1980s, there has yet to be conclusive verification of long-term performance in space. An opportunity was lost in 1994 when the COMET spacecraft, which carried a CPL in its service module,[46] was destroyed on ascent. The next earliest event will be with the EOS-AM,[42] which is slated for launch in 1998.

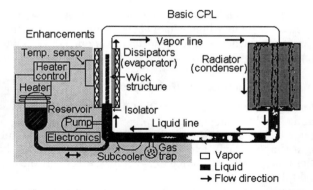

Fig. 6.21 CPL: wick capillary pressure drives the vapor to the radiator where condensation and heat rejection take place; vapor pressure is sufficiently high to loop the liquid back to the evaporator where the process repeats; enhancements improve reliability and provide strict temperature control.

Progress on CPL activities has been periodically assessed in gatherings arranged by NASA Goddard as workshop sessions. An informative package containing presentations from the third CPL workshop, May 19–21, 1992 (no designated number) is available and might be obtained through NASA's thermal departments. Other discussions continue to appear in various technical publications (see, for example, Refs. 42 and 46–50) and Chapter 9 of *Heat Pipe Science and Technology*[27] is devoted to the CPL and the Russian counterpart known as loop heat pipe [(LHP), Fig. 6.22].

Viewed as a specialized heat pipe, the CPL avoids the conflicting requirements of maximum capillary pressure and least resistance to condensate flow by confining

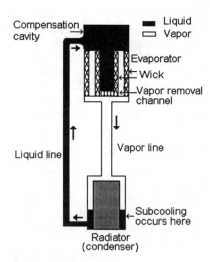

Fig. 6.22 Loop heat pipe (LHP): the compensation cavity is integral to the evaporator and functions as a reservoir.

the wick structure only to the evaporator. Here, capillary pressure drives the vapor along a header to the condenser and continues to force the condensate to loop around back into the evaporator where the cycle repeats. Other than the evaporator section, the CPL flow headers can be made of smooth, low-friction tubing (usually in the range of 1–2 cm diameter). With the latest developments in wick structures, capillary pressures of 4700 N/m^2 have been sustained over a 15-m transport length in ground testing.[42] This compares with less than 100 N/m^2 in 1-m conventional grooved heat pipes.

In theory, only the portion of the system referred to as basic CPL in Fig. 6.21 is needed to collect dissipated heat at the evaporator for transport and rejection at the radiator. Reliability and operation improve when the pressure drop is reduced by sharing the heat load among multiple evaporators and condensers with parallel interconnections, and by incorporating such enhancements as a liquid reservoir, a backup pump, an isolator to prevent vapor backflow, a gas trap to collect non-condensables, and a subcooler to collapse any remaing bubbles in the liquid line. Among these items, the reservoir seems to have become standard in the CPL, as it is found indispensable for injecting liquid into the evaporator (by increasing the pressure with heater power) to initiate the operation and to replenish (reprime) the wick in the event of a burnout. The wick itself is constructed with a central channel to accept the liquid and allow it to migrate radially to soak the capillary pores.

The reservoir can also provide strict temperature control, which may be automated by controlling the heater with thermostats or computer-activated relays. Thus, a pressure change in the evaporator due to a varying heat load is compensated by liquid flow into or out of the reservoir. A somewhat different arrangement of flow lines will, in addition, make the liquid flow from and into the reservoir to compensate for pressure imbalance from heating effects in the condenser. Hence, with sufficient vapor space, heater-induced pressures can maintain the temperature in the evaporator within a narrow range despite large variations in dissipation and environment heating.

Heat transfer from a CPL to an associated heat rejection system has not been fully addressed in the literature. As may be surmised from the analysis of heat pipe radiators, the design of an efficient thermal interface to transfer large amounts of heat between a fluid and a radiator is not simple and presents real engineering and weight problems. In this respect, the CPL may offer a singular advantage as its high capillary pressure permits manifold condensation lines with narrow spacing to maximize the heat rejection per unit radiator area.

The loop heat pipe is schematically shown in Fig. 6.22. It has a lower heat transport capability than the CPL and is different in that it has a liquid section (compensation cavity), which inherently acts as a liquid reservoir. According to Faghri (Ref. 27, p. 621 and Table 9.1), the LHP has been tested successfully in Russia without specialized priming provisions at moderate powers. The action is passive, however, and there is less temperature control with changing power than in a CPL. Presumably, one may consider a more complex system by appending to the compensation cavity a reservoir with vapor space and heater control.

H.   The Variable Conductance Heat Pipe (VCHP)

Another temperature control device with phase change is the passive VCHP described earlier in Sec. IV of this chapter (see Fig. 6.9). A gas-loaded VCHP self-compensates for changes in heat input by the sweep of a noncondensable gas

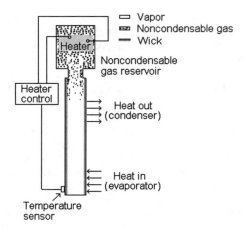

Fig. 6.23 Gas-loaded variable conductance heat pipe with feedback control: heater control reacts to signal from temperature sensor; when heater is on, pressure rise drives noncondensable gas into condenser region cutting back on area for rejecting heat; with heater off, pressure decrease recedes gas into reservoir leaving more area for heat rejection.

front. When heating (in the evaporator) is low, the reduced pressure causes the gas to move from a reservoir into the condenser section, thereby reducing the area of heat transfer to the condensing medium. On increased heating, the front retreats toward the reservoir exposing more area for heat rejection. The action follows the Clausius–Clapeyron equation, which predicts a small temperature band within a large pressure change. This translates into a wide span in active condenser area to accommodate large variations in heating with only a slight change in temperature. Finer and adjustable control can be achieved by adding the heater/sensor system shown in Fig. 6.23, which depicts automatic manipulation of pressure (by applying or removing heater power) in response to a temperature signal.

Another configuration operating along the same principles is the excess-liquid VCHP (second from top in Fig. 6.24). Here, pressure change with heating expands or contracts a fluid bellows immersed in a liquid compartment and forces a flow that adjusts the active area in the condenser. In the vapor-flow modulated VCHP (lowermost in Fig. 6.24), a bellows is fitted with a throttle valve to modulate the amount of vapor into the condenser. These and many other equally inventive devices are described in the literature[25,27] with proposed analytical methods for predicting the position of the gas or liquid front and transient behavior under varying heat loads.

Variable conductance heat pipes in satellites have been limited and appear confined to experiments[51] and perhaps classified military programs. One disadvantage in using them for electronics platform thermal control is loss of real estate for placing components. Much of the radiating area that is active in a maximum heating operation will be at very low temperature during a cold equilibrium condition and probably harmful as a mounting interface for standard electronics. Because this limitation does not generally apply to external radiators free of electronics, the VCHP may prove to be of great benefit in the thermal control of large radiators for high-power satellites.

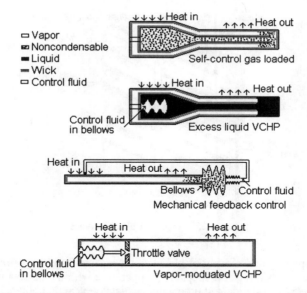

Fig. 6.24 **Variations on the variable conductance heat pipe.**

This last point may be appreciated by referring to the discussion on the heat pipe radiator (Application, Sec. IV of this chapter). If the radiator is oversized to accommodate perturbations in satellite and environment heating, then thermal control during lower heating modes might require, at least intermittently, prohibitively large quantities of electrical heater power to keep the coolant's inlet temperature to the electronics above the allowable limit. An alternative would be to replace some of the conventional heat pipes located toward the header oulet with VCHPs with feedback controls that react to the sensed temperatures in that region. This would effectively adjust the radiating area to produce the desired temperature as the coolant enters the electronics region.

V. Louvers

Louvers are mechanical devices that, in effect, regulate the area of a radiator in response to its temperature. The regulation is preset to accommodate a wide range of heating within a relatively small change in temperature.

The louvers set used most is the venetian blind, shown in Fig. 6.25. It is a framed array of highly reflective blades, with central shafts that fit tightly into the center of bimetallic spring actuators calibrated to expand or contract to various angular positions at prescribed temperatures. The actuators are enclosed within an actuator housing (Fig. 6.26), which is thermally coupled to the radiator but protected and isolated from the external environment by a housing cover. As radiator heating increases, the rise in the temperature within the housing warms up the actuators, which then generate thermal torques that rotate the blades toward their open position leading to an increase in radiation to space. When the radiator temperature decreases, the actuators move the blades toward their closed positions, blocking more of the radiator's view to space and increasing the resistance to radiation. This

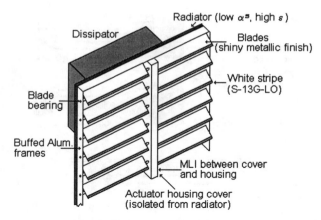

Fig. 6.25   Venetian blind louvers.

automatic opening and closing of blades compensates for changes in dissipation and environment heating and keeps the temperature within a narrow band.

Reference 52 is a standard source with test data on venetian blinds louvers. The following is an updated and expanded version of that work. Brochures and design details may also be obtained from Orbital Sciences Corporation.

A.   Louvers Components

The descriptions in Figs. 6.25 and 6.26 originate from what is generally referred to as the Fairchild louvers. Different sizes have been built, some with single rows

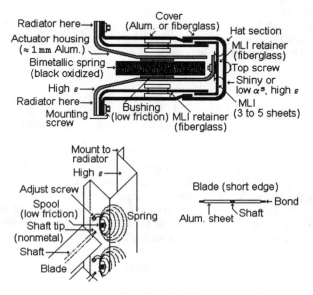

Fig. 6.26   Louvers' actuator and housing assembly.

of blades and others of different aluminum types and thicknesses, but all units have essentially the same components and operate on the same principle.

A standard blade is formed from two stamped, approximately 0.15-mm thick 5052-H38 aluminum sheets cut into 5-cm widths and lengths that depend on the application. The sheets are bonded along their longitudinal edges and to a hollow central aluminum shaft with a solid Vespel (Kapton) tip that fits into a Teflon (or Vespel) spool at the center of the bimetallic (lower left in Fig. 6.26). The other end of the shaft slides into a nonmetallic, low-friction blade bearing assembly with sufficient looseness (end play) to avoid a mechanical bind. The nonmetallic interfaces reduce heat conduction, especially between the blade and actuator.

Before installation within the frame, the blades are polished or buffed to high specularity and, for lower temperature in the sun, a narrow stripe of white paint is laid near the edge to decrease the overall ratio of solar absorptivity to emissivity (Sec. V.D of this chapter). Also, most designs have anodized (high-emissivity) blade interior and the unbonded edges closed by bending an extension in the aluminum sheet. These provisions aid uniformity of blade temperature and reduce thermal distortions.

Polished 2024-T4 aluminum is often used as frame material. The end frames (parallel to the blades rotational axis) support the actuator housing, whereas the side frames (parallel to the housing) tie the actuator assembly and the blades into the framed set. Short stubs on the inside of the side frame prevent the blades from rotating beyond 90 angular deg. In many satellites, neighboring MLI blankets on the structure are made to overlap the frames.

As shown in Fig. 6.26, the actuator housing consists of two separate sections (usually 6061-T6 aluminum, prerferrably 1.0 mm thick or better) tied on two ends by alignment screws. The inside surface is treated for high emissivity (Z306 Chemglaze black has been used) to enhance radiation exchange with the radiator and actuators, and insulating fiberglass washers at the interface with the end frame members are used to reduce their thermal influence. When mounted on a satellite radiator, the louvers set is secured mainly by the housing screws, with good conduction (usually using a filler) along the entire mating length with the radiator. In this way, and by reducing as much as possible the thermal interaction with the frames and the external environment, the actuator housing becomes a radiation enclosure for the actuators at approximately the same temperature as the radiator's. Modern louvers designs and mounting procedures result in less than a 2°C difference between the temperatures of the spring and the facing location on the radiator.

The cover for the actuator housing is not a load-carrying member and may be nonmetallic. It is lined on the inside with MLI (usually three to five sheets of 7.5-μm aluminized Kapton and Dacron mesh interspacers) with retainers to keep the sheets separated but secured. The hat section has its own MLI lining and can be removed (by removing the top screw) for direct accessibility to an adjustment screw used to position the blade at a specified angle for a given temperature. The process is known as louvers calibration and is performed in a temperature controlled air chamber as described in Sec. V.C of this chapter.

The bimetallic in the Fairchild louvers is a GTE 6650 spiral spring manufactured by the Metal Laminates Division of W. M. Chace Company.[53] A chemical treament (oxidizing is common) is usually added to increase the emissivity. A schematic and some characteristics are given in Fig. 6.27. The figure caption points out the

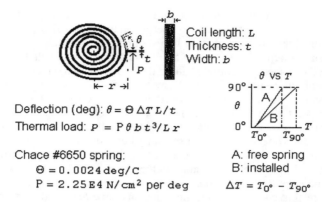

Coil length: L
Thickness: t
Width: b

Deflection (deg): $\theta = \Theta \, \Delta T \, L/t$

Thermal load: $P = P \theta \, b \, t^3/L \, r$

θ vs T

A: free spring
B: installed

$\Delta T = T_{0°} - T_{90°}$

Chace #6650 spring:
$\Theta = 0.0024 \, \mathrm{deg/C}$
$P = 2.25 \, \mathrm{E4} \, \mathrm{N/cm^2}$ per deg

Fig. 6.27 Louvers' spring deflection and thermal load; for standard spring, $L = 68.5$ cm, $r = 1.78$ cm, $b = 0.56$ cm, and $t = 0.25$ mm give $\theta = 90$ deg and $P = 0.15$ N for $\Delta T = 13.7°$C ($19.0 \pm 1.5°$C installed); for large spring, $L = 112.0$ cm, $r = 2.0$ cm, $b = 1.15$ cm, and $t = 0.25$ mm give $\theta = 90$ deg and $P = 0.16$ N for $\Delta T = 8.5°$C ($10.0 \pm 1.0°$C installed).

minuscule forces resulting from induced spring deflection. Hence, any frictional effect on blade motion must be reduced to the lowermost possible level. Improvements in this area have been made by utilizing materials with low coefficients of friction and by forming the spring in a manner that eliminates contact between spirals. But some influence remains, as revealed by observed hysteresis and hesitation in blade motion and in the larger span in temperature (additional 2–5°C) to rotate the spring by 90 deg when installed. These effects are normally incorporated in the input data when doing a worst-case thermal analysis.

Sun shields have been used in some applications to protect louvers parts from the high temperatures associated with specular metallic finishes in the sun. A standard sun shield is made of thin (0.15–0.5 mm) aluminum substrate, welded into a shallow box (about 6 mm deep), which is attached by nonmetallic screws to the frame. Other designs have a bent lip on the side frames along which a flat shield can be slid and secured. The shield's outer, space facing surface is treated with a coating of low solar absorptivity and high emissivity, such as white paint or silvered Teflon (quartz mirrors on a honeycomb substrate have also been used). A high-emissivity coating, such as Chemglaze Z306 black paint, is applied on the inner surface, facing the louvers, to enhance heat exchange with the radiator.

B. Effective Emissivity and Effective Solar Absorptivity

Heat transfer from a louvered radiator at temperature T is characterized by an effective emissivity $\varepsilon_{\mathrm{eff}}$ and an effective solar absorptivity α^s_{eff} that satisfy the following steady-state energy equation in a solar–space environment:

$$q_{\mathrm{louver}} = \varepsilon_{\mathrm{eff}} \, \sigma T^4 - \alpha^s_{\mathrm{eff}} S$$

where q_{louver} is heat radiated per unit radiator area enclosed within (inside) the frame of the louvers set. It is important to note that in this definition the solar input is solar constant S (nominally 1350 W/m^2) not modified by the direction cosines, so that the effect of the multiple reflection of sun rays in the cavities between pairs of blades (or blade and frame) is intrinsic to the value of α^s_{eff}. This is in step with the methods of analysis and test used to determine effective emissivity and solar absorptivity.[54]

Although the preceding characterization does not include influences by external heat sources other than the sun, testing shows that at least ε_{eff} has a role fairly consistent with an equivalent diffuse gray surface emissivity. However, a generalized meaning for α^s_{eff} is much more difficult to advance or confirm mainly because of insufficient test data.

Effective emissivity of an ideal louvered radiator of uniform blade angles and with radiation the only mode of heat transfer may be determined by analysis using the radiation exchange equations described in Chapter 3. The formulation considers the interaction between the radiator and blades with eventual radiation to space. For highly specular and low-emissivity louvers parts, optics geometry indicates that ε_{eff} is nearly proportional to the projected open area through which the radiator views space. Thus, ideally,

$$\varepsilon_{\text{eff}} = (A_o/A)\varepsilon_r$$

where ε_r is radiator emissivity and A_o/A the ratio of the normal area through which the radiator views space to the total area within the frame. From geometry, A_o is proportional to the sine of blade angle θ, which is approximately a linear function of radiator temperature (see Fig. 6.27). Hence, theoretically, effective emissivity has the form

$$\varepsilon_{\text{eff}} = \varepsilon_0 + (\varepsilon_\pi/2 - \varepsilon_0)\sin\theta$$

where ε_0 and $\varepsilon_{\pi/2}$ are the values at blade angles 0 and 90 deg. Using trigonometric relations and the equation of a straight line that relates θ to temperature,

$$\varepsilon_{\text{eff}} = \varepsilon_0, \qquad \text{for } T \leq T_0$$

$$\varepsilon_{\text{eff}} = \varepsilon_0 + (\varepsilon_{\pi/2} - \varepsilon_0)\cos\left(\frac{\pi}{2}\frac{T_{\pi/2} - T}{T_{\pi/2} - T_0}\right), \qquad \text{for } T_0 \leq T \leq T_{\pi/2}$$

$$\varepsilon_{\text{eff}} = \varepsilon_{\pi/2}, \qquad \text{for } T \geq T_{\pi/2}$$

where T_0 and $T_{\pi/2}$ are radiator temperatures corresponding to 0- and 90-deg blade angles.

However, because of some nonspecular parts, nonuniformities in blade positions, nonideal surface conditions, and unavoidable extraneous heat losses, the most useful values of effective emissivity are those obtained from tests in cryogenic vacuum chambers. A typical arrangement is shown in Fig. 6.28. (Sometimes two sets of louvers are tested at the same time with their insulated backs facing each other to eliminate MLI losses.) The effective emissivity is calculated from measured radiator temperatures T corresponding to a series of equilibrium states at different heater power. By proper insulation and the use of guard heaters, radiation from the louvered area can be made equal to the dissipated heat Q^d and

$$\varepsilon_{\text{eff}} = Q^d / \left[A\left(\sigma T^4 - \sigma T_w^4\right)\right]$$

where T_w is chamber wall temperature (usually ≤ 90 K) and A is area within the frame. Some actual data are plotted in Fig. 6.29. Heat transfer by conduction and

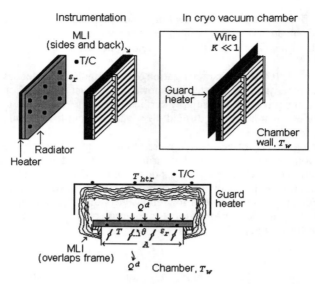

Fig. 6.28 Effective emissivity test: radiator is 0.635-cm-thick aluminum with emissivity ε_r. When thermocouples on guard heater are made to read the same as the radiator's then $\varepsilon_{\text{eff}} = Q^d / [A(\sigma T^4 - \sigma T_W^4)]$.

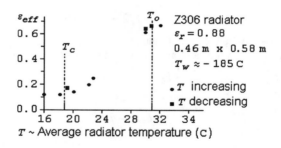

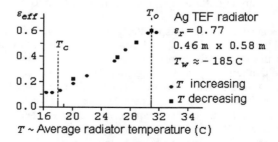

Fig. 6.29 Effective emissivity of unshielded louvered radiators: test data.

from gaps and fin effects, as well as feedback radiation from the frame, blades, and housing cover, are all conveniently included in the test value of ε_{eff}.

Generally, tests show some difference in the effective emissivity between measurements made with increasing and decreasing radiator temperature. This hysteresis is attributed to inconsistent frictional effects on the blade position at it varies with temperature. For most sets, the results in the active regime between closed and open blades can be contained within two straight lines, which are used in analysis to bound the performance. The variation along the bounding lines is given by

$$\varepsilon_{\text{eff}} = \varepsilon_c, \qquad \text{for } T \le T_c$$

$$\varepsilon_{\text{eff}} = \varepsilon_o - \frac{\varepsilon_o - \varepsilon_c}{T_o - T_c}(T_o - T), \qquad \text{for } T_c \le T \le T_o$$

$$\varepsilon_{\text{eff}} = \varepsilon_o, \qquad \text{for } T \ge T_o$$

where ε_c is effective emissivity at radiator temperature T_c, when the blades just begin to open with increasing temperature or first close with decreasing temperature, and ε_o is the value at T_o, when the blades just open fully with increasing temperature or begin to close from fully open position with decreasing temperature. The tolerance selected in performing a thermal analysis is usually biased so that the hot case incorporates a delayed opening whereas the cold case has delayed closure.

A quadratic relation is sometimes found to give a better approximation, especially near the open blades position as the curve approaches an asymptote. In this case,

$$\varepsilon_{\text{eff}} = \varepsilon_c, \qquad \text{for } T \le T_c$$

$$\varepsilon_{\text{eff}} = \varepsilon_o - \frac{\varepsilon_o - \varepsilon_c}{(T_o - T_c)^2}(T_o - T)^2, \qquad \text{for } T_c \le T \le T_o$$

$$\varepsilon_{\text{eff}} = \varepsilon_o, \qquad \text{for } T \ge T_o$$

The theoretical temperature variation with heat generated in the radiator is illustrated in Fig. 6.30. Comparison is made with the performance of passive radiators

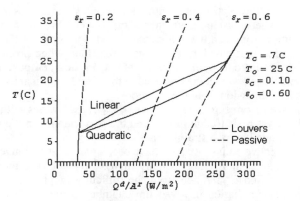

Fig. 6.30 Radiator performance in deep space ($T_\infty = 0$ K), comparison of louvered and unlouvered systems.

of various surface emissivities. The effect of louvers on temperature control can be seen in the flattening of the curves in the active region.

Effective emissivity of radiators with shielded louvers $[\varepsilon_{\text{eff}(s)}]$ is defined in the equation

$$Q^d/A = \varepsilon_{\text{eff}(s)}(\sigma T^4 - \sigma T_\infty^4)$$

where T_∞ is radiation sink temperature. By considering the louvered system to be equivalent to a standard (without louvers) radiator of the same area and temperature and with emissivity ε_{eff}, radiation in terms of transfer to a shield at temperature T_s can be written

$$Q^d/A = \Im_{r-s}(\sigma T^4 - \sigma T_s^4)$$

where $\Im_{r-s}$ may be calculated by the equation of parallel planes in close proximity (Chapter 3, Sec. II.B); that is,

$$\Im_{r-s} = (1/\varepsilon_{\text{eff}} + 1/\varepsilon_i - 1)^{-1}$$

with i indicating interior (facing louvers) shield surface.

An orbital average heat balance on the shield in a space environment gives

$$Q^d/A + \alpha_e^s(q^S + q^A) + \varepsilon_e q^E = \varepsilon_e \sigma T_s^4$$

where e is the exterior (facing space) shield surface. Hence, on eliminating T_s from the equations,

$$\frac{Q^d}{A} = \frac{\varepsilon_e \Im_{r-s}}{\varepsilon_e + \Im_{r-s}} \left\{ \sigma T^4 - \left[\frac{\alpha_e^s}{\varepsilon_e}(q^S + q^A) + q^E \right] \right\}$$

By comparing to the equation defining $\varepsilon_{\text{eff}(s)}$,

$$\varepsilon_{\text{eff}(s)} = \varepsilon_e \Im_{r-s}/(\varepsilon_e + \Im_{r-s})$$

and

$$\sigma T_\infty^4 = (\alpha_e^s/\varepsilon_e)(q^S + q^A) + q^E$$

Testing confirms the relationships within the expected heat losses from fin effects and conduction where the shield attaches to the louvers. In general, the test value of effective emissivity when the blades are closed is approximately the same for both shielded and unshielded systems, indicating that, when properly attached, the shield will not introduce an undue fin effect. In the open position, however, both theory and test show that the presence of the shield significantly increases the resistance to heat emission from the radiator.

As may be seen from Fig. 6.31, profiles of $\varepsilon_{\text{eff}(s)}$ as it varies with temperature show a similar hysteresis to that in ε_{eff}, and the performance in the active region may also be bounded by two straight lines.

Numerous effective emissivity tests have been conducted on louvered radiators of various sizes and emissivities (usually high).[52] By contrast, there have been very few tests to find α_{eff}^s. Aside from the difficulty of producing an accurate simulation of solar radiation in a vacuum chamber, it has not been possible to generalize the test configuration to include solar effects for both in-plane and out-of-plane angles of incidence that would account for differences in area and frame

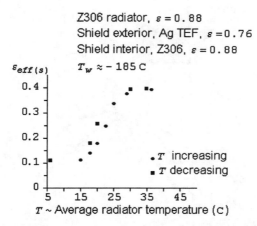

Fig. 6.31 **Effective emissivity of a shielded louvered radiator (test data).**

size. It has been observed, for example, that louvers designed with a relatively deep actuator housing and end frame for accommodating large springs will give reduced absorptivity at shallow sun angles but increased values at higher solar incidence. The specularity of the radiator and the external surface properties of the actuator housing also show measurable effects. But despite these and similar disturbances, the few available data have proven satisfactory as a baseline for making estimates when the radiator is treated with either silvered Teflon or quartz mirrors (data from solar tests with white paints are not available). It should also be mentioned that the general trends observed in testing have been confirmed by ray tracing and radiosity techniques.[54,55]

Solar simulations in a vacuum have employed xenon compact arc type lamps with beam half angle collimation of less than 2 deg. Some data are reproduced in Figs. 6.32 and 6.33. Average values for the case of the solar vector making a complete revolution above the louvers plane are given in Table 6.8. The calculations use predetermined values of ε_{eff} as a function of temperature (or blade angle θ) in the equation

$$\alpha_{\text{eff}}^{s} = \alpha_{\text{eff}}^{s}(\theta, \varphi, \psi) = \left[\varepsilon_{\text{eff}}(\sigma T^4 - \sigma T_w^4) - Q^d/A\right]/S$$

with $\theta = \theta(T)$ and φ and ψ being the aspect of solar vector as described in Figs. 6.32 and 6.33. As noted earlier, simulated solar flux S (usually within 2% of 1350 W/m^2 but wider variations from actual solar spectrum) is not modified by direction cosines.

It is presumed that the values of α_{eff}^{s} and ε_{eff} obtained by these methods can be used in calculations involving transients and heating from sources external to the louvered radiator. It must be recognized, however, that in reality the actuator response to temperature transients in the radiator depends on the type and mass of the spring, and that the blade motion is generally nonuniform and hesitant due to friction. Also, the adding of albedo heating depends on how that radiation is classified with respect to direction of impingement on the louvers. Test information on this interaction is completely lacking and intuition is unavoidable in making judgements. The approach most often adopted is based on assuming diffuse albedo

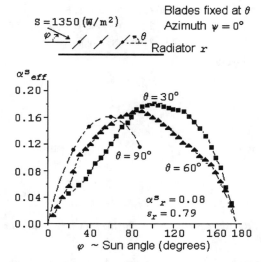

Fig. 6.32 Effective solar absorptivity of louvered radiator, variation with sun angle at fixed azimuth; test data from ATS-6 QUAL (standard spring and actuator housing); quartz mirrors (uncoated) radiator, 0.46 m × 0.58 m; solar vector in plane of paper.

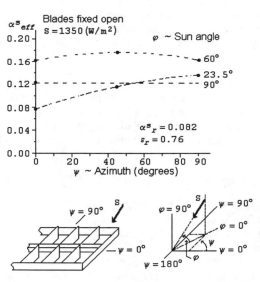

Fig. 6.33 Effective solar absorptivity of louvered radiator, variation with azimuth at fully open blades; test data from INTELSAT CRL (large spring and actuator housing); silvered Teflon radiator, 0.622 m × 0.605 m.

in the same spectrum as solar radiation. The normal component q^A incident on a louvered radiator is then reduced by the effective emissivity (shape factor) and the fraction absorbed made equal to the radiator's solar absorptivity α_r^s. Hence,

$$q^{A,a}_{\text{louvered radiator}} = \alpha_r^s(\varepsilon_{\text{eff}}q^A)$$

and the energy equation of a louvered radiator (dissipation q^d) in space environment with impinging solar, albedo, and Earth heating becomes

$$q^d + \alpha_{\text{eff}}^s S + \alpha_r^s(\varepsilon_{\text{eff}}q^A) + \varepsilon_{\text{eff}}q^E = \varepsilon_{\text{eff}}\sigma T^4$$

or

$$q^d = \varepsilon_{\text{eff}}(\sigma T^4 - \sigma T_\infty^4)$$

with

$$\sigma T_\infty^4 = (\alpha_{\text{eff}}^s/\varepsilon_{\text{eff}})S + \alpha_r^s q^A + q^E$$

Also lacking are data to evaluate the effect of the presence of louvers on the degradation of the radiator's solar absorptivity and the resulting effective solar absorptivity. In determining the degraded value of α_{eff}^s for a louvered silvered Teflon or quartz mirror radiator, the practice has been to keep the ratio $\alpha_{\text{eff}}^s/\alpha_r^s$ constant and equal to the value obtained in test; that is,

$$\text{Degraded } \alpha_{\text{eff}}^s = (\text{test } \alpha_{\text{eff}}^s / \text{test } \alpha_r^s) \text{ degraded } \alpha_r^s$$

Of course, one should expect increasing departure from this relation as degradation increases and specularity changes.

C. Louvers Specifications and Calibration

Louvers performance specifications cite the unit size and the radiator's temperatures and corresponding effective emissivities for open and closed blades. Generally, the requirements are made to agree with historically attainable data, with verification in a vacuum within an achievable tolerance as part of the unit's acceptance criteria.

Specifications for the Fairchild louvers have been guided by data from tests that were performed during NASA's sixth Applications Technology Satellite (ATS-6) and Multimission Modular Spacecraft (MMS) projects, together with research activities funded by NASA GSFC, INTELSAT/COMSAT and Fairchild Space Company (now OSC). Pertinent results are summarized in Table 6.8.

Although calibration for setting blade positions as a function of temperature is performed in air (Fig. 6.34), measurements and observations through portholes in vacuum chambers indicate that, once calibrated, louvers retain reasonable blade uniformity and motion in a cryogenic vacuum. However, there is also observed in a cryogenic vacuum a shift in blade position caused by a difference between radiator and average actuator temperature. The bias is usually quoted in the specifications at $\leq 2°C$, to be verified by analysis and worked into calibration procedures to ensure the required performance in orbit.

Table 6.8   Effective emissivity and solar absorptivity of radiators with Fairchild louvers (test data)

Effective emissivity as function of radiator temperature[a]

Small louvers ($< 20 \times 20$ cm^2):

$\varepsilon_c = 0.154 \pm 0.01$

$\varepsilon_o = 0.65 \pm 0.02$ (Ag TEF radiator, measured $\varepsilon \approx 0.76$)

$\varepsilon_o = 0.68 \pm 0.02$ (quartz mirrors radiator, measured $\varepsilon \approx 0.79$)

$\varepsilon_o = 0.74 \pm 0.01$ (Chemglaze Z306 radiator, measured $\varepsilon \approx 0.88$)

$T_o - T_c = 19$ K ± 1.5 K (nominal spring)

$T_o - T_c = 10$ K ± 1.0 K (large spring, estimated)

Large louvers ($\geq 25 \times 25$ cm^2):

$\varepsilon_c = 0.10 \pm 0.01$

$\varepsilon_o = 0.63 \pm 0.01$ (Ag TEF radiator, measured $\varepsilon \approx 0.76$)

$\varepsilon_o = 0.67 \pm 0.01$ (quartz mirrors radiator, measured $\varepsilon \approx 0.79$)

$\varepsilon_o = 0.72 \pm 0.01$ (Chemglaze Z306 radiator, measured $\varepsilon \approx 0.88$)

$T_o - T_c = 19$ K ± 1.5 K (nominal spring)

$T_o - T_c = 10$ K ± 1.0 K (large spring)

For IR input, adjust by ε_{eff} (that is, absorbed IR = ε_{eff} times normal impinging IR)

For shielded louvers ($\geq 25 \times 25$ cm^2, Chemglaze Z306 on shield interior and Ag TEF on the exterior)

ε_{eff}, ε_c, and ε_o are labeled $\varepsilon_{\text{eff}(s)}$, $\varepsilon_{c(s)}$, and $\varepsilon_{o(s)}$

$\varepsilon_{c(s)} = 0.10 \pm 0.01$

$\varepsilon_{o(s)} = 0.39 \pm 0.006$ (Chemglaze Z306 radiator, measured $\varepsilon \approx 0.88$)

$T_o - T_c = 18$ K ± 2.0 K (nominal spring)

$T_o - T_c = 10$ K ± 1.0 K (large spring, estimated)

Effective solar absorptivity

Quartz mirrors radiator, 45.75×58.2 cm^2, nominal spring, nominal actuator housing and frame depths

Measured radiator $\alpha_r^s \approx 0.08$, $\varepsilon_r \approx 0.79$

Blades fully open ($\theta = 90$ deg), solar vector normal to blade shaft (azimuth $\psi = 0$ deg), sun angle φ off louvers plane:

φ, deg:	23.5,	40,	60,	90
α_{eff}^s:	0.112,	0.146,	0.162,	0.114

Solar vector coning a complete revolution around azimuth ψ, sun angle φ off louvers plane:

φ deg:	0,	23.5,	30,	60
average α_{eff}^s:	≈ 0.0,	0.10 (estimated),	0.117,	0.160

Blades closed: $\alpha_{\text{eff}}^s \approx 0.0$ or slightly negative

Ag TEF radiator, 62.2×60.5 cm^2, large spring, increased actuator housing and frame depths

Measured radiators $\alpha_r^s \approx 0.082$, $\varepsilon_r \approx 0.76$

Blades fully open ($\theta = 90$ deg), solar vector normal to blade shaft (azimuth $\psi = 0$ deg), sun angle φ off louvers plane:

φ, deg:	23.5,	60,	90
α_{eff}^s:	0.078,	0.163,	0.126

(cont.)

Table 6.8   (Continued)

Blades fully open, scalar vector coning a complete revolution around azimuth ψ, sun angle off louvers plane:

φ, deg: 0,   23.5,   60
average α_{eff}^s: ≈ 0.0,   0.109,   0.178

Blades closed:

α_{eff}^s: ≈ 0.0 or slightly negative

α_{eff}^s for other α_r^s, new α_{eff}^s = (old α_{eff}^s /old α_r^s) new α_r^s, with maximum α_{eff}^s = 1.0
For albedo input, adjust f (albedo factor) by ε_{eff} (that is, new $f = \varepsilon_{\text{eff}} f$) and use α^s of
    radiator.

[a]Closed and open c and o; $|T_{\text{radiator}} - T_{\text{actuator}}|$ generally less than 2°C in cryogenic vacuum chambers; all temperatures refer to radiator; optical properties are measured as-applied values; $\varepsilon_{\text{eff}} = \varepsilon_c$(const), $T \leq T_c$; $\varepsilon_{\text{eff}} = \varepsilon_o - [(\varepsilon_o - \varepsilon_c)/(T_o - T_c)](T_o - T)$, $T_c \leq T \leq T_o$; $\varepsilon_{\text{eff}} = \varepsilon_o$(const), $T \geq T_o$.

Typical specifications for selecting and calibrating actuators to be used for thermal control between about 0 and 20°C may read as follows:

1) For selection of actuator springs (preinstallation with steady temperature in liquid bath), verify the same angular displacement ±1.5 deg for intervals from −5 to 5°C, 5 to 15°C, and 15 to 25°C; and 90 deg (+1.5, −0) angular displacement for change from 5 to 19°C.

2) For angular uniformity in air of installed blades, verify that all blades are closed at 0°C ± 0.3°C, all blades are partially opening within ±1.5 deg with temperature increase to equilibrium level 3°C ± 0.3°C, all blade angles are within ±1.5 deg with temperature increase to equilibrium level 10°C ± 0.3°C, all blades are fully open with temperature increase to equilibrium level 21°C ± 0.3°C, and all blades are partially closing within ±1.5 deg with temperature decrease to equilibrium level 17°C ± 0.3°C.

3) For calibration air chamber, temperature control will be from −20 to 100°C with ±0.3°C uniformity.

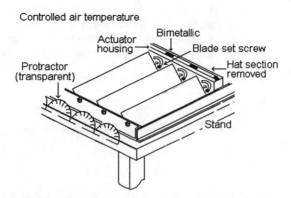

Fig. 6.34 Louvers' calibration: set is placed on a test stand in air chamber with the hat section removed and spring tension released; at prescribed temperatures, the tension is readjusted and secured to align the blades along the desired angles.

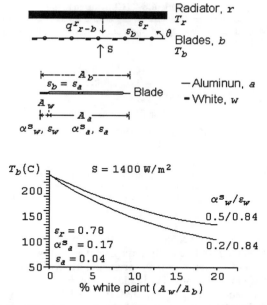

Fig. 6.35 Effect of white stripe on louvers' blade temperature, average properties of painted blade are $\alpha_{av} = (\alpha_a A_a + \alpha_w A_w)/A_b$, $\varepsilon_{av} = (\varepsilon_a A_a + \varepsilon_w A_w)/A_b$.

D. Louvers in the Sun–Blade Temperature

It is the high specularity and low emissivity of louvers blades that make them efficient reflectors of the dissipation when they are open and a barrier to radiation when closed. Spectral measurements on 5052-H38 aluminum (polished sheets) used in Fairchild louvers yield the following optical properties: specularity (percent of reflection with angle of incidence equals angle of reflection) $= 0.93 \pm 0.02$, $\alpha^s = 0.15 \pm 0.02$, $\varepsilon = 0.04 \pm 0.01$, and $\alpha^s/\varepsilon = 4.25 \pm 0.25$.

The high ratio of α^s/ε translates into a high temperature in sunlight. A simplistic but revealing evaluation can be made by considering the case of normal solar impingement on a plane of closed louvers. The situation is depicted at top in Figure 6.35. Without an enhancement coating, a heat balance on the blade gives

$$\varepsilon_b \sigma T_b^4 = \alpha_b^s S + \Im_{r-b}(\sigma T_r^4 - \sigma T_b^4)$$

$$T_b = \left\{ [1/(\varepsilon_b \sigma + \Im_{r-b}\sigma)][\alpha_b^s S + \Im_{r-b}\sigma T_r^4] \right\}^{1/4}$$

with subscripts r and b referring to radiator and blade, respectively, and $\Im_{r-b}$ approximately

$$\Im_{r-b} = (1/\varepsilon_r + 1/\varepsilon_b - 1)^{-1}$$

Using $\varepsilon_r = 0.76$ (silvered Teflon), $\alpha_b^s = 0.17$, $\varepsilon_b = 0.04$, and $S = 1400$ W/m^2 (high value solar flux) gives

$$\Im_{r-b} = (1/0.76 + 1/0.04 - 1)^{-1} = 0.0395$$

and, at radiator temperature of, say, 20°C,

$$T_b = \{\{1/[(0.04 + 0.0395)(5.67E-8)]\}\{0.17(1400)$$
$$+ (0.0395)(5.67E-8)(293)^4\}\}^{1/4} = 487\,\text{K}$$

or nearly 215°C.

More complicated computations based on the ideas in Chapter 3 are needed to find the temperature at other solar incident angles and blade angles.[54] Also, to be truly representative, the analysis should refer to the particular hardware being studied, including the properties and dimensions of the frames and actuator housing cover. In this respect, computer modeling using radiosity and ray tracing techniques has been fairly successful in showing trends, but insufficient knowledge on specular properties and behavior along incident and reflected angles often renders the numerical results doubtful and requires serious judgement on the tolerances and biases to be used.

In any event, a high blade temperature in the sun has been confirmed in solar simulation chambers with instrumented blades not cooled with an auxiliary surface finish.[55] Some material softening and warping were evidenced, and there was a measurable effect on actuator tracking attributed to increased heat conduction from the hot blades. But perhaps most crucial was the observation that high temperatures aggravate the outgassing rates of the adhesives used to bond the blade sections together and to the shaft. During the late 1970s, various compounds were assessed at Fairchild Space Company, but none were conclusively judged adequate for meeting both strength and the long-term outgassing and contamination requirements. Further studies in the 1980s considered eliminating adhesives altogether and replacing the aluminum with gold-plated stainless steel with a laser seam weld as the method for forming the blade.

However, the commonly adopted solution to high blade temperature in modern louvers remains the laying of a narrow stripe of white paint near the outer longitudinal edge that is exposed when the louvers are closed. The average solar absorptivity to emissivity ratio is thus reduced, leading to cooler temperatures, as indicated in the curves in Fig. 6.35. Generally, the region with white paint is kept under 15% of the surface area (one side) to avoid exessive disruption in surface specularity.

Figure 6.36 shows test verification of the effect of employing the white stripe. The data were obtained on an early louvers design instrumented and tested specifically to determine the effect.[55] The dashed lines shown connecting the test points follow trends obtained from a computer model applied to that particular set. Occurrence of maximum temperature at a blade angle 60 deg and sun incidence also near 60 deg (note the angular directions in Fig. 6.36) seems to be a consistent characteristic of louvers mounted to radiators with specular properties in the solar radiation spectrum. The effective emissivity is seen to suffer slightly provided the amount of white paint does not exceed some 15% of the blade surface area and is confined to the outermost edge in the open position.

E. Heat Rejection Capability: Unshielded versus Shielded Louvers

Protecting louvers with sun shields introduces substantial resistance to heat rejection from the radiator. On the other hand, a sun shield removes direct solar impingement that, as noted in Fig. 6.32 and the table of properties, magnifies solar

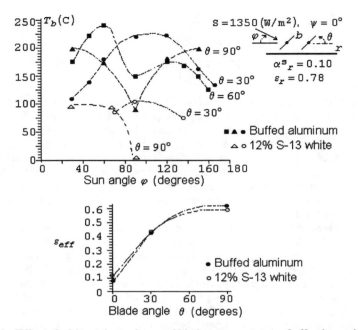

Fig. 6.36 **Effect of white stripe on louvers' blade temperature and effective emissivity, test data from ATS-F test unit 2, with nominal spring and actuator housing, 0.482 m × 0.565 m silvered Teflon radiator.**

heating at increased incident angles. For example, although the solar absorptivity of an OSR radiator may be only 0.08, the effective solar absorptivity with open blades under one solar constant can exceed 0.1 at zero azimuth and 23.5 deg sun incidence. The natural question is whether there are circumstances in which shielding against the sun sufficiently counters the reduction in heat emission.

The two situations with identical dissipation and environment heating are described in Fig. 6.37. The orbital average equations for heat rejection capability at given radiator temperature T are

$$(Q^d/A)_u = \varepsilon_{\mathrm{eff}}\{\sigma T^4 - [(\alpha^s_{\mathrm{eff}}/\varepsilon_{\mathrm{eff}})S + \alpha^s_r q^A + q^E]\}$$

and

$$(Q^d/A)_s = \varepsilon_{\mathrm{eff}(s)}\{\sigma T^4 - [(\alpha^s_e/\varepsilon_e)(q^S + q^A) + q^E]\}$$

with u and s implying unshielded and shielded, respectively. Maximum heat rejection (blades fully open) can, thus, be compared in orbits and orientations that define the incident fluxes.

The comparison is more meaningful when both systems are of the same construction and dimensions, with each designed for maximum heat rejection from the radiator. OSR treatments are common on exposed radiators and shields' exteriors. In a shielded system, the solar absorptivity of the interior has no bearing on the result, and high-emissivity black paints are commonly used on the radiator and the inner surface of the shield. Typical properties are, for unshielded louvers, quartz

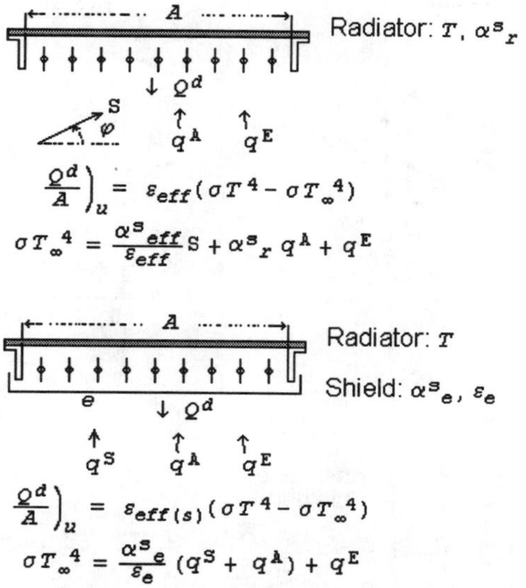

Fig. 6.37 Heat balance on unshielded and shielded louvered radiators.

mirrors radiator: $\alpha_r^s/\varepsilon_r = 0.08/0.79$, $\varepsilon_{\text{eff}} = 0.67$, and α_{eff}^s (average value with sun coning a full revolution at 23.5 deg off the louvers plane) $= 0.10$. For shielded louvers, Z306 radiator and shield interior, and silvered Teflon shield exterior, the properties are $\alpha_e^s/\varepsilon_e = 0.082/0.76$ and $\varepsilon_{\text{eff}(s)} = 0.39$.

It is clear from the relative values of effective emissivity that an unshielded louvered radiator has a correspondingly higher heat rejection rate in a wholly IR surrounding, including the case when radiation is emitted to space with no external heating. Dependence on temperature for this case is shown in Fig. 6.38a. Theoretically, based on the preceding values the unshielded system is over 70% more effective at all temperatures for which the rejection is positive; this is concluded from

$$(Q^d/A)_u/(Q^d/A)_s = \varepsilon_{\text{eff}}/\varepsilon_{\text{eff}(s)} = 1.71$$

Figure 6.38b gives results from a comparison of the performance during summer and winter solstice on the north and south radiators of a satellite in GEO (Chapter 4, Sec. V.D). Assuming direct solar impingement as the only source of environment heating, the equations of heat rejection capability become

$$(Q^d/A)_u = \varepsilon_{\text{eff}}\sigma T^4 - \alpha_{\text{eff}}^s S$$

and

$$(Q^d/A)_s = \varepsilon_{\text{eff}(s)}\left[\sigma T^4 - \left(\alpha_e^s/\varepsilon_e\right)q^S\right]$$

Maximum normal incident flux on either the north or south radiator is

$$q^S = S \sin 23.5 \text{ deg}$$

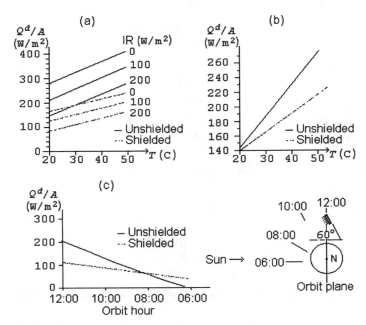

Fig. 6.38 Heat rejection capability (blades fully open) of unshielded and shielded louvered radiators: a) IR surroundings, b) NS radiator in GEO with sun vector coning at 23.5 deg off the louvers' plane ($S = 1350$ W/m^2, $\alpha^s_{\text{eff}} = 0.1$), and c) low Earth polar orbit, with solar vector at 60 deg to the louvers' plane and radiator temperature of 30°C.[52]

and the temperatures at which the unshielded system yields higher rejection rates are found from

$$\varepsilon_{\text{eff}}\sigma T^4 - \alpha^s_{\text{eff}}S > \varepsilon_{\text{eff}(s)}\left[\sigma T^4 - \left(\alpha^s_e/\varepsilon_e\right)q^S\right]$$

For the given data and $S = 1350$ W/m^2, the temperature must be greater than

$$\{(1/5.67E-8)[(0.1)(1350) - (0.39)(0.082/0.76)$$

$$\times(1350\sin 23.5\text{ deg})]/(0.67 - 0.39)\}^{\frac{1}{4}} = 290\text{ K}$$

or 17°C. The results in Fig. 3.38b are based on the assumption that the louvers will be fully open when the radiator temperature is 17°C or higher. Many electronics panels during hot case operation are at temperatures somewhat higher than 20°C, which suggests that louvered radiators on the north and south panels of a geosynchronous satellite will be more effective without a shield.

Comparisons in LEOs require knowing the changing azimuth, as well. as the angle of solar incidence. These data are contained in flux programs usually used to obtain normal incident fluxes in intervals of orbit time as well as integrated averages (Chapter 4). For exposed louvered radiators, a matrix may be added to the input file to define the values of effective solar absorptivity in terms of surface aspect (ψ, φ), where ψ is azimuth and φ sun angle. A listing of $\alpha^s_{\text{eff}}(\psi, \varphi, \omega t)$, where ωt is time or position (dimensionless) in orbit, can then be obtained for a set of blade angles. In addition, the output can be made to yield the integrated average of the absorbed solar flux at the specified blade angle based on the formula

$$\left(\alpha_{eff}^s S\right)_{\text{orbital average}} = (1/\text{orbit period}) \int_{\text{orbit}} \alpha_{eff}^s[\psi(\omega t), \varphi(\omega t), \omega t] S(\omega t) \, d(\omega t)$$

The explicit functional dependence in $S(\omega t)$ is a reminder that S is the solar constant when the surface is in sunlight and zero otherwise.

Results from Sec. 5.3 of Ref. 52 are reproduced in Fig. 6.38c. The analysis deals with a low-Earth polar orbit (685 km) in which the louvers are fully open and the sun is at 60 deg off the louvers' plane. The combination gives a relatively high value of total absorbed flux leading to a situation where shielding might improve heat rejection capability. It has been noted, however, that for most common orbits and optimized radiator orientations, an unshielded louvered system offers higher rejection rates, especially at the temperatures when the louvers are fully open.

Example: Thermal Analysis of an Attitude and Data Handling System

To illustrate the method of analysis of louvered radiators, consider an attitude and data handling system (ADS) for a spacecraft in GEO. The equipment is packaged in a module that is to be isolated as much as is practical from the rest of the satellite and oriented with its primary radiator facing northward. It is known that the dissipation will vary significantly, depending on the different phases and operational activities of the mission, and a network of heat pipes has been recommended to be embedded in the radiator to increase the effectiveness of distributing and radiating the dissipated heat. This, combined with the large mass of the components, justifies conducting a preliminary thermal analysis based on steady-state isothermal conditions and negligible heat exchange with the interior.

The ADS is shown in Fig. 6.39. It is estimated that electronics waste heat will vary between 200 and 450 W (with 10% margins to be included in the analysis), and the specifications require that the radiator temperature be maintained between 0°C and 45°C with a 20-W heater power budget. The dimensions have been tentatively assigned as shown, and it has been agreed that quartz mirrors are to be used for radiator treatment. Negotiations with the customer established that the data in Table 6.9 be used for hot case and cold case analysis.

From the given dimensions, the total area available for radiation on the module's north side is

$$A_{\text{available}}^r = (1.65)^2 = 2.72 \text{ m}^2$$

It must now be determined if this area is sufficiently large to maintan 45°C or lower in a hot-biased environment with a dissipation of 495 W. Hot case data include maximum solar input, maximum view to the solar array at 55°C and no heat losses to the satellite interior. A steady-state heat balance on the radiator in an enclosure with the solar array sa and space gives

$$Q^d/A + \alpha^s q^S = (F_E F_A)_{\text{ADS-sa}}\left(\sigma T^4 - \sigma T_{\text{sa}}^4\right) + (F_E F_A)_{\text{ADS-space}}(\sigma T^4 - 0)$$

Q^d/A is dissipation per unit radiator area at temperature T. Substituting the hot

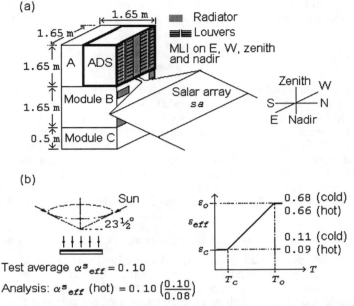

Fig. 6.39   Attitude and data handling system (ADS) in GEO: a) location on the satellite and radiator orientation and b) louvers characteristics, hot case α_{eff}^s is the test value adjusted by the ratio of degraded radiator α_{eff}^s to original test value.

case data with $T = 45°C$ and using the relations

$$F_A)_{\text{ADS}-\text{sa}} + F_A)_{\text{ADS}-\text{space}} \approx 1.0$$

$$(F_E F_A)_{1-2} \approx \varepsilon_1 \varepsilon_2 (F_A)_{1-2}$$

$$(F_E F_A)_{i-\text{space}} = \varepsilon_i (F_A)_{i-\text{space}}$$

$$Q_{\text{hot}}^d / A + (0.10)(522.4) = [(0.76)(0.78)(0.05) + (0.76)(0.95)][(5.67E-8)$$

$$\times (45 + 273)]^4 - (0.76)(0.78)(0.05)(5.67E-8)(55 + 273)^4$$

$$Q_{\text{hot}}^d / A + 52.24 = 4.2618E-8(45 + 273)^4 - 19.45$$

or

$$Q_{\text{hot}}^d / A = 364.12 \, \text{W/m}^2$$

For the available area, the maximum dissipation that can be tolerated passively without exceeding 45°C is

$$Q_{\text{hot}(2.72 \, \text{m}^2, 45°\text{C})}^d = (364.12)(2.72) = 990.4 \, \text{W}$$

Therefore, there is significantly more area (twice) than needed to accommodate 495 W at 45°C.

Table 6.9   Data for hot and cold case analysis, Fig. 6.39

Hot case
ADS dissipation (includes 10% positive margin)
    Continuous 495 W
Environment flux (no Earth heating)
    Summer solstice $S = 1310$ W/m^2, solar vector coning at 23.5 deg off the
    plane of the radiator
Normal incident flux
    $q^S = 1310 \sin 23.5 = 522.4$ W/m^2
External heating input
    From solar array: $T_{sa} = 55°C$, $\varepsilon_{sa} = 0.78$, $F_{ADS-sa} = 0.05$
    None from module A: $T_A = 40°C$, $K_{ADS-A} = 0$
Insulation effectiveness
    $\varepsilon_{ins} = 0$
Radiator properties (uncoated quartz mirrors)
    α^s/ε (end of 5 yr) $= 0.10/0.77$
    With unshielded louvers (open blades and sun at 23.5 deg): $\alpha^s_{eff} = 0.13$,
        $\varepsilon_c = 0.09$, $\varepsilon_o = 0.66$
Cold case
ADS dissipation (includes 10% negative margin)
    Continuous 180 W
Normal incident flux
    $q^S = 0$
External heating input
    Negligible
Insulation effectiveness
    $\varepsilon_{ins} = 0.02$
Radiator properties
    α^s/ε (beginning of life) $= 0.08/0.80$
    With unshielded louvers (no solar input): $\varepsilon_c = 0.11$, $\varepsilon_o = 0.68$

If all 2.72 m^2 are used for passive thermal control, then a hot case calculation gives

$$T_{hot\,(2.72\,m^2,495\,W)} = \{(1/4.2618E-8)[(495/2.72) + 52.24 + 19.45]\}^{1/4}$$

$$= 277.8\,K\,(4.8°C)$$

and for cold case with no external heating

$$T = (Q^d/\sigma\varepsilon A)^{1/4}$$

or (using cold case data)

$$T_{cold\,(2.72)\,m^2,180\,W} = \{180/[(5.67E-8)(0.80)(2.72)]\}^{1/4} = 195.4\,K\,(-77.6°C)$$

which is well below the acceptable limit of 0°C, even before including extraneous heat losses.

Heater power could be used to bring the temperature up to 0°C, but according to cold case assumptions, this calculation must include emission and heat losses through the insulation Q_{ins} to deep space (0 K). Insulation losses may be considered to be through the three MLI surfaces in the vicinity of the ADS radiator; that is,

$$Q_{\text{ins}} = \varepsilon_{\text{ins}} A_{\text{ins}} \sigma T^4$$

with

$$A_{\text{ins}} \approx (3)(1/2)(1.65)(1.65) = 4.08 \, \text{m}^2$$

Thus, using a worst insulation effectiveness of 0.02 and an ADS temperature of 0°C,

$$Q_{\text{ins}} \approx (0.02)(4.08)(5.67E-8)(273)^4 = 25.7 \, \text{W}$$

Therefore, a conservative estimate of the heater power to keep the temperature at 0°C is found from the energy balance

$$Q_{\text{heater}} + Q_{\text{cold}}^d - Q_{\text{loss}} = A\left(\varepsilon \sigma T_{0°C}^4\right)$$

where cold case values are used; that is,

$$Q_{\text{heater(cold: 2.72 m}^2\text{,45 W)}} = 2.72(0.80)(5.67E-8)(273)^4 + 25.7 - 180 = 531 \, \text{W}$$

This number is prohibitively higher than the specified allowable of 20 W.

Heater power may be reduced by covering enough of the radiator surface with MLI to allow the temperature in the hot case to be at the specifications' upper limit of 45°C. The radiator size in this case can be calculated from the earlier equation

$$Q_{\text{hot}}^d / A = 364.12 \, \text{W/m}^2$$

in which $Q_{\text{hot}}^d = 495$ W. Hence,

$$A_{(45°C, \text{ hot case})} = 495/364.12 = 1.36 \, \text{m}^2$$

The insulation loss in the cold case is now increased to

$$Q_{\text{ins}} \approx 25.7 + (0.02)(2.72 - 1.36)(5.67E-8)(273)^4 = 34.3 \, \text{W}$$

and the heater power for the new configuration in the cold case is

$$Q_{\text{heater(cold:1.36 m}^2\text{,180 W)}} = 1.36(0.80)(5.67E-8)(273)^4 + 34.3 - 180 = 197 \, \text{W}$$

which is an improvement, but still considerably higher than the allocated 20 W.

Because further reduction in radiating area would lead to a hot case temperature exceeding the 45°C limit, the option of thermal control by louvers must be considered. Hence, with A_l representing the area covered by louvers, and assuming the rest of the radiator to have MLI through which no heat is lost, the hot case heat balance becomes

$$Q^d / A_l + \alpha_{\text{eff}}^s S = (F_E F_A)_{l(\text{ADS-sa})}\left(\sigma T^4 - \sigma T_{\text{sa}}^4\right) + (F_E F_A)_{l(\text{ADS-space})}\sigma T^4$$

with subscript l indicating louvers. Treating this as an imaginary surface with emissivity ε_{eff},

$$(F_E F_A)_{l(\text{ADS-sa})} \approx \varepsilon_{\text{eff}} \varepsilon_{\text{sa}} F_{A(\text{ADS-sa})}$$

$$(F_E F_A)_{l(\text{ADS-space})} \approx \varepsilon_{\text{eff}} F_{A(\text{ADS-space})}$$

and, hence (assuming the louvers will be open during hot case at 45°C),

$$495/A_l + (0.13)(1310) = [(0.66)(0.78)(0.05)$$
$$+ (0.66)(0.95)](5.67E-8)(45 + 273)^4$$
$$- (0.66)(0.78)(0.05)(5.67E-8)(55 + 273)^4$$

or

$$A_l = 2.588 \, \text{m}^2$$

which can be accommodated within the available space.

The cold case equates cold case dissipation $Q^d_{\text{cold}(l)}$ to radiation through the louvered radiator and losses from adjacent insulation; that is,

$$Q^d_{\text{cold}(l)} = (\varepsilon_{\text{eff}} A_l + \varepsilon_{\text{ins}} A_{\text{ins}}) \sigma T^4_{\text{cold}}$$

where

$$A_l = 2.588 \, \text{m}^2$$

and

$$\varepsilon_{\text{ins}} A_{\text{ins}} = 0.02[4.08 + (2.72 - 2.588)] = 0.084 \, \text{m}^2$$

It is specified that the louvers shall be calibrated so that it is certain they will be fully open when the radiator in vacuum is at 45°C and closed at 0°C. Because the nominal calibration range for standard actuators is 19°C, both ends would be met if the louvers are calibrated in air to be fully open at 25°C (hence, guaranteed open at 45°C in a vacuum) and fully closed at 6°C (hence, guaranteed closed at 0°C in a vacuum). Other choices are, of course, possible, but it is important to have an early opening as the temperature increases toward the hot end and an early closure with decreasing temperature toward the closed end.

Temperature as function of dissipated power is obtained by assuming the linear profile of effective emissivity recommended in Table 6.8. Assuming a 2°C difference in the actuator temperature between air and vacuum performance, the radiator's effective emissivity is given by

$$\varepsilon_{\text{eff}} = 0.11, \qquad \text{for } T < 4°C$$

$$\varepsilon_{\text{eff}} = 0.66 - [(0.66 - 0.11)/19](23 - T), \qquad \text{for } 4°C < T < 23°C$$

$$\varepsilon_{\text{eff}} = 0.66, \qquad \text{for } T > 23°C$$

Assigning ε_{eff} to each T, Table 6.10 is generated.

Table 6.10   Temperature versus dissipation for louvered radiator

T_{cold}, °C	ε_{eff}	$Q^d_{\text{cold, }(l)}$, W
0	0.110	116.1
4	0.110	123.1
6	0.168	178.2
6.1	0.171	181.2
7.0	0.197	207.0

Hence, covering 2.588 m^2 of the radiator with louvers and insulating the remaing 0.132 m^2 would result in a cold case temperature of just under 6.1°C when the dissipation is 180 W, whereas the hot case temperature (495 W dissipation) will not exceed 45°C.

This is a substantial improvement on passive control in terms of the cold case temperature and the fact that no heater power is consumed. Furthermore, in this example, the cold case margin (6.1°C vs allowable 0°C) and the availability of up to 20 W in heater power should prompt investigating means for reducing the hot case temperature (presently 45°C) by allowing the cold case to drop nearer to 0°C and, if necessary, using heater power to ensure it does not fall below that limit. Along this direction, an approach often practiced by thermal engineers is to supplement louvers control by passive radiation, which helps overcome some of the blockage and trapped solar radiation due to the presence of the louvers.

The iterations for the hybrid system are usually based on maximizing the louvered area and, hence, sizing the passive portion to just satisfy the selected hot case temperature. Cold case calculations are then performed to determine the need and amount of heater power to maintain at least 0°C.

Assuming the radiator remains isothermal, the hot case heat balance now reads

$$Q^d + \alpha^s_{\text{eff}} A_l S + \alpha^s_p A_p q^S$$
$$= \left[(F_E F_A)_{l(\text{ADS-sa})} A_l + (F_E F_A)_{p(\text{ADS-sa})} A_p \right] \left(\sigma T^4 - \sigma T^4_{\text{sa}} \right)$$
$$+ \left[(F_E F_A)_{l(\text{ADS-space})} A_l + (F_E F_A)_{p(\text{ADS-space})} A_p \right] \sigma T^4$$

where A_l and A_p are louvered and passive areas. The exchange factors in the hot case with the louvers open are calculated as

$$(F_E F_A)_{l(\text{ADS-sa})} = (0.66)(0.78)(0.05) = 0.026$$
$$(F_E F_A)_{p(\text{ADS-sa})} = (0.77)(0.78)(0.05) = 0.030$$
$$(F_E F_A)_{l(\text{ADS-space})} = (0.66)(0.95) = 0.627$$
$$(F_E F_A)_{p(\text{ADS-space})} = (0.77)(0.95) = 0.731$$

Thus, in the hot case,

$$495 + (0.13)(1310)A_l + (0.1)(522.4)A_p$$
$$= (0.026A_l + 0.030A_p)(5.67E-8)[T^4 - (55+273)^4)]$$
$$+ (0.626A_l + 0.731A_p)(5.67E-8)T^4$$

Rearranging,

$$[0.652(5.67E-8)T^4 - 187.4]A_l + [0.761(5.67E-8)T^4 - 71.9]A_p = 495$$

with the constraint

$$A_l + A_p = 2.588 \, \text{m}^2$$

Table 6.11 Heater power versus temperature for hybrid radiator

T_{cold}, °C	ε_{eff}	$180 + Q_{heater}$, W	Q_{heater}, W
0	0.110	192.8	12.8
2	0.110	198.5	18.5
3	0.110	201.4	21.4
4	0.110	204.4	24.4
5	0.139	229.3	49.4

Therefore, each selected hot case temperature will result from one unique combination of active and passive areas. For example, if the desired T_{hot} is 40°C then

$$167.42A_l + 342.24A_p = 495$$
$$A_l + A_p = 2.588 \, m^2$$

which gives

$$A_l = 2.235 \, m^2$$
$$A_p = 0.353 \, m^2$$

Corresponding temperatures and heater powers (if needed) in the cold case can be found by iterating on

$$Q_{cold}^d + Q_{heater} = (\varepsilon_{eff}A_l + \varepsilon A_p + \varepsilon_{ins} A_{ins})\sigma T_{cold}^4$$

where, as before,

$$\varepsilon_{ins} A_{ins} = 0.084 \, m^2$$

and ε_{eff} is a known function of temperature. Hence

$$180 + Q_{heater} = [\varepsilon_{eff}(2.235) + (0.80)(0.353) + 0.084](5.67E-8)T_{cold}^4$$

and the results for various temperatures are listed in Table 6.11.

It is concluded that when 2.235 m² of louvered area are supplemented by 0.353 m² of passive radiators, 20 W of heater power will maintain the cold case at about 2.5°C. In thermal control terms, a thermostatically controlled heater (Sec. VI of this chapter) of sufficient rating (say, 50 W) and nominal set point at about 2.5°C would, in the cold case, operate at a duty cycle that will result in an average orbital consumption of 20 W. The analysis also shows that there will be no heater power consumed during a hot case orbit when the temperature is 40 °C.

Although there is a clear advantage in using louvers to reduce heater power, the accompanying penalty from trapped solar energy and reduced effective radiation area must not be underestimated. A review of the analysis will show that a minor increase in passive area to compensate for nonisothermal distributions or to counter a more serious degradation of surface properties will radically increase the heater power needed to maintain tolerable temperatures in the cold case. This is described in Fig. 6.40 in terms of a reduced maximum temperature for a given fixed total area. A more informative statement is contained in thermal performance curves, which relate the temperature to dissipated heat considered as a variable in the heat

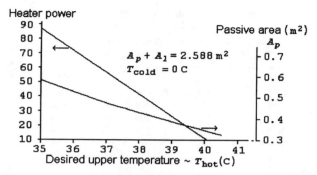

Fig. 6.40   Passive area and heater power to keep the temperature of ADS hybrid louvered/passive radiator between 0°C and maximum hot case value.

balance equations. Additional or reduced heating from any effect (including heater power) can then be translated into an equivalent dissipation and the corresponding temperature read off directly from the plots. Results from this example are given in Fig. 6.41. The data are found from the equations

$$Q_{\text{hot}} = \left\{ \left[(F_E F_A)_{l(\text{ADS}-\text{sa})} + (F_E F_A)_{l(\text{ADS}-\text{space})} \right] A_l + \left[(F_E F_A)_{p(\text{ADS}-\text{sa})} \right. \right.$$

$$\left. \left. + (F_E F_A)_{p(\text{ADS}-\text{space})} \right] A_p \right\} \sigma T^4$$

or

$$Q_{\text{hot}} = (0.653 A_l + 0.761 A_p) \sigma T^4$$

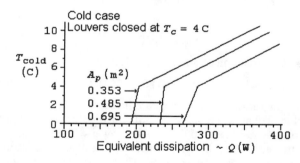

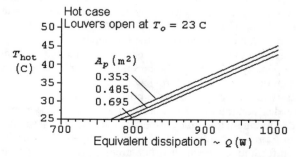

Fig. 6.41   Thermal performance of hybrid louvered/passive radiator.

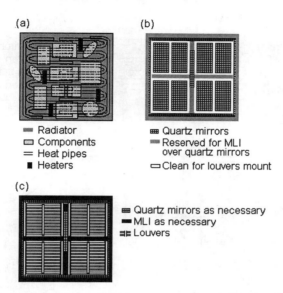

(a)

(b)

■ Radiator
▫ Components
= Heat pipes
■ Heaters

▦ Quartz mirrors
▦ Reserved for MLI
   over quartz mirrors
▭ Clean for louvers mount

(c)

▦ Quartz mirrors as necessary
▬ MLI as necessary
▤ Louvers

Fig. 6.42   Design concept for the ADS radiator: a) components, heat pipes and heaters layout, b) radiator's exterior surface treatment, and c) louvers in place.

and

$$Q_{cold} = (\varepsilon_{eff} A_l + \varepsilon A_p + \varepsilon_{ins} A_{ins}) \sigma T_{cold}^4$$

$$Q_{cold} = [\varepsilon_{eff} A_l + (0.80) A_p + 0.084] \sigma T_{cold}^4$$

A design concept is given Fig. 6.42. A layout of the electronics with heat pipes and heaters is shown in Fig. 6.42a and Fig. 6.42b shows the radiator (looking from the outside) treated with quartz mirrors except on strips reserved for mounting the louvers frames and actuator housings. Also shown in Fig. 6.42b are the regions for auxiliary passive thermal control. The passive area just inside the rim may be changed at any time before flight by adding or removing MLI. In some cases, the MLI can be an extension from thermal blankets on other sides of the module.

A typical configuration with the louvers in place is shown in Fig. 6.42c. Tooling and manufacturing procedures are simplified by making all louvers sets the same shape and size. In this example, four units, each measuring about 74.75 × 74.75 cm² provide the desired 2.235 m² total louvered area. A larger number of smaller units may be more accessible and, as it often happens, uniformity of size and symmetry may have to give way to accommodating a radiator with penetrations and protruding instruments and sensors.

The reader may want to investigate the increase in heat rejection capability if louvered radiators are protected from direct sun while maintaining an open view to space. This could require the design of open shades (perhaps attached to each louvers frame) of sufficient height to block off the sun at maximum angle of incidence. The interior surface of the shade should have high reflectivity and be tilted at least slightly outward to reduce the effects of radiation exchange with the radiator.

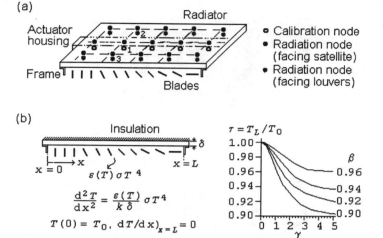

(a)

(b)

Fig. 6.43 Thermal modeling of louvered radiators: a) numerical thermal model, temperature of node 1 defines the emissivity of nodes 2 and 3 and b) analytical model (one-sided radiator), implicit solution in terms of parameters $\beta = [\varepsilon_o T_c - \varepsilon_c T_o]/[(\varepsilon_o - \varepsilon_c)T_o]$ and $\gamma = [(\varepsilon_o - \varepsilon_c)L^2 \sigma T_o^4/k\delta(T_o - T_c)]^{\frac{1}{2}}$.

F. Thermal Modeling of Louvered Radiators

The louvers' effect in thermal modeling is obtained by placing radiator nodes along the strip facing the actuator housing. Normally, the interaction of these calibration nodes with the actuator itself is not included, but in the computer execution their temperatures set the values of emissivity, in accordance with a given logic, of all other nodes radiating through the louvers.

One scheme[56] uses the temperature of each calibration node to set the emissivity of the louvers-facing nodes along the line perpendicular to the actuator housing and passing through the calibration node. For example, the temperature of node 1 in Fig. 6.43a sets the same value of emissivity to nodes 2 and 3 according to the temperature-effective emissivity matrix in the input file. When details on louvers size and placement are known, the calibration nodes may be positioned facing the actuators themselves. The average temperature of two adjacent calibration nodes is then used to set the emissivity of all other nodes between them and all regions between the corresponding blades. Similar procedures are written for effective solar absorptivity as function of sun angles (in-plane ψ and elevation φ) and temperature (or blade angle θ).

For an unshielded louvered radiator, the quasi-steady-state equation

$$Q/A = \varepsilon_{\text{eff}}[T(\theta)](\sigma T^4 - IR) - \alpha_r^s \varepsilon_{\text{eff}} q^A - \alpha_{\text{eff}}^s[\psi, \varphi, \theta(T)]S$$

is used at each designated (ψ, φ) in an orbit position to generate a list of Q as function of $T(\theta)$. The values are tabulated as bivariate arrays in the computer input file and a program (for example, SINDA) performs a first interpolation using orbit position (or time) to select for each θ the corresponding Q and T array. A second interpolation using sun angles gives the particular Q that enters the heat balance of each node that includes heat exchanges with surrounding nodes. Because double interpolations in large models can be lengthy and time-consuming, consideration should always be given to performing orbital average analyses.

Shielded louvered radiators are similarly modeled, except that here there is no direct solar input to nodes located on the radiator. The nodes on the shield are usually given a one-to-one radiation coupling to opposing nodes on the radiator and, as in the case of exposed systems, the actuator enters the model only in the form of a variable properties matrix.

G. Lateral Temperature Distribution in Louvered Panels

Generally, the effective properties of louvered radiators used in thermal modeling are the same profiles obtained from tests conducted under isothermal conditions, with all the blades in nearly parallel positions. The input file merely supplies a property–temperature matrix from which iterations are made corresponding to the temperature of the calibration node, regardless of the relative positions of the blades.

The validity of this approach has been verified[57] in a study comparing test data with results of the analytical model described by Fig. 6.43b. The linear form of the effective emissivity, as obtained from isothermal definitions, was used to predict the temperature distribution due to nonuniform heating.

Assuming one-dimensional, steady-state conduction in a louvered fin of uniform thickness δ, length L, and conductivity k, the energy equation with radiation to deep space is written in the standard form

$$\frac{d^2 T}{dx^2} = [\varepsilon_{\text{eff}}(T)/k\delta]\sigma T^4$$

with imposed edge ($x = 0$) and thin end ($x = L$) conditions

$$T(0) = T_0, \qquad \frac{dT}{dx}\bigg|_{x=L} = 0$$

The effective emissivity as function of temperature is given by

$$\varepsilon_{\text{eff}}(T) = \varepsilon_c(\text{const}), \qquad \text{for } T \leq T_c$$

$$\varepsilon_{\text{eff}}(T) = aT - b, \qquad \text{for } T_c \leq T \leq T_o$$

$$\varepsilon_{\text{eff}}(T) = \varepsilon_o(\text{const}), \qquad \text{for } T \geq T_c$$

where

$$a = \frac{(\varepsilon_o - \varepsilon_c)}{(T_o - T_c)}$$

$$b = \frac{(\varepsilon_o T_c - \varepsilon_c T_o)}{(T_o - T_c)}$$

Substituting

$$T^4 \approx T_m^4 + 4T_m^3(T - T_m)$$

the equation (in dimensionless quantities) in the active region becomes

$$\frac{d^2\tau}{d\xi^2} = B^2(\tau - \beta)(\tau - 0.75\tau_m)$$

with

$$\tau(0) = 1 \quad \text{and} \quad \left.\frac{d\tau}{d\xi}\right|_{\xi=1} = 0$$

and

$$\tau = T/T_0, \quad \xi = x/L, \quad B^2 = 4(\varepsilon_o - \varepsilon_c)L^2\sigma T_0^3\tau_m^3/[(\tau_o - \tau_c)k\delta]$$

$$\beta = (\varepsilon_o\tau_c - \varepsilon_c\tau_o)/(\varepsilon_o - \varepsilon_c)$$

The solution (keeping τ_m a constant) is represented by an elliptic integral of the first kind (see, for example, Ref. 58) and is given as

$$\sqrt{\frac{2}{3}}B(1 - \xi) = \int_{\tau_1}^{\tau} \frac{d\eta}{[(\eta - \tau_1)(\eta - \tau_2)(\eta - \tau_3)]^{0.5}}$$

where τ_1 (dimensionless temperature at $\xi = 1$), τ_2 and τ_3 are the roots of the cubic

$$(\tau - \beta)^3 + 1.5(\beta - 0.75\tau_m)(\tau - \beta)^2 = (\tau_1 - \beta)^3 + 1.5(\beta - 0.75\tau_m)(\tau_1 - \beta)^2$$

To obtain the best estimate for τ_1, the elliptic integral is evaluated at $\tau(0) = 1$ and the derivative with respect to τ_m taken on both sides of the equation with the result evaluated for $\partial\tau/\partial\tau_m = 0$ (for details, see Ref. 57). Numerical iterations on temperature at fin's end [$T(L)$] give the results shown graphically in Fig. 6.43b. The parameter

$$\gamma = \{(\varepsilon_o - \varepsilon_c)L^2\sigma T_0^4/[(T_o - T_c)k\delta]\}^{\frac{1}{2}}$$

combines the louvers' characteristics with radiator properties and the imposed temperature at the edge [$T(0)$].

Confirmation was obtained in a vacuum test of a special louvered radiator [advanced louver system (ALS) 0.58×0.46 m^2] that was meticulously assembled for reduced hysteresis ($T_o - T_c < 14°$C in a vacuum) and calibrated to less than one angular degree blade uniformity. The instrumentation is described in Fig. 6.44. Applying heat in a region near the edge of the radiator was intended to generate a sufficiently large longitudinal temperature drop in the panel to ascertain the influence as the blades assume different angular positions.

The specimen was suspended in the chamber by thin steel wires with its insulated back facing the insulated surface of another louvered assembly undergoing independent testing. In this arrangement, extraneous heat losses were considerably reduced, and the amount of heat conducted at $x = 0$ was very nearly the measured EI (and I^2R) heater power.

Various levels of heater power aimed at establishing specific steady-state values of edge temperature were applied. The case where edge heating produced a temperature profile spanning nearly exactly the active range (T_o–T_c) is as shown. A heater power value of 33.3 W produced an average edge temperature of 31.9°C (which is equal to T_o). In this instance the end temperature was measured at 18.7°C. Averages from other thermocouple readings are also indicated on the curve in Fig. 6.44.

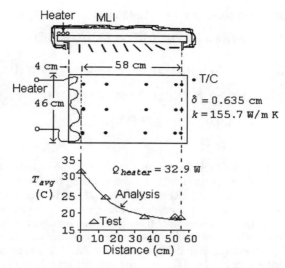

Fig. 6.44   Lateral conduction in louvered radiators.[57]

Multiplying both sides of the energy equation by $d\tau/d\xi$ and substituting the linear expression for effective emissivity result in

$$d\left(\frac{d\tau}{d\xi}\right)^2 = \left(2aL^2\sigma T_0^4/k\delta\right)(\tau - \beta)\tau^4\,d\tau$$

Because

$$Q_{\text{heater}} = -k\delta w\left(\frac{dT}{dx}\right)_{x=0}$$

where w is radiator width, a first integration from $x = 0$ to $x = L$ gives

$$Q_{\text{heater}}L/k\delta wT_0 = \gamma\left[(1/3)\left(1 - \tau_1^6\right) - (2\beta/5)\left(1 - \tau_1^5\right)\right]^{\frac{1}{2}}$$

From the curves or tables of τ_1 vs γ, an imposed T_0 and known louvered radiator characteristics yield a unique value τ_1 and corresponding T_L. Substituting into the last equation will yield the heater power needed to establish these conditions. Using the data in Fig. 6.44 with $T_0 = 31.9 + 273 = 304.9$ K $\left(\sigma T_0^4 = 490 \text{ W/m}^2\right)$,

$$k\delta = (155.7)(0.635/100) = 0.989 \text{ W/K}$$

$$\beta = (\varepsilon_o\tau_c - \varepsilon_c\tau_o)/(\varepsilon_o - \varepsilon_c)$$

$$= \{[(0.707)(291) - (0.115)(304.9)]/[(0.707 - 0.115)304.9]\} = 0.9456$$

$$\gamma = \{(\varepsilon_o - \varepsilon_c)L^2\sigma T_0^4/[(T_o - T_c)k\delta]\}^{1/2}$$

$$= \{(0.592)(0.58)^2(490)/[13.9(0.989)]\}^{1/2} = 2.664$$

$$\tau_1 \text{ (from the curves for } \beta = 0.9456 \text{ and } \gamma = 2.664) \approx 0.955,$$

$$\text{or} \qquad T_L = (0.955)(304.9) = 291.2(18.2°C)$$

$$Q_{\text{heater}} = (k\delta w T_0/L)\gamma\left[(1/3)\left(1 - \tau_1^6\right) - 0.4\beta\left(1 - \tau_1^5\right)\right]^{1/2}$$

$$= [0.989(0.46)(304.9)(2.664)/0.58][(1/3)(1 - 0.955^6)$$

$$-0.4(0.9456)(1 - 0.955^5)]^{\frac{1}{2}}$$

$$= 33.0 \text{ W}$$

It should be pointed out that the radiator thickness (0.635 cm) was selected in anticipation of developing this particular profile during the test.

The very close correlation between analysis and test in this example could be due to the meticulous and unique process of assembling and calibrating this particular louvers set. But similar results have also been obtained in thermal tests of satellites equipped with standard louvers, built and calibrated in accordance with less demanding procedures than those used on the ALS. It is suggested that using the effective emissivity obtained from isothermal definitions in modeling is valid when louvers carry a sufficiently large number of blades or when temperature variations are not too large as to make the angular position of one blade much different from an adjacent one. In actual practice, strength and weight considerations lead to minimizing the height of the relatively heavy louvers frame, which in turn restricts the width of a blade to avoid physical interference with the mounting panel, especially during vibration. Hence, a large louvers set usually carries many blades spaced within short distances. On the other hand, smaller panels with fewer blades inherently exhibit small temperature gradients.

H. Transients in Louvered Panels

Satellite louvers are mostly mounted to massive equipment platforms that undergo relatively small temperature changes during an orbit. Because recommended mounting procedures (which often include a thermal filler between the actuator housing and the radiator) are directed toward enhanced thermal coupling between the radiator and the actuator housing, the actuator temperature also varies slightly and will remain within about 2°C from the radiator temperature. Thus, in a normal orbit, including eclipse, the blades generally fluctuate by a small angle off an average position defined by the average orbital temperature.

Significant transients could conceivably occur if louvers were to be used for the thermal control of panels having low thermal masses. More commonly, the transient problem is of importance during prolonged cool-down periods, when the satellite is forced into a low-power mode due to delayed deployment or a specialized fail-safe maneuver. As a matter of fact, because louvers can suppress the cool-down rate, they provide substantial margins to riding out anomalies that otherwise could result in very cold temperatures.

Cool-down at uniform temperature is given by the equations

$$Mc\frac{dT}{dt} = -\varepsilon(T)\sigma T^4$$

$$T(0) = T_i$$

where the initial temperature T_i can be within any range of louvers performance depending on the calibration levels. This is accounted for by specifying the effective emissivity within the limits of integration. For example, if $T_i > T_o$ then the time t_f it takes for the temperature to fall from T_i to a specified value T_f is determined as follows.

$T_f > T_o$:

$$t_f = \int_{T_i}^{T_f} \frac{\varepsilon_o}{Mc} \sigma T^4 \, \mathrm{d}T$$

$T_o > T_f > T_c$:

$$t_f = \int_{T_i}^{T_o} \frac{\varepsilon_o}{Mc} \sigma T^4 \, \mathrm{d}T + \int_{T_o}^{T_f} \frac{aT - b}{Mc} \sigma T^4 \, \mathrm{d}T$$

$T_c > T_f$:

$$t_f = \int_{T_i}^{T_o} \frac{\varepsilon_o}{Mc} \sigma T^4 \, \mathrm{d}T + \int_{T_o}^{T_c} \frac{aT - b}{Mc} \sigma T^4 \, \mathrm{d}T + \int_{T_c}^{T_f} \frac{\varepsilon_c}{Mc} \sigma T^4 \, \mathrm{d}T$$

Transients in actual louvered panels do not appear to deviate appreciably from theoretical predictions when the rate of temperature change is on the order usually encountered in flight and satellite vacuum testing. Under most normal operations the actuator temperature follows the radiator's closely with the blades moving in a slow, rather smooth fashion. But the few available data from unit tests with imposed rapid temperature change show distinct deviations and inconsistencies that cannot be predicted or explained solely on thermal basis. Typical heat-up and cool-down profiles (after unpublished results from tests conducted at Fairchild Space Company in December 1980) are shown in Fig. 6.45. The delay in louvers action and the overshoots and undershoots in temperature are indicative of nonuniform and generally nonrepeatable frictional effects. Whether these are deserving of further study is not clear, but a complete characterization of louvers transients will not be possible without much additional analysis and testing.

I. Systems Evaluation of Louvers

Louvers are potentially the most reliable compensators for reduced spacecraft power and defficiencies in other thermal hardware. Therefore, using them on satellites can be greatly rewarding even when their need appears marginal. The arguments are stated as follows:

1) Because louvers are calibrated to respond to specified temperatures, the blades will automatically tend to close to counterbalance reductions in available power. Beyond the benefit of reduced heater power, the action becomes critically important for maintaining adequate temperatures in cases of prolonged spacecraft misalignment prior to solar array deployment and other emergency situations with long periods of battery discharge.

2) Predicted temperatures in louvered satellites are less dependent on uncertainties in power and hardware properties. In particular, louvers will offset heat losses due to an underestimated MLI effectiveness. This fact, plus the inevitable changes in the satellite power budget during the course of a program, provides a pursuasive position in favor of maximizing the louvered radiator area on a satellite.

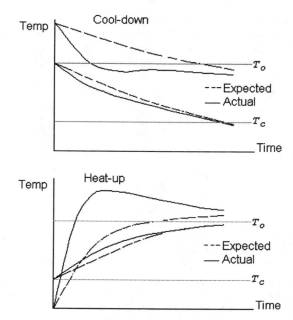

Fig. 6.45   Transients in louvered radiators: comparisons between actual and observed behavior.

A distinct disadvantage in using louvers is that their very presence reduces the active radiating area. By narrowing the frames and the actuator housing cover, modern designs limit the blockage to about 18% of the total area. Still, any reduction in radiation capability is a drawback, especially when only a limited area is available in preferred locations. As a solution, thermal engineers often design with combinations of louvered and nonlouvered radiators.

The apparent simplicity of construction and operation of louvers inevitably invites urges for modifications. Whether further improvements on present designs are worth the additional effort is constantly being debated by engineers and technicians of all disciplines. Before committing to making changes that appear attractive on paper, it may be beneficial to review the following:

1) As in all thermal hardware, the lowest risk in using louvers is to resort to parts and constructions equivalent to those verified in test and flight. The overwhelming success of venetian blind louvers logically makes them the more attractive choice.[59]

2) Reducing louvers' weight is a legitimate target. [Global Positioning System (GPS) louvers average about 3.5 kg/m^2.] It must be noted, however, that current venetian blind louvers are themselves the result of numerous, excruciating efforts to cut back on weight. Hence, to claim equivalent integrity, modified units must undergo similar, if not the same extensive analyses and qualification testing. Along these lines, it should be pointed out that a major concern in weight reduction is a reduced frame rigidity, which could lead to distortions and the binding or hindering of blades' motion.

3) A similar discussion applies to upgrading the radiation characteristics, particularly the effective emissivity. Current designs employ a number of features that were added following a study program that concentrated on precisely this kind of

enhancement.[60] Nonetheless, it may be useful to investigate further reductions in radiation blockage by using frames of different streamlines or by placing the actuators (with the housing and housing cover) along a frame and away from the main view of the radiator. In fact, some louvers with actuator housings serving as frame members have been built and successfully flown. They are normally the smaller size with rows of single blades instead of the standard double-blade system.

4) Another attractive excursion is the study of different actuating systems that would narrow the temperature range between open and closed blades. For spring-actuated louvers, no further progress has been made in this area since the introduction in the early 1980s of a rather large, heavy spring that reduced the range to about 10 K (vs 18 K for the standard size). More improvement may be possible by using newer, high-tech bimetallics. But resorting to a larger spring will undoubtedly lead to an increase in the width and height of the actuator housing, thus introducing additional blockage. Hence, any benefit that might be gained from narrowing the temperature span must be weighed against the potential of operating at a higher temperature.

5) Unshielded louvers offer greater net heat rejection in most orbital situations, even when there is considerable exposure to the sun. On the other hand, the shield eliminates the problem of elevated temperatures on louvers' elements. It also removes uncertainties due to the scarcity of flight and test data on exposed louvered radiators in a solar and, especially, albedo environment.

6) Because of its low mass, a shield can undergo wide temperature cycling in orbit. This could cause cracking and a deterioration of adhesive film coatings (paticularly silvered Teflon) used for surface treament. At least part of the aggravation may be due to residual air bubbles trapped between the substrate and the film, and this may be avoided by applying adhesive tapes in a vacuum. Vacuum-packing bags have been successfully used in this process.

7) The cost of a shield vs the added cost to make a louvers set operate in the sun has not generally been a deciding factor. The choice has usually been a matter of weight and the availability of radiator area for disposing dissipated power.

Surprisingly, studies on shielding in the form of surrounding shades of sufficient height to block off solar impingement while maintaining the louvers in full view of space do not seem to have been reported. At its face value, a deployable, nonload-carrying conical frame made of flexible material with an aluminized interior (toward the louvers) would appear to offer a significant improvement. But one must assume that such additions would be complicated by questions of reliability and would undoubtedly require extensive mechanical and thermal testing.

VI.   Heaters and Heater Control

Heaters are resistor elements that generate heat ($I^2 R$ dissipation) when an electric current passes through them. They have been a standard fixture in satellite thermal systems since the early days of the space program, and it is not unusual that experimenters and subsystem engineers supply them as integral elements of their equipment. Their main function is to maintain components' temperatures above critical values, but they are also used to warm up dormant electronics prior to activation, to control temperature differences for enhanced dimensional stability, and as dissipators of excess satellite power.

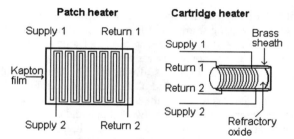

Fig. 6.46   Satellite heaters with two independent elements.

Patch heaters are the kind used most often. They consist of etched foil (electrical resistance elements, such as Nichrome) bonded within two thin sheets of a flexible insulator (75–125 μm Kapton is common), which can be made into various shapes and sizes, including long narrow strips. A typical patch is shown in Fig. 6.46. The wiring is often two or more independent elements for redundancy or varied heating levels. Standard installations use film adhesives to bond the heater on the intended surface, and sometimes clamps or screws are added to secure the corners against lifting.

Shunt heaters, in the shape of cylindrical cartridges of wound resistors encased within metallic housings (usually about 0.5-cm diameter and various lengths), are sometimes used in regions of high temperatures, where adhesive bonds are not reliable. The cartridge (on right in Fig. 6.46) is normally potted into a hole drilled into a component chassis or support bracket. Shunt heaters are common as dissipators of excess solar array power. They are also standard in the catalyst bed chamber of a hydrazine thruster, where they are used to raise the temperature to the minimum generally required before firing.

Control of a heater operation (current off and on) remains mostly through the snap action of a bimetallic thermostat (mechanical switch) that automatically opens and closes the electric circuit at prescribed temperatures (called set points). However, the use of solid-state controllers (electronic thermostats) is becoming more frequent for faster and narrower control. The heaters in these systems are controlled by relays operated by a power switch unit that receives (through an interface unit) signals from temperature sensors (thermistors) and makes comparisons with stored data of the various set points. At programmed time intervals, incoming temperature values are read and the relay's status updated to conform to the set points. Besides speed and accuracy (less than 0.1°C dead band), the set point of a solid-state thermostat can be reprogrammed while in orbit.

Mechanical thermostats and thermistors for the solid-state system are mounted and secured with good thermal contact (bonds, fillers, mechanical attachments, etc.) to the location where the temperature is to be sensed. When it is not possible to place a unit on the intended area, an analysis must be performed to determine the necessary adjustments on the set points needed to compensate for the temperature difference in the new location.

Heater–thermostat wiring configurations depend mainly on redundancy requirements.[61] An ideal design would have two independent circuits, one from each separate spacecraft power bus, with each circuit carrying its own redundancy (parallel wiring) in heater elements and thermostats (see Fig. 6.47), with options for ground override. Such elaborate schemes have been used where temperatures are

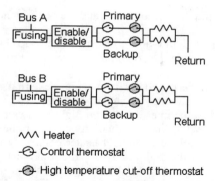

Fig. 6.47   Multiredundant heater–controller system.

exceedingly critical, as in hydrazine propellant systems (Fig. 6.48). In most other applications, however, single-point failures can be avoided with moderately uncomplicated circuitry.

Information and design data on heaters and controllers are usually found in manufacturers catalogs.[62–64]

A.   Heaters Classification

Heaters are classified in accordance with their intended use. But the classification may seem arbitrary as the same set of heaters is often configured for more than one function. Some designations include the following:

1) Thermostatically controlled primary and secondary (backup) thermal control heaters (also called zone heaters) are bonded to the mounting panels of electronics and instruments and other structures requiring temperature control.

2) Primary and secondary component heaters (also known as trim heaters) are an integral part of the components (including instruments) and are assembled and tested as part of the component. They are usually off when the component is not active.

3) Makeup heaters and associated backups supplement or substitute for component heaters but are considered satellite system elements. They are usually placed

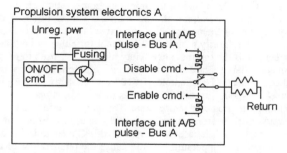

Fig. 6.48   Typical heater control of the catalyst bed compartment of a hydrazine propulsion system, identical redundant control through electronics B from satellite bus B.

externally in the vicinity of a component and may or may not be thermostatically controlled.

4) Survival heaters with low-power ratings are activated by low-temperature set points and used to maintain the components at temperatures above the minimum nonoperating limit.

5) Hard-line heaters derive power from sources external to the satellite and are used during launchpad and other ground operations.

6) Test heaters are sometimes bonded on flight hardware but used only during ground testing. Their extensions are generally cut and removed before flight.

Directions for a reliable and efficient heater control system often include requirements for reducing the multiplicity of parts. But the degree of simplification has to be commensurate with available margins, both in the heater system and the other thermal hardware. In this regard, thermal control system specifications frequently contain guidelines toward minimizing the number of individual heater patches and heater power levels. In particular, an assessment should be made of a common system that combines the functions of the makeup and survival heaters with that of the thermal control heaters.

B. Heater–Controller Performance Characterization and Specifications

A heater–controller system is characterized by the heater power and the controller set points. Heater power usually refers to the integrated average rate of energy consumption over a sufficiently long period in which the heater operates at some duty cycle, which is defined as the percent of the time the heater in on. The duty cycle is expected to be lower for a higher power rating, which is the power the heater is designed to deliver when it is continuously on.

Military Specification MIL-STD-1540B, *Test Requirements for Space Vehicles*[65] includes instructions for setting margins on heater power. Paragraph 3.45 of that document reads as follows:

> A thermal uncertainty margin is included in the thermal analysis of space vehicles to account for uncertainties in parameters such as complicated view factors, surface properties, contamination, radiation environments, joint conduction, and inadequate ground simulation. For components that have no thermal control, or have passive thermal control, the maximum predicted component temperatures should be at least 11 deg C above the maximum temperature estimated for each component based on measurements and analysis, and the minimum temperature should be at least 11 deg C below the minimum temperature estimated for each component based on measurements and analysis. The 11 deg C is the thermal uncertainty margin for the component. For active thermal control subsystems, a remaining control authority of at least 25 percent for either or both hot and cold limits is specified as the thermal uncertainty margin. It is used to provide a control margin equivalent to the 11 deg C uncertainty margin specified for passively controlled components. For example, if a 100 watt capacity proportional control heater is used, it should operate at 80 watts or less to maintain the component above the minimum predicted temperature. The duty cycle should be less than 80 percent for an on-off heater. A control authority margin in excess of 25 percent should be demonstrated in cases where an 11 deg C change in the analytically predicted component temperatures would cause the temperature of any part of the actively controlled component to exceed the acceptable temperature limit.

The reference to a heater control authority in relation to the 11°C margin seems open to different interpretations. For thermal control heaters, an acceptable sub-stitute statement that has been negotiated into some NASA contract specifications states that: the electrical resistance of the thermal control heaters shall be such that the power rating (heater continuously on) at minimum satellite voltage will main-tain the temperature under cold case conditions at least 11°C above the minimum allowable.

As the set point of a heater–controller system is determined based on temperature rather than power limits, the result of having an oversized heater is a shorter duty cycle netting the same average power consumption for maintaining the set temperatures.

Specifications for margins become less clear when heaters are used as "insur-ance" thermal hardware. Although they may be enabled (vs disabled), these heaters will come into operation only by ground command or through thermostatic ac-tion under extreme or unusual conditions. Heater and thermostat operations in these designs can be verified at the satellite level during thermal vacuum testing (Chapter 7, Sec. II.D) by creating cold conditions with reduced satellite power and lower surrounding temperatures.

Other specifications pertain to redundancy and override capability. They might include the following:

1) All heater switching circuits shall be subject to ground override.

2) All heater controllers shall be redundant and each with capability of inde-pendent operation.

Set points of redundant thermostats enabled through a common supply must be offset to avoid overlapping their operation and reducing reliability. Because of their wide dead band, the offsets in mechanical thermostats could force the backup unit to operate during a primary mode, leading to higher power consumption. By comparison, the substantially larger available range for settings with electronic thermostats offers a decided advantage.

C.   Predictions with Heater Thermostatic Control

Computer routines (for example, THRMST for SINDA) are available for in-corporating into the thermal model the operation of a themostatic switch. But predictions in a satellite that carries a large number of automatically controlled heaters can be tricky and the numerical results misleading if all of the heaters are continually operating off and on.

The difficulty is evident mostly when a number of heater-controller systems are being used to restrain the thermal distortion of a light structure. The require-ment sometimes translates into maintaining temperature differences on the order of fraction of a degree among key locations. Constraints of this magnitude are easily overwhelmed by small variations in the values of thermal and physical properties, tolerances on thermostat set points, inherent uncertainties in computer numerical calculations, and drifts (overshoot and undershoot) in temperature that are characteristic of light coupled systems (see example on thermal flutter at end of Chapter 5).

To some thermal engineers, designs having this characteristic are not viable because they would require extensive testing and dedicated provisions for on-orbit adjustments of set points and heater power levels. An alternative has been to increase the thermal mass and replace the requirement of a highly uniform spacial temperature distribution with limits on localized variations. The theory is that the effect of large differences in the istantaneous temperature profile become unimportant if small variations (say, within $\pm0.1°C$ by using solid-state relays) can be maintained on an known, calibrated temperature map. The procedure would then require calibrations to be completed just before initiating observations or measurements, which are frequently performed a few times in the same segment (location) of consecutive orbits.

Composites with negligible coefficients of expansion (Chapter 2, Sec. III and Fig. 2.4a) provide a direct approach for limiting thermal distortions without strict control of temperature differences. The dimensional stability of NASA's Far Ultraviolet Spectroscopic Explorer (FUSE) Instrument, slated for launch in 1998, is achieved by a thermal design that combines solid-state heater control with a low coefficient of expansion composite of graphite fiber in cyanate ester.

Predictions in massive systems with heater thermostatic control can be managed by iterating on the results of steady-state computer runs with estimated constant power values imposed on individual heaters. If needed, the computations are repeated with updated powers until the calculated temperatures of the nodes locating the thermostats or temperature sensors are consistent with the value of the set point. In most cases, two or three iterations are found sufficient to complete the assessment. The heaters are then sized with the required margins and, traditionally, a computer run is made with time-varying heating to verify the duty cycles in at least one orbital case.

D. Systems Evaluation of Controlled Heaters

There is an unavoidable tolerance about each set point in the bimetallic thermostat that can usually be limited to about $\pm3°C$. If the dead band between the open and closed temperatures is confined to less than about $6°C$, the overlapping of tolerances could cause dithering with a drift in the set point as the thermostat begins to rapidly open and close. This problem does not exist with solid-state controllers (even though the dead band is a fraction of a degree), and for this and reasons relating to accuracy and longer life, the solid-state device and direct relay control by onboard computers are now generally preferred.

Redundancy must not be jeopardized by the physical proximity of paired heater elements or controllers components. A failure of a heater wire in a Kapton patch has been known to cause damage and failure in an adjacent element. Such occurrences are reduced by adequate spacing between wires within the limits of not seriously affecting heating uniformity. Similarly, in the mounting of thermostats and sensing elements to conductive interfaces, common bonds must be avoided when a local danage could affect neighbouring units.

Caution must be exercised when heaters are to be covered by MLI. An MLI treatment rarely has uniform effectiveness and, therefore, if the thermal conductivity of the insulated surface is not sufficiently high, a nominal insulation effectiveness in the area of sensed temperatures might cause a heater to remain on while other regions of high effectiveness rise to excessive temperatures. The possibilty of this happening with heated hydrazine propellant lines is discussed in an example in Chapter 5, Sec. IV.B.

The number of heaters and sensors in a satellite can be reduced if heat pipes are used. Heat pipes by their nature tend to isothermalize a platform, which makes it possible to establish fewer representatives locations for the temperature. Thus,

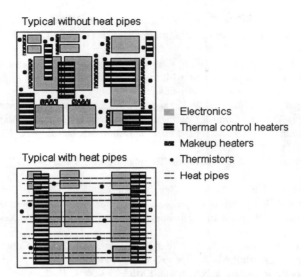

Fig. 6.49 Number and arrangement of heaters and thermistors on panels with and without heat pipes.

a single heater–controller system and backup, with the heater patch spread over strategic regions on a mounting platform, could provide control over all components on that platform that have the same temperature requirements (see Fig. 6.49).

References

[1]Gilchrist, W. H., "GSFC Procedures for Layout and Cleaning of Thermal Blankets," NASA Document 905-G-407, 1977.

[2]NASA-CR-72605; also Lockheed Missile & Space Co. Rep. LMSC-A903316, 1971.

[3]Stimpson, L. D., and Jaworski, W., "Effects of Overlaps, Stitches, and Patches on Multilayer Insulation," AIAA Paper 72-285, 1972.

[4]Lin, E. I., Stultz, J. W., and Reeve, R. T., "Effective Emittance for Cassini Multilayer Insulation Blankets and Heat Loss near Seams," *Journal of Thermophysics and Heat Transfer*, Vol. 10, No. 2, 1996, pp. 357–364.

[5]Stein, B., "LDEF Materials Overview," *Long Duration Exposure Facility (LDEF)—Sixty-Nine Months in Space Second Post-Retrieval Symposium*, edited by A. S. Levine, NASA CP-3144, 1992, pp. 741–789.

[6]Hedin, A. E., "MSIS-86 Thermopheric Model," *Journal of Geophysical Research (Atmosphere)*, Vol. 92, No. A5, 1987, pp. 4649–4662.

[7]Levine, A. S. (ed.), *Long Duration Exposure Facility (LDEF)—Sixty-Nine Months in Space First Post-Retrieval Symposium*, NASA CP-3134, 1991, pp. 643–661, 763–779, 831–845, 899–917, 935–944, 975–987, 1073–1107.

[8]Levine, A. S. (ed.), *Long Duration Exposure Facility (LDEF)—Sixty-Nine Months in Space Second Post-Retrieval Symposium*, NASA CP-3144, 1992, pp. 741–789, 1061–1072, 1099–1110.

[9]Levine, A. S. (ed.), *Long Duration Exposure Facility (LDEF)—Sixty-Nine Months in Space Third Post-Retrieval Symposium*, NASA CP-3275, 1993.

[10]Triolo, J. J., and Ousley, G. W., "Returned Solar Max Hardware Degradation Study Results," *Space Environment Effects on Material Workshop*, NASA CP-3035, 1988, pp. 133–159.

[11]de Groh, K. K., and Banks, B. A., "Atomic-Oxygen Undercutting of Long Duration Exposure Facility Aluminum Kapton Multilayer Insulation," *Journal of Spacecraft and Rockets*, Vol. 31, No. 4, 1994, pp. 656–664.

[12]Hemminger, C. S., Stuckey, W. K., and Uht, J. C., "Space Environmental Effects on Silvered Teflon Thermal Control Coatings," *Long Duration Exposure Facility (LDEF)–Sixty-Nine Months in Space First Post-Retrieval Symposium*, edited by A. S. Levine, NASA CP-3134, 1991, pp. 831–845.

[13]"Specifications for Application of Thermal Control Paint S-13G," NASA MSFC 10MO1835.

[14]Hall, D. F., and Fote, A. A., "α_s/ε_H Measurements of Thermal Control Coatings over Four Years at Geosynchronous Altitude," *Spacecraft and Contamination: Sources and Prevention*, Vol. 91, Progress In Astronautics and Aeronautics, AIAA, New York, 1984, pp. 215–234.

[15]Hall, D. F., and Fote, A. A., "10-Year Performance of Thermal Control Coatings at Geosynchronous Altitude," AIAA Paper 91-1325, 1991.

[16]*Solar Max Repair Mission*, NASA 408-SMRM-79-0001, 1985.

[17]Henninger, J. H., "Solar Absorptance and Thermal Emittance of Some Common Spacecraft Thermal-Control Coatings," NASA RP-1121, 1984.

[18]Hall, D. F., and Fote, A. A., "Thermal Control Coatings Performance at near Geosynchronous Altitude," *Journal of Thermophysics and Heat Transfer*, Vol. 6, No. 4, 1992, pp. 665–671.

[19]Wales, R. O. (ed.), *ATS-6 Final Engineering Performance Report. Vol. I—Program and Systems Summaries; Mechanical and Thermal Details*, NASA RP-1080, 1981.

[20]Stultz, J. W., "Solar Absorptance of Second Surface Mirrors for High Angles of Incidence," AIAA Paper 74-670, 1974.

[21]Drolen, B. L., "Bidirectional Reflectance and Specularity of Twelve Spacecraft Thermal Control Materials," *Journal of Thermophysics and Heat Transfer*, Vol. 6, No. 4, 1992, pp. 672–679.

[22]Brosmer, M. A., Fischer, W. D., and Hall, D. F., "Thermal Analysis of Flight Calorimeter Instrument Designs and Calibration Test Methods," AIAA Paper 87-1622, 1987.

[23]Karam, R., and Braun, C., "Thermal Aspects of Precision Orbit Determination (POD)," AIAA Paper 90-1703, 1990.

[24]Meshishnek, M. J., Gyetvay, S. R., and Jaggers, C. H., "Long Duration Exposure Facility Experiment Deintegration/Findings and Impacts," *Long Duration Exposure Facility (LDEF)—Sixty-Nine Months in Space First Post-Retrieval Symposium*, edited by A. S. Levine, NASA CP-3134, 1991, pp. 1073–1107.

[25]Chi, S. W., *Heat Pipe Theory and Practice*, Hemisphere, Washington, DC, 1976, pp. 43–47, 79–84.

[26]Dunn, P., and Reay, D. A., *Heat Pipes*, 2nd ed., Pergamon, Oxford, England, UK, 1982.

[27]Faghri, A., *Heat Pipe Science and Technology*, Taylor and Francis, Washington, DC, 1995, Chap. 9, p. 621.

[28]Ivanovsky, M. N., Sorokin, V. P., and Yagodkin, I. V., *The Physical Principles of Heat Pipes* (translated from Russian), Clarendon Press, Oxford, England, UK, 1982.

[29]Schlitt, K. R., Kirkpatrick, J. P., and Brennan, P. J., "Parametric Performance of Extruded Axial Grooved Heat Pipes from 100 K to 300 K," AIAA Paper 74-724, 1974.

[30]Brennan, P. J., Kroliczek, E. J., and Jen, H., "Axially Grooved Heat Pipes: 1976," AIAA Paper 77-747, 1977.

[31]Berger, M. E., and Kelly, W. H., "Application of Heat Pipes to the ATS-F Spacecraft," American Society of Mechanical Engineers, Paper 73-ENAs-45,1973.

[32]Cotter, T. P., "Theory of Heat Pipes," Los Alamos Scientific Laboratory Report LA-3246-MS, 1965; also USAEC Report LA-3246, 1965.

[33]Winter, E. R. F., and Barsch, W. O., "The Heat Pipe," *Advances in Heat Tansfer,* edited by T. F. Irvine, Jr., and J. P. Hartnett, Vol. 7, Academic, New York, 1971, pp. 219–320g.

[34]Kemme, J. E., "High Performance Heat Pipes," *Proceedings: 1967 Thermionic Conversion Specialist Conference,* 1967.

[35]Carey, V. P., *Liquid-Vapor Phase-Change Phenomena,* Hemisphere, Washington, DC, 1992, pp. 169–214.

[36]Nishikawa, K., and Fujita, Y., "Boiling Heat Transfer and its Augmentation," *Advances in Heat Transfer,* edited by T. F. Irvine, Jr., and J. P. Hartnett, Vol. 20, Academic, New York, 1990, pp. 1–82.

[37]Marcus, B. D., "Theory and Design of Variable Conductance Heat Pipe," NASA-CR-2018, 1972.

[38]van Stralen, S. J. D., Zijl, W., and de Vries, D. A., "Bubble Growth Rates in Aqueous Binary Systems at Subatmospheric Pressures," *Heat and Mass Transfer Source Book: First all-Union Conference,* Scripta, Washington, DC, 1976, pp. 188–296.

[39]Cox, R. L., and Leach, J. W., "Flexible Deployable–Retractable Space Radiators," Vol. 60, Progress in Astronautics and Aeronautics, AIAA, New York, 1977, pp. 243–262; also AIAA Paper 77-764, 1977.

[40]Feig, J., "Radiator Concepts for High Power System in Space," AIAA Paper 84-0055, 1984.

[41]Masumoto, H., "Development of a VCHP/FCHP Radiator System for 3-Axis Stabalized Geostationary Satellite Application," AIAA " Paper 85-1012, 1985.

[42]Butler, D., and Hoang, T., "The Enhanced Capillary Pumped Loop Flight Experiment: A Prototype of the EOS Platform Thermal Control System," AIAA Paper 91-1377, 1991.

[43]Karam, R. D., and Hwangbo, H., "Thermal Interaction between a Heat Pipe Radiator and a Coolant Fluid Header," *Advances in Heat Pipe Technology: Proceedings of the Fourth International Heat Pipe Conference,* edited by D. A. Reay, Pergamon, Oxford, England, UK, 1981, pp. 493–503.

[44]Balaji, C., Sri Jayaram, K., and Venkateshan, S. P., "Thermodynamic Optimization of Tubular Space Radiators," *Journal of Thermophysics and Heat Transfer,* Vol. 10, No. 4., 1996, pp. 705–707.

[45]Raab, B., "A Thermally Integrated Spacecraft Design Approach Using Nuclear Dynamic Power Systems," *Proceedings of the Thirteenth Intersociety Energy Conversion Engineering Conference,* 1978, pp. 150–157.

[46]Roukis, S. L., Kroliczek, E., and Hall, G., "COMET Service Module Thermal Control System Design Using a Capillary Pumped Loop," Society of Automotive Engineers, Paper 921367, 1992.

[47]Kroliczek, E. J., Ku, J., and Ollendorf, S., "Design, Development, and Test of a Capillary Pump Loop Heat Pipe," AIAA Paper 84-1720, 1984.

[48]Ku, J., Kroliczek, E., Butler, D., and McIntosh, R., "Functional and Performance Tests of Two Capillary Pumped Loop Engineering Models," AIAA Paper 86-1248, 1986.

[49]Ku, J., Kroliczek, E., Taylor, W. J., Schweickart, R., and McIntosh, R., "Capillary Pumped Loop—GAS and Hitchhiker Flight Experiments," AIAA Paper 86-1249, 1986.

[50]Ku, J., "Overview of Capillary Pumped Loop Technology," *Proceedings of the American Society of Mechanical Engineers National Heat Transfer Conference*, 1993 .

[51]Kirkpatrick, J. P., and Brennan, P. J., "Advanced Thermal Control Flight Experiment," *ATS-6 Final Engineering Performance Report, Vol. I—Program and Systems Summaries; Mechanical and Thermal Details*, edited by R. O. Wales, NASA RP-1080, 1981, pp. 201–223.

[52]Eby, R. J., and Karam, R. D., "Louvers for Spacecraft Temperature Control," AIAA Paper 87-1567, 1987.

[53]CHACE Design Catalogue, GTE Metal Laminates Div., Reidsville, NC.

[54]Furukawa, M., "Analytical Studies on Design Optimization of Movable Louvers for Space Use," *Journal of Spacecraft and Rockets*, Vol. 16, No. 6, 1979, pp. 412–425.

[55]Michaleck, T. J., Stipandic, E. A., and Coyle, M. J., "Analytical and Experimental Studies of an All-specular Thermal Control Louver System in a Solar Environment," AIAA Paper 72-268, 1972.

[56]Hwangbo, H., Hunter, J. H., and Kelly, W. H., "Analytical Modeling of Spacecraft with Heat Pipes and Louvers," AIAA Paper 73-773, 1973.

[57]Karam, R. D., "Temperature Distribution in Louvered Panels," *Journal of Spacecraft and Rockets*, Vol. 16, No. 2, 1979, pp. 92–97.

[58]Byrd, P. F., and Friedman, M. D., *Handbook of Elliptic Integrals for Engineers and Physicists.*, Springer-Verlag, Berlin, 1954, Formulas 710.07 and 730.00.

[59]Hardt, B. E., Karam, R. D., and Eby, R. J., "Louvers," *Satellite Thermal Control Handbook*, edited by D. G. Gilmore, The Aerospace Corp. Press, El Segundo, CA, 1994, pp. 4-99–4-121.

[60]Kelly, W. H., Reisenweber, J. H., and Flieger, H. W., "High-Performance Thermal Louver Development," AIAA Paper 76-460, 1976.

[61]Gilmore, D. G., "Heaters, Thermostats, Solid State Controllers," *Satellite Thermal Control Handbook*, edited by D. Gilmore, The Aerospace Corp. Press, El Segundo, CA, 1994, pp. 4-87–4-95.

[62]"MINCO Thermofoil Heaters," MINCO Bulletin HS-200, Minco Products Inc., Minneapolis, MN, 1982.

[63]"Elmwood Sensors: Elmwood Sensors 3000 Series," Elmwood Sensors Inc., Pawtucket, RI, 1990.

[64]"Sundstrand Thermal Switches: Sundstrand 550 Series Precision Thermal Switches," Sundstrand Data Control Inc., Redmond, WA, 1985.

[65]"Military Standard, Test Requirements for Space Vehicles," MIL-STD-1540B-U.S. Air Force, Oct. 1982, Paragraph 3.45, pp. 16–17.

Chapter 7

Thermal Verification Tests

I. Introduction

THERMAL verification tests are of two varieties, those dedicated to confirming the validity of thermal control and those devised to affirm components' integrity and workmanship as individual units and at the satellite level. In the validation test, the input is a predetermined heating load (usually satellite power and simulated orbital fluxes) with temperature as output to be compared and evaluated relative to predictions. The input in the workmanship test is a temperature level (and, consequently, a thermal stress) that usually exceeds the expected value in normal operation, and the output is performance and survivability under short-term but extreme conditions. Workmanship tests are intended to screen out flawed components and project reliability in normal long-term orbital operation.

Test requirements vary, depending on the particular program and acceptable risks. But there is general agreement that thermal testing is more revealing when done in a vacuum, as the absence of heat transfer by convection is likely to lead to results similar to those that would be encountered in space.

For standards on definitions, methods, and levels of heating and cooling, Military Standard Specifications MIL-STD-1540-B (Ref. 1) and companion document MIL-STD-340 (Ref. 2) are considered a starting point, although many of the documents' requirements relate to a time when lesser quality parts and materials were being used and serious discrepencies were frequently observed between predictions and actual performance. For many NASA programs, the 1540-B is viewed more as a guideline than a set of rules demanding strict adherence. NASA's procedures for testing depend more on the history of the component and the margins in the thermal and mechanical designs.[3]

II. Thermal Balance Test

As stated in Chapter 1, the initial guide in satellite thermal control is a temperature matrix to which the satellite must conform if it is to function effectively throughout its mission. In the design phase, the thermal engineer invokes various analyses to determine the necessary hardware which, on paper, results in temperatures that do not exceed assigned specifications even under worst tolerance and heating conditions. Sometimes, intermediary development tests are performed during this phase to determine or confirm hardware properties and performance. Final validation before flight is through a thermal balance test, which is conducted to verify that the methods of analysis are legitimate and the temperature of the assembled satellite will indeed be within the confines of mission requirements. It

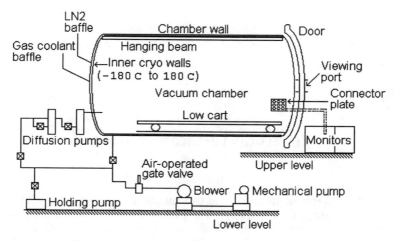

Fig. 7.1   Schematic of a thermal vacuum chamber.

is expected at this point that any discrepencies would be minor and, if necessary, may be removed by slight modifications on the design and by correlation of the thermal model with test results.

Thermal balance tests are performed in vacuum chambers (air pressure in the 10^{-6} torr range) equipped with heating sources and cryogenic surroundings for simulating orbital fluxes and space background. Most chambers are also fitted with contamination detectors and condensable collectors for chemical analysis. A connector pin plate serves as an interface for transmitting data to external monitors. A schematic showing the major parts of a typical vacuum chamber is given in Fig. 7.1.

Normally, only the main electronics canister of a satellite is tested, but it is often tested with nonflight attachments designed to be heated or cooled to recover the thermal effects of deployed structures such as solar arrays and antennas. Guard heaters, reflective shields, and isolators are also used wherever necessary to nullify heat exchanges that are characteristic of the test configuration but absent in orbit. The stationing of the satellite in the chamber is depicted in Fig. 7.2.

In many programs, the thermal balance test occupies only a segment of an extended thermal vacuum test conducted to verify the functional performance

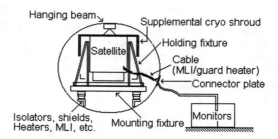

Fig. 7.2   Typical placement of a satellite electronics canister in a vacuum chamber.

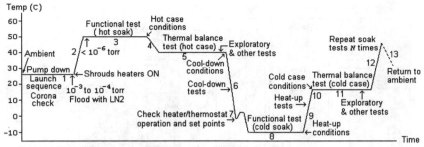

1. Chamber evacuated at ambient temperature. Corona and launch functional tests
2. Shrouds flooded with LN2 while heaters on high. Satellite on high power to hasten hot soak approach
3. Satellite functional tests with platforms at high acceptance level temperature
4. Shrouds temperatures and satellite power set to thermal balance hot case test conditions
5. Hot case thermal balance test followed by exploratory tests
6. Satellite power on low (some components off) and heaters OFF on shrouds to hasten cold soak approach
7. Check satellite heaters and thermostat set points
8. Satellite functional tests at low acceptance level temperatures (possible disabled heater-thermostat system)
9. Satellite power on high and shrouds at maximum heating to hasten temperature rise
10. Shrouds temperatures and satellite power set to thermal balance cold case conditions
11. Cold case thermal balance test followed by exploratory tests
12. Heat-up for second hot soak functional test

Fig. 7.3   Thermal vacuum tests profile, the validity of the thermal design is assessed from data in the period 4–11.

of the assembled satellite while it is exposed to various levels of temperature. A common profile is given in Fig. 7.3. It shows, in addition to thermal balance, various checkout procedures and functional operations that emulate orbital scenarios but at imposed elevated and reduced temperatures. The values marked on the ordinate are an example of the relative levels at which the tests are conducted. Usually, however, not all satellite parts have the same temperature, and provisions must be made to insure that critical components do not exceed their specified limits.

A. Instrumentation

Test temperatures are monitored through readouts from thermocouples placed, usually with adhesive tape to avoid damaging the thermal finish, in regions where the temperatures are representative of the thermal performance and may be used to make comparisons with predictions. Supplementary data are also gathered from flight thermistors. The electrical power (I^2R and EI) is tapped across the satellite or its separate modules and often across individual components of special interest. Data are retrieved along connections leading to recorders and monitors external to the vacuum chamber. Normally, wires are collected into as few bundles as possible and wrapped with MLI. Guard heaters are placed over the insulation wraps and exercised to limit fin effects.

The number of thermocouples depends on the size of the satellite and also the complexity of the thermal system. The count is constrained only by accessibility and the availability of monitoring channels, and sometimes it may have to be in the hundreds to make a detailed assessment. Figure 7.4 is an example derived from a section of the TOPEX/Poseidon instrument module.[4] Essential locations common to most satellites are listed in Table 7.1.

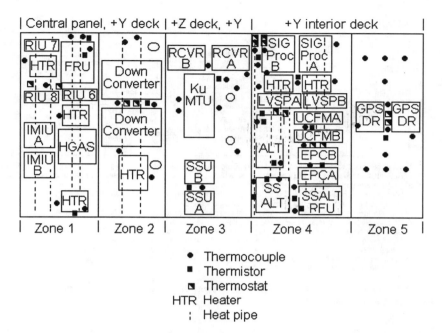

Fig. 7.4   Instrumentation of NASA/JPL TOPEX/Poseidon instrument module for the thermal balance test.

B.   Definition of Equilibrium

Most thermal balance tests are designed to evaluate average orbital behavior defined by equilibrium or quasi-equilibrium temperature states. These correspond to constant profiles of average external heating flux and constant or uniformly periodic operational power. As noted before on a number of occasions, averaging is an adequate description when the satellite is sufficiently massive that its internal temperature does not vary significantly in the course of an orbit.

Achieving a state of equilibrium or quasi-equilibrium is essential as even the smallest monotonic rate of change in temperature can eventually build up to a few degrees, which may be crucial to proving the performance. Generally, a state where only the minutest change in temperature is observed everywhere could take considerable (and unaffordable) test time to attain. Equilibrium, however, can be inferred by observing thermocouple and thermistor readings following forced, momentary increases or decreases in temperature. Thus, when a sensor indicates a slow approach to equilibrium from cooler to warmer levels, a deliberate short-term increase in heating (manipulating component power, heaters, or sink temperature) will cause the temperature to overshoot its equilibrium value, but then begin to converge back as original test conditions are restored. Similarly, when equilibrium is being approached from a higher temperature, a short undershoot in temperature (momentarily turning down power) will confirm the equilibrium value if it begins to rise again. Sometimes a series of small, stepped changes may be necessary to avoid excessive drifts from near equilibrium.

Table 7.1   Thermocouples locations and data reduction in the thermal balance test

Item	Thermocouple location	Data reduction
Electronics components	Chassis, covers, and components baseplates	Check calibration of components flight thermistors
Structure	Near components flight thermistors	Check calibration of structure thermistors; use in thermal model correlation
Structure	Near heaters	Track heaters activity and effectiveness
Structure	Near thermostats	Monitor and verify thermostat set points
Components mounting platforms	Vicinity of components	Compare with components thermistors for effectiveness of component mounting interface
Interior mounting platforms and radiators	Interior surfaces corresponding to collection of nodes in thermal model	Use in thermal model correlation
Exterior (space-facing) radiators	Exterior surfaces, including louvered planes, corresponding to collection of nodes in thermal model	Use in thermal model correlation; use with interior thermocouple data to verify ΔT across mounting panel
Mounting platforms	On platforms along heat pipe runs, including both ends	Thermal model correlation; verify heat pipe performance
Cables, wiring, stands, stimulators, appendages	Across penetrations and on guard heaters	Negate heat loss and gain by controlling ΔT
Cold plates, shrouds	Uniform distribution	Control of temperature input

In many cases the arithmetic average of the readings from a selected set of thermocouples serves as the standard indication of equilibrium. Clearly, however, a judgement must be made to distinguish sets requiring independent monitoring and manipulation and to sort out, or even discard, what may be considered nonessential or maverick data.

Examples based on actual test data are shown in Figs. 7.5 and 7.6. As stable conditions were being approached, the average temperature in zone 2 in Fig. 7.5 was following a slow but steady rise. The set point of the heater thermostat in that region was considerably below the expected temperature, and the simplest option for speeding up the trend and proving equilibrium was to override the thermostat circuit and manually operate the heater for short periods at rated value. During about that same time frame, zone 4 was showing a steady decrease in its temperature. The drop was momentarily hastened by turning off one of the experiments. Equilibrium was then ascertained by noting that in each case when conditions returned to normal operation there was a convergence to the same temperature from above and from below.

Figure 7.6 describes a quasi-equilibrium state. The data were from a thermocouple located in the coolest region of an insulated hydrazine propellant line thermally

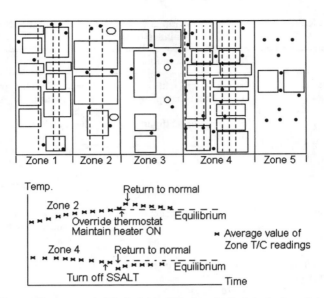

Fig. 7.5   Thermal balance equilibrium; equilibrium is declared when the same T/Cs reading (average) is approached from above and from below (test results from the instrument module of NASA/JPL PTOPEX/Poseidon).

controlled by a heater–thermostat system. Although there was definite cyclic activity confirming a stable situation, the cycle's profiles were not exactly identical and continued in this fashion throughout 12 days of tests. But despite the nonrepetitive variation, the behavior was considered acceptable and declared to fall within the quasi-equilibrium criterion after making allowances for inconsistensies in the thermostat dead band.

Tolerances may be included in the definitions of equilibrium in thermal balance tests as follows:

1) In approaching equilibrium from higher to lower temperature, the value must be confirmed to within ±1.0°C by a momentary forced decrease in temperature followed by a spontaneous temperature rise.

2) In approaching equilibrium from lower to higher temperature, the value must be confirmed to within ±1.0°C by a momentary forced increase in temperature followed by a spontaneous temperature decrease.

3) Quasi-equilibrium is confirmed when, under stable or regularly periodic heating, the temperature vascilltes at least three times about an average value.

C.   Flux Simulation: Heating Shrouds

Orbital heating fluxes are commonly simulated by the emitted energy from heated shrouds placed facing the satellite's exposed radiators. Although the process is wholly in the IR regime and, therefore, not a true simulation of energy that includes the solar spectrum, it has proven convenient and adequate for most applications. Occasionally, situations requiring knowledge of behavior in an actual solar environment have been handled by solar simulation with a spectrum of

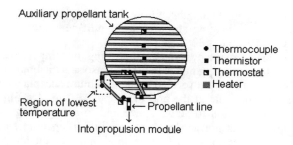

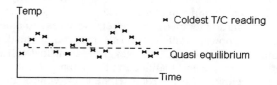

Fig. 7.6 Quasi-equilibrium is declared when a stable cyclic variation in temperature is observed (test results from the auxiliary propulsion system of NASA/JPL TOPEX/Poseidon).

xenon or quartz–iodine lamps at about 3000 K. Usually, however, these tests are conducted on specific local regions or relatively small components. Flux simulation with IR lamps and skin heaters bonded directly to a surface (mostly on the inside) are becoming less common as experience has increased in designing and operating shrouds.

Shrouds are coated with high emissivity and placed as close as practical in front of the radiator. The gap around the perimeter is normally closed by a reflective material, such as MLI with an inner lining of aluminized Kapton slightly slanted outward to effectively make the shape factor from radiator to shroud approximately 1.0.

The schematic shown in Fig. 7.7 represents a common shroud design in which the target temperature is established by thermostatic control of bonded heaters while liquid nitrogen continuously flows through an integrated pipe system. The tubing layout must result in acceptable uniformity of shroud temperature, and manifold constructions to upgrade uniformity are sometimes made to contend

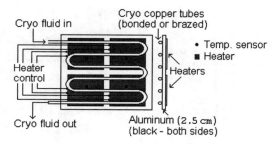

Fig. 7.7 Heating shroud for simulating orbital fluxes and space background in a vacuum chamber.

with large areas and extended lengths of flow lines. A typical thermal specification for the design of a test heating shroud may require temperature control from $-180°C\,(-15°C, +10°C)$ to $180°C\,(\pm5°C)$.

The lower, cryogenic range pertains to controlling the temperature of radiators with negligible or no impinging fluxes in orbit. It also sets the boundary condition at which satellite operations are verified at the minimum acceptance (soak) limit. As described in the example at the end of chapter 4 (see also Fig. 4.11b), the difference between radiation to actual space and to surroundings at up to 100 K is not too significant at normal levels of satellite dissipation, and therefore a large tolerance on temperature may be permitted in the cryogenic range. This is an important advantage as sensitivity to heat variations at cryogenic temperatures makes it difficult to maintain a tight tolerance.

Upper values of shroud temperature are relatively easy to attain because of almost limitless heater power available from electrial sources on the ground. Generally, there is ample margin to establish a background sink that will allow verification at the maximum acceptance or qualification temperature limit.

Less effective techniques for shroud temperature control include cooling by gaseous nitrogen or combined parallel flows with glycol, or by using direct radiation to surrounding cryogenic chamber walls. On rare occasions, the temperature has been varied by changing the rate of flow of the coolant. In all cases, however, the minimum achievable temperature is restricted by the limitations of commercially available cryogenic fluids.

The equivalence principles discussed in Chapter 4, Sec. V.B (also Chapter 6, Secs. II.A and V.B) can be used to translate the total absorbed heating by a radiator in space into a facing shroud temperature. The connection is shown in Fig. 7.8.

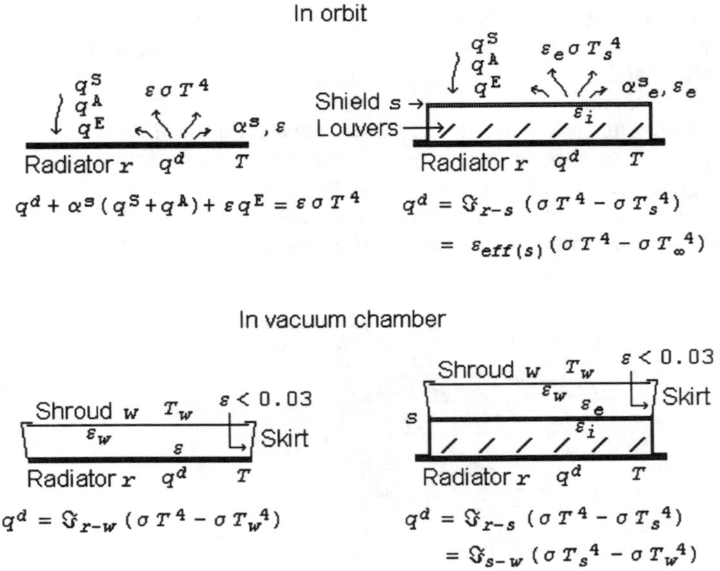

Fig. 7.8 Models for determining the shroud temperature that will simulate the absorbed external heating in orbit.

Table 7.2 Equivalent shroud temperatures for simulating orbital fluxes[a,b]

Standard radiator (temperature T, emissivity ε, and solar absorptivity α^s):

$$\sigma T_w^4 = (\varepsilon/\Im_{r-w})\big[\sigma T_\infty^4 - (1 - \Im_{r-w}/\varepsilon)\sigma T^4\big]$$

$$\Im_{r-w} = [1/\varepsilon + 1/\varepsilon_w - 1]^{-1}$$

$$\sigma T_\infty^4 = (\alpha^s/\varepsilon)(q^S + q^A) + q^E$$

Insulated radiator [temperature T, $\varepsilon_{\text{ins}} < \approx 0.05$, $\varepsilon_{\text{outer cover}} > \approx 0.50$, $\alpha_{\text{ins}}^s \equiv \varepsilon_{\text{ins}}(\alpha^s/\varepsilon)_{\text{outer cover}}$]:

$$\sigma T_w^4 = (\varepsilon_{\text{ins}}/\Im_{\text{ins}})\big[\sigma T_\infty^4 - (1 - \Im_{\text{ins}}/\varepsilon_{\text{ins}})\sigma T^4\big]$$

$$\Im_{\text{ins}} = [1/\varepsilon_{\text{ins}} + 1/\varepsilon_w - 1]^{-1}$$

$$\sigma T_\infty^4 = (\alpha_{\text{ins}}^s/\varepsilon_{\text{ins}})(q^S + q^A) + q^E$$

Unshielded louvers (see Chapter 6, Sec. V.E) (α^s is radiator solar absorptivity; T is radiator temperature):

$$\sigma T_w^4 = (\varepsilon_{\text{eff}}/\Im_{r-w})\big[\sigma T_\infty^4 - (1 - \Im_{r-w}/\varepsilon_{\text{eff}})\sigma T^4\big]$$

$$\Im_{r-w} = [1/\varepsilon_{\text{eff}} + 1/\varepsilon_w - 1]^{-1}$$

$$\sigma T_\infty^4 = (\alpha_{\text{eff}}^s/\varepsilon_{\text{eff}})S + \alpha^s q^A + q^E$$

Shielded louvers (Chapter 6, Sec. V.E) (shield external emissivity ε_e, internal ε_i, T is radiator temperature):

$$\sigma T_w^4 = [(1 - \varepsilon_{\text{eff}(s)}/\varepsilon_e)(1 + \Im_{r-s})/\Im_{s-w}]\sigma T_\infty^4$$

$$- [(1 - \varepsilon_{\text{eff}(s)}/\varepsilon_e)(\Im_{r-s}/\Im_{s-w}) - \varepsilon_{\text{eff}(s)}/\varepsilon_e]\sigma T^4$$

$$\Im_{r-s} = [1/\varepsilon_{\text{eff}(s)} + 1/\varepsilon_i - 1]^{-1}$$

$$\Im_{s-w} = [1/\varepsilon_e + 1/\varepsilon_w - 1]^{-1}$$

$$\sigma T_\infty^4 = (\alpha_{\text{eff}}^s/\varepsilon_{\text{eff}})S + \alpha^s q^A + q^E$$

The calculations are based on a shield temperature consistent with

$$\sigma T_s^4 = (\alpha_e^s/\varepsilon_e)(q^S + q^A) + q^E + q^d/\varepsilon_e$$

[a] Equations for exchange factor $\Im$ assume effective shape factor from radiator to corresponding shroud equal to 1.0.

[b] T_w is shroud (wall) temperature and ε_w emissivity of shroud surface facing the radiator.

By combining energy conservation with the condition that the temperature will be the same in space as in the vacuum chamber when the dissipation is q^d, the results in Table 7.2 are obtained. The radiator's absorptivity and emissivity are replaced by α_{ins}^s and ε_{ins} when dealing with insulated radiators and by α_{eff}^s and ε_{eff} when louvers are used. The equations are, of course, predicated on the assumption that surface and effective properties, together with the individual values of absorbed orbital fluxes, are known. Two sets of shroud temperatures are usually calculated corresponding to hot case and cold case conditions.

D. Hot Case, Cold Case, Cool-Down, and Exploratory Tests

Traditionally, the thermal balance test is conducted under conditions that represent the hot case and cold case which bracket the range of operating temperatures in which the thermal design was based. The hot case and cold case were defined earlier (Chapter 5, Sec. III.C) in terms of the values of properties and parameters to be used in analysis. In testing, however, it is not possible to impose tolerances and deviations on the actual design or material and hardware properties, but there is considerable flexibility in selecting operational scenarios in terms of components dissipation and the amount of external heating (shroud temperature). A test hot case, for example, may consist of a steady-state, maximum component activity at high bus voltage, with shroud temperatures calculated based on EOL values of surface properties with positive margins on the impinging fluxes. A cold case, on the other hand, may relate to the thermal performance at minimum satellite activity with most of the shrouds set at the lowest temperature.

Other, exploratory and special tests may also be planned and conducted on the balance configuration with minimal or no modifications. One is a cool-down test in which most of the components are turned off (or put on stand-by) while surrounding shrouds are driven as cold as possible. The heater–thermostat circuits are enabled during this test and their function verified as the temperature drops to the set point level. In addition to checking out the thermostats, the recorded cool-down rates provide the data needed to confirm calculations involving thermal mass and transient parameters.

Suspected deficiencies in the thermal design can be scanned and scrutinized by varying the temperatures of specific shrouds or manipulating satellite heaters, guard heaters, and the operation of individual components. A common example pertains to using heating shrouds opposite MLI blankets for the purpose of judging the MLI effectiveness by noting the influence of a hot or cold shroud on the insulated surface. Another exercise is to increase the heat load at one end of a heat pipe radiator to determine its ability to transport heat to the other end. Satellite heater–thermostat systems invariably have versatile backup and override capabilities that can be utilized for this and other diagnostic testing.

E. Planning and Performing the Thermal Balance Test

Among the considerations that enter into planning and performing a thermal balance test are the following:

1) The necessity or extent of a thermal balance test must be reviewed when the thermal design resembles or is a modification of previous, successfully flown satellites. It is possible to agree that if similar units exhibit sufficient thermal margins (heaters' duty cycle below predictions, temperatures well within the allowables, etc.) then there may be justification to forgo a costly thermal balance test. The risk is a function of the margins exhibited in previous performances plus those imposed in screening, accepting, and qualifying individual or collective components.

2) Historically, hydrazine propulsion systems have always required a thermal balance test as part of the subsystem acceptance criteria. Hydrazine systems are thermally critical because the propellant must be kept at some margin (usually 10°C) above its freezing temperature (1.5°C) at all locations, including thin and straddling lines with few points of attachment. It must also be guaranteed that temperatures remain below an upper limit (usually 35 to 40°C) to be consistent

with the requirements of neighboring electronics. This reduced permissible range in temperature excursions usually invites complicated redundancies and, in many cases, intricate design techniques whose consistency can best be confirmed by repeating the thermal balance test for each unit.

3) It is sometimes argued that because the thermal balance test cannot reveal the long-term degradation of thermal hardware, substitutes for thermal flight components should be used whenever possible to avoid exposure and handling during the test. This applies mainly to MLI blankets and radiator coatings with sensitive solar absorptivity. In the case of MLI blankets, the use of test or spare hardware must depend on the complexity and repeatability of fabrication and assembly. With regard to coatings, decisions must take at least two factors into account: One is whether there was sufficient testing at the subsystem level to prove integrity in a vacuum under assembled conditions, and the other is ensuring that the emissivity of the substitute coating is identical (or very nearly so) to that of the flight component.

4) Maintaining heat pipes in a horizontal position during thermal testing minimizes gravity effects. This is especially significant when grooved heat pipes are used, and some programs may even require that a negative leveling bias be imposed as a guarantee against aiding capillary flow. Test fixtures must, therefore, be designed with leveling provisions and must be such that thermal distortions that occur in the chamber will not have a positive, advantageous effect on heat pipe performance. But as was noted in Chapter 6, Sec. IV.F (item 5), it should be possible to waive some of the requirements on testability with heat pipes if a superior thermal design is otherwise compromised. It remains, however, that any reduced ground performance must be within predictions, and an analysis must confirm enhanced operations in space.

F. Thermal Model Correlation

Correlating the thermal model with the results of the thermal balance test is a requirement intended to secure a reliable analytical tool for evaluating orbital performance, especially in cases of irregularities or in conducting special maneuvers and operations. It should be noted, however, that model correlation is a separate and independent process from verifying the thermal performance. A satellite that does not meet the specified temperature during the thermal balance test can conceivably be correlated with a modified model but must be considered a failure and treated accordingly.

An ideal correlation would make the mathematical representation an exact reflection of the physical entity. But this may not be entirely possible, nor are its benefits necessarily overly significant. A serious limitation happens to be an intrinsic property of a good and forgiving thermal design, in which, when a component fails to perform as required, another, independent component compensates by exceeding its own intended performance. Hence, approximately the same correlation would result from adjustments on the parameters of either component. An obvious example on this is when louvers compensate for MLI deficiency as the blades move toward their closed position. Other contradictions include, among others, the inability to distinguish among the proportions of internal heat conduction, radiation, and the distribution of the I^2R loss from connectors and harnesses.

A test configuration that could pinpoint all subtleties would be an extraordinary challenge to design and implement. Moreover, test times would undoubtedly become long and expensive. The practical approach is to consider correlation in terms of the adequacy of the thermal design as a whole, and to maintain a flexibile attitude toward discrepencies that have small or irrelevant bearing on the overall performance.

The following guidelines have proven useful in defining the scope of correlation:

1) Local and total dissipated power must be known within a small tolerance ($\pm 2\%$ is not unreasonable) to carry out a legitimate correlation. In most thermal designs, the temperature dependence on power is considerably stronger than on other parameters, which makes accurate monitoring of power an essential requirement in instrumentation design and layout.

2) Correlation is more relevant at the electronics and experiments mounting platforms and where thermal distortion is of particular concern. Temperatures of surrounding walls, including structure and MLI surfaces, become pertinent only to the extent of affecting the platform temperature or the distortion issue.

3) When faced with an option, it may be better to effect correlation by making adjustments on conduction rather than radiation couplings. Although heat exchange by conduction is more predictable, it takes lesser variations on the conductance than on script F to produce the same deviation in the calculated temperature. A reference analysis comparing test results with a computer run in which at least some internal radiation couplings are disconnected may also prove useful.

4) Experience indicates that a correlation goal within $\pm 3°C$ at relevant locations is achievable in most designs with active thermal hardware. A general criterion of correlation within $\pm 5°C$ in all locations should be negotiable. Differences greater than $6°C$ or negative and positive deviations in the same general region should be explained and recommendations made for corrections. But in any event, final decisions must depend on systems evaluation of the discrepency in terms of risks and mission margins.

G. Passed and Failed Thermal Balance Tests

The thermal design is said to pass (vs fail) the thermal balance test if the measured temperatures fall within the specifications for the conditions expected in orbit. A more general definition requires, in addition, that all thermal components function in accordance with their intended capability. Hence, a system that consumes considerably more heater power than predicted to maintain the minimum allowable temperature may be viewed as a failed or at least marginal design. Such is also the case for temperatures remaining nominal only because of excessive heat loss from a faulty MLI blanket or actual operating power being lower than was estimated.

Response to failures and the course of corrective action will have to depend on the magnitude of the problem and the willingness for accepting risks. Minor discrepencies are often readily corrected without need for further evaluation. For example, a biased discrepency traced to a somewhat less than acceptable MLI performance (say, 0.03 effective emissivity instead of the maximum 0.02 permitted) might easily be overcome by reducing the effective radiating area with additional MLI or by the application of tapes with appropriate coatings. Conversely, a consistent hot bias might be removed by exposing more radiator area. Other recourses include adjustments on louvers and thermostat set points.

Gross discrepencies must be handled at the systems level. Major corrective action could cause considerable delay, and decisions at this juncture will have to consider the possibility of long-term operation near acceptance or qualification temperatures with reduced performance or higher risk of early failure. The reputation that thermal engineers have of being too meticulous and too conservative probably stems from their deep-rooted awareness of the dire consequences of a failed thermal balance test.

III. Testing with Temperature Margins

One of the difficult functions in satellite engineering is formulating criteria for thermal stress tests that will project adequate long-term operation. A serious complication is devising rules by which testing will reveal workmanship flaws and design defects but will not be excessive to the point of causing a failure that is improbable in normal circumstances. Although the involvement of thermal control in this area is peripheral, they are nevertheless instrumental in establishing the temperature levels on which the margins for verification are added.

Military Standard Specification *MIL-STD-1540-B* is often considered the source for satellite test margins.[1] However, as noted before, the recommended levels and durations in 1540-B are considered excessive in some circles, and waivers are frequently granted based on historical evidence and intended usage. On the other hand, the categories and, to a large extent, the methods of testing are uniform in much of the industry. These may be summarized as follows:

1) Parts thermal tests serve as a selection process for primitive parts (chips, fuses, relays, screws, pieces of aluminum, bare circuit boards, etc.) that make up spacecraft components. These items are generally free of soldered joints, connecting wires, and loose or bent parts, and their integrity is judged by dimensional tolerances and stability and material adequacy. Thermal verification for these items can be quite severe and might include exposures to temperatures ranging from oven levels to cryogenics.

2) Component thermal cycling tests, which are part of the components acceptance criteria prior to their integration into a satellite, are performed at acceptance level for flight and spare units and at the qualification level for protoflight or qualification units. These tests are conducted with the component mounted to a temperature-controlled plate (referred to as a cold plate) in exactly the same fashion as on a spacecraft platform. The component is then operated for extended periods at plate temperatures that cycle between two plateaus which exceed the allowables by predetermined margins. Testing is most appropriately performed in a vacuum chamber at ambient wall temperature and with MLI around the component cover. The defining features of the component cycling test are the temperature levels, the dwell time at each temperature plateau, the transition rate from one temperature to the other and the number of cycles. A cold start is also demonstrated at a cold startup temperature.

3) Component burn-in tests are extensions of the thermal cycling test designed to accumulate test time under stressed conditions. Levels and manner of the burn-in tests vary depending on the heritage of the component, but a common procedure is to operate at an elevated temperature for a minimum of 100 h. Sometimes the test is run in ambient air for longer hours and at temperatures that exceed the upper qualification limit.

4) Satellite level component cycling tests are often conducted in a vacuum under the arrangement of the thermal balance test (phases indicated by 3 and 8 in Fig. 7.3). In these tests, shrouds heating (controlled by feedback from temperature sensors

on the radiators) is continually adjusted to maintain the components mounting platforms at the desired temperature (usually acceptance level) while the satellite is being operated through functional tests in actual mission scenarios. As in the case of component level testing, the factors defining this test are temperature levels, dwell times, transition rates from the hot to cold plateau, and the number of cycles.

IV. Typical Verification Tests

Examples of verification tests performed as part of some of NASA's 5-year and 7-year mission programs are given in Table 7.3. The data derive from projects ranging from the NIMBUS series (1960s) through ATS-6 (1975) and TOPEX/ Poseidon (1992).

Current practices often combine component level acceptance and qualification under one test, but solar arrays' workmanship continues to be verified by temperature cycling in a vacuum at qualification levels. Satellite functional tests in a vacuum remain standard, with at least three cycles of hot and cold soak. However, in some cases the thermal balance test is being abbreviated into a single go/no-go check point.

Table 7.3   Satellite thermal tests

Component	Analysis, °C	Testing
Heritage electronics		Flight components thermal cycling
General		(vacuum, chamber walls at 22°C and MLI on
Platform	0 to 40	cover):
Batteries		Individual components mounted on con-
Platforms	0 to 22	trolled cold plate exactly as on the spacecraft.
Cell differential	5	Functional tests conducted with cold plate at -10°C for 6 h and 55°C (35°C for batteries) for 6 h. Cycle is repeated eight times. Transition rate limited to 3°C/h.
		Qualification components thermal cycling (same arrangement as flight components): Functional tests conducted with cold plate at -15°C (-10°C for batteries) for 12 h and 61°C (35°C for batteries) for 12 h. Cycle repeated 24 times. Transition rate limited to 3°C/h.
		Spacecraft level thermal vacuum (components integrated into satellite): Functional tests conducted with mounting radiators at 50°C (35°C for batteries) then -10°C. Cycle repeated three times at 12-h dwells and once (soak test) at 72-h dwell. Transition rate measured at 1.8°C/h.
		Thermal balance test: Functional performance in hot case, cold case, and contingency low-power case. Temperature drop rate in cool-down test: 1.2°C/h.

(cont.)

Table 7.3   (Cont.)

Component	Analysis, °C	Testing
Nonheritage electronics		Flight components thermal cycling:
Platform		24 cycles at −20°C and 75°C with functional
Interface unit	0 to 35	tests in 12-h dwells.
Propulsion unit	10 to 40	Qualification components thermal cycling:
		3 cycles at −30°C and 85°C with functional
		tests in 72-h dwells.
		Spacecraft level thermal vacuum and balance
		(components integrated into satellite):
		Same conditions as heritage components.
Earth sensor		Flight components thermal cycling:
Platform	3 to 45	24 cycles at −20–65°C with 12-h dwells.
Bolometer	−7 to 50	Functional tests in vacuum:
		With solar simulation (three incident angles)
		at cold plate 0°C (cold case) and 45°C (hot
		case).
Solar array panels		Flight panels vacuum cycling
Local	−85 to 84	(IR lamps/cryogenic walls):
Average	−80 to 84	Panels attached to framed stringers simulating
		actual mechanical boundaries; 50 cycles (150
		h) from −82 to 105°C. Panels periodically
		monitored for circuit continuity. Two chamber
		breaks for visual inspection. Test repeated for
		100 additional cycles.
Solar array mechanisms		Flight array:
(At deployment)		Deployment verification at ambient condi-
Hinges	−23 to 50	tions; 19 deployments including 5 with one
Dampers	−45 to 54	(of two) unoperating (failed) damper.
		Thermal/structural model (TSM)
		(vacuum, simulated solar panel interface):
		Hinge motion (partial deployment) demon-
		strated at −53 and 80°C. Temperature differ-
		ence across hinge measured at 8°C.
		Dampers:
		Dampening rates confirmed in vacuum at
		−55, 20, and 64°C. No fluid leak in 12-h soak
		at 70°C. Test conducted three times on same
		set of nonflight dampers.
		Miscellaneous:
		Harnessing, attachments, and tapes qualified
		using representative sample panel; 50 cycles
		from −82 to 105°C.
Louvers and sun shields		Qualification unit thermal cycling
Exposed blade	−50 to 125	(vacuum, solar simulation):
Shield (silvered	−73 to 30	Blades cycled 50 times at −55 to 155°C.
Teflon on Al)		Shield cycled 50 times at −82 to 50°C.
		Flight units thermal performance tests
		(vacuum, cryogenic walls):

(cont.)

Table 7.3   (Cont.)

Component	Analysis, °C	Testing
		Effective emissivity verification.
		Flight units thermal vacuum tests
		(units on satellite):
		Performance verification during satellite thermal cycle/thermal balance tests.
Heat pipes (grooved)	0 to 40	Flight component acceptance (ambient):
		Proof pressure at 101 atm and 120°C. 60 W-m at 65°C with 5-mm evaporator tilt. Leak rate less than $1.0E-8$ scc/s helium at 15.3 atm.
		Flight units thermal vacuum tests
		(units bonded into satellite radiators):
		Performance verification during satellite thermal cycle/thermal balance tests.
Propulsion (hydrazine) system		Parts tests:
Hydrazine elements	10 to 40	Valve seats verified at LN2 and
Electronics	10 to 45	155°C. Catalyst bed and coatings verified in oven at 1200°C.
		Flight system (module) acceptance tests:
		Thermal balance in vacuum with cold case defined by zero external heating flux and minimum bus voltage.
		Flight system thermal vacuum tests
		(module integrated into satellite):
		Thermal control verification during satellite thermal cycle/thermal balance tests.

References

[1] "Military Standard, Test Requirements for Space Vehicles," U.S. Air Force, MIL-STD-1540B, Oct. 1982, pp. 1–83.

[2] "Application Guidelines for MIL-STD-1540B, Test Requirements for Space Vehicles," U.S. Air Force, MIL-STD-330, July, 1985.

[3] Welch, J. W., "Thermal Testing," *Satellite Thermal Control Handbook*, edited by D. G. Gilmore, The Aerospace Corp. Press, El Segundo, CA, 1994, Chap. 9.

[4] Cleveland, P. E., and Braun, C. E., "Thermal Design of the TOPEX/Poseidon Instrument Module," *Proceedings of the Twentieth Intersociety Conference on Environmental Systems*, 1990, pp. 29–37.

Index

PROGRESS IN ASTRONAUTICS AND AERONAUTICS
SERIES VOLUMES

*1. Solid Propellant Rocket Research (1960)
Martin Summerfield
Princeton University

*2. Liquid Rockets and Propellants (1960)
Loren E. Bollinger
Ohio State University
Martin Goldsmith
The Rand Corp.
Alexis W. Lemmon Jr.
Battelle Memorial Institute

*3. Energy Conversion for Space Power (1961)
Nathan W. Snyder
Institute for Defense Analyses

*4. Space Power Systems (1961)
Nathan W. Snyder
Institute for Defense Analyses

*5. Electrostatic Propulsion (1961)
David B. Langmuir
Space Technology Laboratories, Inc.
Ernst Stuhlinger
NASA George C. Marshall Space Flight Center
J. M. Sellen Jr.
Space Technology Laboratories, Inc.

*6. Detonation and Two-Phase Flow (1962)
S. S. Penner
California Institute of Technology
F. A. Williams
Harvard University

*7. Hypersonic Flow Research (1962)
Frederick R. Riddell
AVCO Corp.

*8. Guidance and Control (1962)
Robert E. Roberson
Consultant
James S. Farrior
Lockheed Missiles and Space Co.

*9. Electric Propulsion Development (1963)
Ernst Stuhlinger
NASA George C. Marshall Space Flight Center

*10. Technology of Lunar Exploration (1963)
Clifford I. Cumming
Harold R. Lawrence
Jet Propulsion Laboratory

*11. Power Systems for Space Flight (1963)
Morris A. Zipkin
Russell N. Edwards
General Electric Co.

*12. Ionization in High-Temperature Gases (1963)
Kurt E. Shuler, Editor
National Bureau of Standards
John B. Fenn,
Associate Editor
Princeton University

*13. Guidance and Control–II (1964)
Robert C. Langford
General Precision Inc.
Charles J. Mundo
Institute of Naval Studies

*14. Celestial Mechanics and Astrodynamics (1964)
Victor G. Szebehely
Yale University Observatory

*15. Heterogeneous Combustion (1964)
Hans G. Wolfhard
Institute for Defense Analyses
Irvin Glassman
Princeton University
Leon Green Jr.
Air Force Systems Command

*16. Space Power Systems Engineering (1966)
George C. Szego
Institute for Defense Analyses
J. Edward Taylor
TRW Inc.

*17. Methods in Astrodynamics and Celestial Mechanics (1966)
Raynor L. Duncombe
U.S. Naval Observatory
Victor G. Szebehely
Yale University Observatory

*18. Thermophysics and Temperature Control of Spacecraft and Entry Vehicles (1966)
Gerhard B. Heller
NASA George C. Marshall Space Flight Center

*19. Communication Satellite Systems Technology (1966)
Richard B. Marsten
Radio Corporation of America

---

*Out of print.

*40. **Aerodynamics of
Base Combustion (1976)**
S. N. B. Murthy, Editor
J. R. Osborn,
Associate Editor
Purdue University
A. W. Barrows
J. R. Ward,
Associate Editors
*Ballistics Research
Laboratories*
ISBN 0-915928-04-3

41. **Communications
Satellite Developments:
Systems (1976)**
Gilbert E. LaVean
*Defense Communications
Agency*
William G. Schmidt
CML Satellite Corp.
ISBN 0-915928-05-1

*42. **Communications
Satellite Developments:
Technology (1976)**
William G. Schmidt
CML Satellite Corp.
Gilbert E. LaVean
*Defense Communications
Agency*
ISBN 0-915928-06-X

*43. **Aeroacoustics: Jet
Noise, Combustion and
Core Engine Noise (1976)**
Ira R. Schwartz, Editor
*NASA Ames Research
Center*
Henry T. Nagamatsu,
Associate Editor
*General Electric Research
and Development Center*
Warren C. Strahle,
Associate Editor
*Georgia Institute of
Technology*
ISBN 0-915928-07-8

*44. **Aeroacoustics: Fan
Noise and Control; Duct
Acoustics; Rotor Noise
(1976)**
Ira R. Schwartz, Editor
*NASA Ames Research
Center*
Henry T. Nagamatsu,
Associate Editor
*General Electric Research
and Development Center*
Warren C. Strahle,
Associate Editor
*Georgia Institute of
Technology*
ISBN 0-915928-08-6

*45. **Aeroacoustics: STOL
Noise; Airframe and
Airfoil Noise (1976)**
Ira R. Schwartz, Editor
*NASA Ames Research
Center*
Henry T. Nagamatsu,
Associate Editor
*General Electric Research
and Development Center*
Warren C. Strahle,
Associate Editor
*Georgia Institute of
Technology*
ISBN 0-915928-09-4

*46. **Aeroacoustics:
Acoustic Wave
Propagation; Aircraft
Noise Prediction;
Aeroacoustic
Instrumentation (1976)**
Ira R. Schwartz, Editor
*NASA Ames Research
Center*
Henry T. Nagamatsu,
Associate Editor
*General Electric Research
and Development Center*
Warren C. Strahle,
Associate Editor
*Georgia Institute of
Technology*
ISBN 0-915928-10-8

*47. **Spacecraft Charging
by Magnetospheric
Plasmas (1976)**
Alan Rosen
TRW Inc.
ISBN 0-915928-11-6

48. **Scientific
Investigations on the
Skylab Satellite (1976)**
Marion I. Kent
Ernst Stuhlinger
*NASA George C. Marshall
Space Flight Center*
Shi-Tsan Wu
University of Alabama
ISBN 0-915928-12-4

49. **Radiative Transfer
and Thermal Control
(1976)**
Allie M. Smith
ARO Inc.
ISBN 0-915928-13-2

*50. **Exploration of the
Outer Solar System (1976)**
Eugene W. Greenstadt
TRW Inc.
Murray Dryer
*National Oceanic and
Atmospheric Administration*
Devrie S. Intriligator
*University of Southern
California*
ISBN 0-915928-14-0

*51. **Rarefied Gas
Dynamics, Parts I and II
(two volumes) (1977)**
J. Leith Potter
ARO Inc.
ISBN 0-915928-15-9

*52. **Materials Sciences in
Space with Application to
Space Processing (1977)**
Leo Steg
General Electric Co.
ISBN 0-915928-16-7

*85. Entry Vehicle Heating and Thermal Protection Systems: Space Shuttle, Solar Starprobe, Jupiter Galileo Probe (1983)
Paul E. Bauer
McDonnell Douglas Astronautics Co.
Howard E. Collicott
The Boeing Co.
ISBN 0-915928-74-4

*86. Spacecraft Thermal Control, Design, and Operation (1983)
Howard E. Collicott
The Boeing Co.
Paul E. Bauer
McDonnell Douglas Astronautics Co.
ISBN 0-915928-75-2

*87. Shock Waves, Explosions, and Detonations (1983)
J. R. Bowen
University of Washington
N. Manson
Universite de Poitiers
A. K. Oppenheim
University of California at Berkeley
R. I. Soloukhin
Institute of Heat and Mass Transfer, BSSR Academy of Sciences
ISBN 0-915928-76-0

88. Flames, Lasers, and Reactive Systems (1983)
J. R. Bowen
University of Washington
N. Manson
Universite de Poitiers
A. K. Oppenheim
University of California at Berkeley
R. I. Soloukhin
Institute of Heat and Mass Transfer, BSSR Academy of Sciences
ISBN 0-915928-77-9

*89. Orbit-Raising and Maneuvering Propulsion: Research Status and Needs (1984)
Leonard H. Caveny
Air Force Office of Scientific Research
ISBN 0-915928-82-5

90. Fundamentals of Solid-Propellant Combustion (1984)
Kenneth K. Kuo
Pennsylvania State University
Martin Summerfield
Princeton Combustion Research Laboratories, Inc.
ISBN 0-915928-84-1

91. Spacecraft Contamination: Sources and Prevention (1984)
J. A. Roux
University of Mississippi
T. D. McCay
NASA Marshall Space Flight Center
ISBN 0-915928-85-X

92. Combustion Diagnostics by Nonintrusive Methods (1984)
T. D. McCay
NASA Marshall Space Flight Center
J. A. Roux
University of Mississippi
ISBN 0-915928-86-8

93. The INTELSAT Global Satellite System (1984)
Joel Alper
COMSAT Corp.
Joseph Pelton
INTELSAT
ISBN 0-915928-90-6

94. Dynamics of Shock Waves, Explosions, and Detonations (1984)
J. R. Bowen
University of Washington
N. Manson
Universite de Poitiers
A. K. Oppenheim
University of California at Berkeley
R. I. Soloukhin
Institute of Heat and Mass Transfer, BSSR Academy of Sciences
ISBN 0-915928-91-4

95. Dynamics of Flames and Reactive Systems (1984)
J. R. Bowen
University of Washington
N. Manson
Universite de Poitiers
A. K. Oppenheim
University of California at Berkeley
R. I. Soloukhin
Institute of Heat and Mass Transfer, BSSR Academy of Sciences
ISBN 0-915928-92-2

96. Thermal Design of Aeroassisted Orbital Transfer Vehicles (1985)
H. F. Nelson
University of Missouri-Rolla
ISBN 0-915928-94-9

97. Monitoring Earth's Ocean, Land, and Atmosphere from Space—Sensors, Systems, and Applications (1985)
Abraham Schnapf
Aerospace Systems Engineering
ISBN 0-915928-98-1

98. Thrust and Drag: Its Prediction and Verification (1985)
Eugene E. Covert
Massachusetts Institute of Technology
C. R. James
Vought Corp.
William F. Kimzey
Sverdrup Technology AEDC Group
George K. Richey
U.S. Air Force
Eugene C. Rooney
U.S. Navy Department of Defense
ISBN 0-930403-00-2

99. Space Stations and Space Platforms—Concepts, Design, Infrastructure, and Uses (1985)
Ivan Bekey
Daniel Herman
NASA Headquarters
ISBN 0-930403-01-0

100. Single- and Multi-Phase Flows in an Electromagnetic Field: Energy, Metallurgical, and Solar Applications (1985)
Herman Branover
Ben-Gurion University of the Negev
Paul S. Lykoudis
Purdue University
Michael Mond
Ben-Gurion University of the Negev
ISBN 0-930403-04-5

101. MHD Energy Conversion: Physiotechnical Problems (1986)
V. A. Kirillin
A. E. Sheyndlin
Soviet Academy of Sciences
ISBN 0-930403-05-3

102. Numerical Methods for Engine-Airframe Integration (1986)
S. N. B. Murthy
Purdue University
Gerald C. Paynter
Boeing Airplane Co.
ISBN 0-930403-09-6

103. Thermophysical Aspects of Re-Entry Flows (1986)
James N. Moss
NASA Langley Research Center
Carl D. Scott
NASA Johnson Space Center
ISBN 0-930430-10-X

*104. Tactical Missile Aerodynamics (1986)
M. J. Hemsch
PRC Kentron, Inc.
J. N. Nielson
NASA Ames Research Center
ISBN 0-930403-13-4

105. Dynamics of Reactive Systems Part I: Flames and Configurations; Part II: Modeling and Heterogeneous Combustion (1986)
J. R. Bowen
University of Washington
J.-C. Leyer
Universite de Poitiers
R. I. Soloukhin
Institute of Heat and Mass Transfer, BSSR Academy of Sciences
ISBN 0-930403-14-2

106. Dynamics of Explosions (1986)
J. R. Bowen
University of Washington
J.-C. Leyer
Universite de Poitiers
R. I. Soloukhin
Institute of Heat and Mass Transfer, BSSR Academy of Sciences
ISBN 0-930403-15-0

*107. Spacecraft Dielectric Material Properties and Spacecraft Charging (1986)
A. R. Frederickson
U.S. Air Force Rome Air Development Center
D. B. Cotts
SRI International
J. A. Wall
U.S. Air Force Rome Air Development Center
F. L. Bouquet
Jet Propulsion Laboratory, California Institute of Technology
ISBN 0-930403-17-7

*108. Opportunities for Academic Research in a Low-Gravity Environment (1986)
George A. Hazelrigg
National Science Foundation
Joseph M. Reynolds
Louisiana State University
ISBN 0-930403-18-5

109. Gun Propulsion Technology (1988)
Ludwig Stiefel
U.S. Army Armament Research, Development and Engineering Center
ISBN 0-930403-20-7

---

*Out of print.

135. Numerical Approaches to Combustion Modeling (1991)
Elaine S. Oran
Jay P. Boris
Naval Research Laboratory
ISBN 1-56347-004-7

136. Aerospace Software Engineering (1991)
Christine Anderson
U.S. Air Force Wright Laboratory
Merlin Dorfman
Lockheed Missiles & Space Company, Inc.
ISBN 1-56347-005-0

137. High-Speed Flight Propulsion Systems (1991)
S. N. B. Murthy
Purdue University
E. T. Curran
Wright Laboratory
ISBN 1-56347-011-X

138. Propagation of Intensive Laser Radiation in Clouds (1992)
O. A. Volkovitsky
Yu. S. Sedenov
L. P. Semenov
Institute of Experimental Meteorology
ISBN 1-56347-020-9

139. Gun Muzzle Blast and Flash (1992)
Günter Klingenberg
Fraunhofer-Institut für Kurzzeitdynamik, Ernst-Mach-Institut
Joseph M. Heimerl
U.S. Army Ballistic Research Laboratory
ISBN 1-56347-012-8

140. Thermal Structures and Materials for High-Speed Flight (1992)
Earl. A. Thornton
University of Virginia
ISBN 1-56347-017-9

141. Tactical Missile Aerodynamics: General Topics (1992)
Michael J. Hemsch
Lockheed Engineering & Sciences Company
ISBN 1-56347-015-2

142. Tactical Missile Aerodynamics: Prediction Methodology (1992)
Michael R. Mendenhall
Nielsen Engineering & Research, Inc.
ISBN 1-56347-016-0

143. Nonsteady Burning and Combustion Stability of Solid Propellants (1992)
Luigi De Luca
Politecnico di Milano
Edward W. Price
Georgia Institute of Technology
Martin Summerfield
Princeton Combustion Research Laboratories, Inc.
ISBN 1-56347-014-4

144. Space Economics (1992)
Joel S. Greenberg
Princeton Synergetics, Inc.
Henry R. Hertzfeld
HRH Associates
ISBN 1-56347-042-X

145. Mars: Past, Present, and Future (1992)
E. Brian Pritchard
NASA Langley Research Center
ISBN 1-56347-043-8

146. Computational Nonlinear Mechanics in Aerospace Engineering (1992)
Satya N. Atluri
Georgia Institute of Technology
ISBN 1-56347-044-6

147. Modern Engineering for Design of Liquid-Propellant Rocket Engines (1992)
Dieter K. Huzel
David H. Huang
Rocketdyne Division of Rockwell International
ISBN 1-56347-013-6

148. Metallurgical Technologies, Energy Conversion, and Magneto-hydrodynamic Flows (1993)
Herman Branover
Yeshajahu Unger
Ben-Gurion University of the Negev
ISBN 1-56347-019-5

149. Advances in Turbulence Studies (1993)
Herman Branover
Yeshajahu Unger
Ben-Gurion University of the Negev
ISBN 1-56347-018-7

150. Structural Optimization: Status and Promise (1993)
Manohar P. Kamat
Georgia Institute of Technology
ISBN 1-56347-056-X

151. **Dynamics of Gaseous Combustion (1993)**
A. L. Kuhl
Lawrence Livermore National Laboratory
J.-C. Leyer
Universite de Poitiers
A. A. Borisov
USSR Academy of Sciences
W. A. Sirignano
University of California
ISBN 1-56347-060-8

152. **Dynamics of Heterogeneous Gaseous Combustion and Reacting Systems (1993)**
A. L. Kuhl
Lawrence Livermore National Laboratory
J.-C. Leyer
Universite de Poitiers
A. A. Borisov
USSR Academy of Sciences
W. A. Sirignano
University of California
ISBN 1-56347-058-6

153. **Dynamic Aspects of Detonations (1993)**
A. L. Kuhl
Lawrence Livermore National Laboratory
J.-C. Leyer
Universite de Poitiers
A. A. Borisov
USSR Academy of Sciences
W. A. Sirignano
University of California
ISBN 1-56347-057-8

154. **Dynamic Aspects of Explosion Phenomena (1993)**
A. L. Kuhl
Lawrence Livermore National Laboratory
J.-C. Leyer
Universite de Poitiers
A. A. Borisov
USSR Academy of Sciences
W. A. Sirignano
University of California
ISBN 1-56347-059-4

155. **Tactical Missile Warheads (1993)**
Joseph Carleone
Aerojet General Corporation
ISBN 1-56347-067-5

156. **Toward a Science of Command, Control, and Communications (1993)**
Carl R. Jones
Naval Postgraduate School
ISBN 1-56347-068-3

157. **Tactical and Strategic Missile Guidance Second Edition (1994)**
Paul Zarchan
Charles Stark Draper Laboratory, Inc.
ISBN 1-56347-077-2

158. **Rarefied Gas Dynamics: Experimental Techniques and Physical Systems (1994)**
Bernie D. Shizgal
University of British Columbia
David P. Weaver
Phillips Laboratory
ISBN 1-56347-079-9

159. **Rarefied Gas Dynamics: Theory and Simulations (1994)**
Bernie D. Shizgal
University of British Columbia
David P. Weaver
Phillips Laboratory
ISBN 1-56347-080-2

160. **Rarefied Gas Dynamics: Space Sciences and Engineering (1994)**
Bernie D. Shizgal
University of British Columbia
David P. Weaver
Phillips Laboratory
ISBN 1-56347-081-0

161. **Teleoperation and Robotics in Space (1994)**
Steven B. Skaar
University of Notre Dame
Carl F. Ruoff
Jet Propulsion Laboratory, California Institute of Technology
ISBN 1-56347-095-0

162. **Progress in Turbulence Research (1994)**
Herman Branover
Yeshajahu Unger
Ben-Gurion University of the Negev
ISBN 1-56347-099-3

163. **Global Positioning System: Theory and Applications, Volume I (1996)**
Bradford W. Parkinson
Stanford University
James J. Spilker Jr.
Stanford Telecom
Penina Axelrad,
Associate Editor
University of Colorado
Per Enge,
Associate Editor
Stanford University
ISBN 1-56347-107-8

164. **Global Positioning System: Theory and Applications, Volume II (1996)**
Bradford W. Parkinson
Stanford University
James J. Spilker Jr.
Stanford Telecom
Penina Axelrad,
Associate Editor
University of Colorado
Per Enge,
Associate Editor
Stanford University
ISBN 1-56347-106-X